SOLUTIONS MANUAL

Physical Chemistry

FOURTH EDITION

Solutions Manual

Physical Chemistry

Fourth Edition

KEITH J. LAIDLER
University of Ottawa

JOHN H. MEISER
Ball State University

BRYAN SANCTUARY
McGill University

B. RAMU RAMACHANDRAN
Louisiana Tech University

HOUGHTON MIFFLIN COMPANY   BOSTON   NEW YORK

Executive Editor: Richard Stratton
Editorial Associate: Marisa R. Papile
Production Coordinator: Jill Haber
Senior Manufacturing Coordinator: Florence Cadran
Senior Marketing Manager: Katherine Greig
Editorial Assistant: May Jawdat

Printed in the U.S.A.

ISBN: 0-618-12342-3

23456789-POO-06 05 04 03

Contents

CHAPTER 1 The Nature of Physical Chemistry and the Kinetic Theory of Gases 1

CHAPTER 2 The First Law of Thermodynamics 23

CHAPTER 3 The Second and Third Laws of Thermodynamics 48

CHAPTER 4 Chemical Equilibrium 72

CHAPTER 5 Phases and Solutions 95

CHAPTER 6 Phase Equilibria 116

CHAPTER 7 Solution of Electrolytes 136

CHAPTER 8 Electrochemical Cells 157

CHAPTER 9 Chemical Kinetics I. The Basic Ideas 178

CHAPTER 10 Chemical Kinetics II. Composite Mechanisms 208

CHAPTER 11 Quantum Mechanics and Atomic Structure 232

CHAPTER 12 The Chemical Bond 251

CHAPTER 13 Foundations of Chemical Spectroscopy 261

CHAPTER 14 Some Modern Applications of Spectroscopy 285

CHAPTER 15 Molecular Statistics 291

CHAPTER 16 The Solid State 310

CHAPTER 17 The Liquid State 332

CHAPTER 18 Surface Chemistry and Colloids 340

CHAPTER 19 Transport Properties 352

1. THE NATURE OF PHYSICAL CHEMISTRY AND THE KINETIC THEORY OF GASES

■ Classical Mechanics and Thermal Equilibrium

1.1.   Work, $w = \int_{l_0}^{l} F(l) \cdot dl = \frac{1}{2} mu^2 - \frac{1}{2} mu_0^2$

Speed = 88 km h^{-1} = 88(km h^{-1}) × (1/3600 s h^{-1}) × (10^3 m km^{-1}) = 24.4 m s^{-1}

Civic: $w = \frac{1}{2}(1000 \text{ kg}) (24.4 \text{ m s}^{-1})^2 = 298 \text{ kJ}$

Taurus: $w = \frac{1}{2}(1600 \text{ kg}) (24.4 \text{ m s}^{-1})^2 = 476 \text{ kJ}$

The work required is directly proportional to the mass of the car.

1.2.   If t^u is the value of the temperature,

$$28.1 = 27.5(1 + 0.160 \times 10^{-4}t^u + 0.100 \times 10^{-7}t^{u2})$$

which reduces to

$$(0.100 \times 10^{-7})t^{u2} + (0.160 \times 10^{-4})t^u - 0.0218 = 0$$

The solution is $t^u = 879$; therefore

$$t = 879 \text{ °C}$$

1.3. $m_1 = 1 \times 10^{-24}$ g $u_1 = 500$ m s^{-1}

$m_2 = 1 \times 10^{-23}$ g $u_2 = 0$

Conservation of momentum: $m_1 u_1 + m_2 u_2 = m_1 u_1' + m_2 u_2'$ (1)

Conservation of energy: $\frac{1}{2} m_1 u_1^2 + \frac{1}{2} m_2 u_2^2 = \frac{1}{2} m_1 u_1'^2 + \frac{1}{2} m_2 u_2'^2$ (2)

$u_2 = 0$;   from Eq. 1: $u_1' = u_1 - \dfrac{m_2}{m_1} u_2'$

Substitute into Eq. 2 and solve for u_2' :

$$u_2' = \frac{2u_1}{1 + \frac{m_2}{m_1}} = \frac{2 \times 500 \text{ m s}^{-1}}{1 + \frac{1 \times 10^{-23} \text{ g}}{1 \times 10^{-24} \text{ g}}} = 90.9 \text{ m s}^{-1}$$

$$\text{Kinetic energy} = \frac{1}{2} m_2 u_2'^2 = \frac{1}{2}(1 \times 10^{-26} \text{ kg}) \times (90.9)^2 \text{ m}^2 \text{ s}^{-2} = 4.13 \times 10^{-23} \text{ J}$$

1.4. $\text{Power} = \dfrac{dw}{dt} = \dfrac{8000 \times 10^3 \text{ J}}{24 \text{ h} \times 3600 \text{ s h}^{-1}}$

$\qquad\qquad = 92.6 \text{ J s}^{-1} = 92.6 \text{ W.}$

1.5.   a.   Mass is extensive, depending upon the amount present.

      b.   Density is intensive, the ratio of two extensive quantities.

      c.   Temperature is intensive

      d.   Gravitational field is intensive.

■ Gas Laws and Temperature

1.6.   The difference in height of the two columns is

      34.71 cm – 10.83 cm = 23.88 cm

Since the arm is open to the atmosphere, this pressure must be added to the barometric pressure. The definition 1 mmHg = 1 Torr applies. Therefore, 23.88 cmHg = 238.8 Torr.

      238.8 Torr + 738.4 Torr = 977.2 Torr

1.7.   a.   A column of mercury 1 m^2 in cross-sectional area and 0.001 m in height has a volume of 0.001 m^3 and a mass of 0.001 m$^3 \times$ 13 595.1 kg m^{-3}. Then

$$1 \text{ mmHg} = 0.001 \text{ m}^3 \times 13\,595.1 \text{ kg m}^{-3} \times 9.806\,65 \text{ m s}^{-2} = 133.322\,3874 \text{ Pa}$$

By definition, 1 atmosphere = 101 325 Pa.

$$1 \text{ Torr} = \frac{1}{760}(101\,325 \text{ Pa}) = 133.322\,3684 \text{ Pa}$$

Thus

$$1 \text{ mmHg} = \frac{133.322\,3874}{133.322\,3684} = 1.000\,000\,14 \text{ Torr}$$

The torr is now defined as 1 mmHg.

b. From $PV = nRT$, and since $n = \dfrac{N}{L}$

$$PV = \frac{NRT}{L} \quad \text{and the number density is} \quad \frac{N}{V} = \frac{PL}{RT}$$

For $P = 10^{-6}$ Torr:

$$N/V = \frac{10^{-6} \text{ Torr (1 atm/760 Torr)}}{8.3145 \text{ J K}^{-1} \text{ mol}^{-1} \ 298.15 \text{ K}} \times 101\ 325 \text{ (Pa atm}^{-1}\text{) } 6.022 \times 10^{23} \text{ mol}^{-1}$$

$$= 3.24 \times 10^{16} \text{ m}^{-3}$$

For $P = 10^{-15}$ Torr:

$$N/V = \frac{10^{-15} \text{ Torr (1 atm/760 Torr)}}{8.3145 \text{ (J K}^{-1} \text{ mol}^{-1}\text{) } 298.15 \text{ K}} \times 101\ 325 \text{ (Pa atm}^{-1}\text{) } 6.022 \times 10^{23} \text{ mol}^{-1}$$

$$= 3.24 \times 10^{7} \text{ m}^{-3}$$

This is still a substantial number.

1.8. A column of mercury 1 m^2 in cross-sectional area and exactly 0.760 m in height has a volume of 0.760 m^3 and a mass of 0.760 m^3 × 13 595.1 kg m^{-3}. Mass times the gravitational acceleration is a weight or force. The column's weight on the unit area gives a pressure

$$0.760 \text{ m}^3 \times 13\ 595.1 \text{ kg m}^{-3} \times 9.806\ 65 \text{ m s}^{-1} = 101\ 325 \text{ kg m s}^{-2}$$

Since 1 Pa = 1 kg m s^{-2}, the pressure is 101.325 kPa

1.9. The pressure of two different liquids of the same volume at the same temperature is proportional to their densities. Assume a 1 mm height of liquid of diameter to accommodate 1 cm^{-3}. Then

$$\frac{P_{DBP}}{P_{Hg}} = \frac{1.047 \text{ g cm}^{-3}}{13.595 \text{ g cm}^{-3}}$$

or

$$P_{DBP} = 1 \text{ Torr} \times \frac{1.047}{13.595} = 0.077 \text{ Torr}$$

Thus, 1 mm of DBP is equivalent to 0.077 Torr. This can be stated such that 1 Torr is equivalent to millimetres of DBP:

$$1 \text{ Torr}/0.077 \text{ (Torr mm}^{-1}\text{)} = 12.99 \text{ mm DBP}$$

1.10. $P_1V_1 = P_2V_2$

$$697 \text{ Torr} \times 0.251 \text{ dm}^3 + 0.0104 \text{ Torr} \times V_2 = 287(V_2 + 0.251)$$

$$174.95 + 0.0104 \ V_2 = 72.037 + 287 \ V_2$$

$$102.91 = 286.99 \ V_2$$

$$V_2 = 0.359 \text{ dm}^3$$

1.11. $P_1V_1 = P_2V_2$

$$V_2 = \frac{1.80 \times 10^5 \, Pa \times 0.300 \, dm^3}{1.15 \times 10^5 \, Pa} = 0.470 \, dm^3$$

1.12. $\dfrac{V_1}{T_1} = \dfrac{V_2}{T_2}$

$$V_2 = \frac{0.300 \, dm^3}{330 \, K} (550 \, K) = 0.500 \, dm^3$$

1.13. Concentration $= n/V = P/RT$. Since $P = 1.013\,25 \times 10^5 \, Pa = 1 \, atm$

a. $\dfrac{n}{V} = \dfrac{1.01325 \times 10^5 \, Pa}{8.3145 \, J \, K^{-1} mol^{-1} \times 298.15 \, K}$

   $= 40.87 \, mol \, m^{-3}$ since $J/Pa = m^3$

   $= 0.0409 \, mol \, dm^{-3}$

Number of molecules per unit volume

   $= 0.0409 \, (mol \, dm^{-3}) \, 6.022 \times 10^{23} \, mol^{-1}$

   $= 2.46 \times 10^{22} \, dm^{-3}$

b. $\dfrac{n}{V} = \dfrac{1.00 \times 10^{-4} \, Pa}{8.3145 \, J \, K^{-1} mol^{-1} \times 298.15 \, K}$

   $= 4.03 \times 10^{-8} \, mol \, m^{-3} = 4.03 \times 10^{-11} \, mol \, dm^{-3}$

   $= 2.43 \times 10^{13} \, dm^{-3}$

1.14. The temperature is held constant, so Boyle's law applies.

$P_2 = P_1 \, (V_1/V_2)$ where P_2 is the pressure.

Since h, the height of the mercury column on the trapped air side, is proportional to the volume of a uniform tube, we can write

$P_2 = P_1 \times 100 \, cmHg/(100 - h) \, cmHg$

where h is the final height in centimetres of Hg in the long arm. Since

$P_2 = 40 - h + P_1$

elimination of P_2 gives

$$\frac{P_1 \times 100}{100 - h} = 40 - h + P_1$$

Substituting the value of 76 cmHg (760 Torr) for P_1 and solving for h gives

$h^2 - 216 \, h + 4000 = 0$

From the quadratic equation we find:

$h = 195.5$ cmHg   or   20.5 cmHg

The value of h cannot be 195.5 cmHg since the tube length is only 100 cm. Therefore,

$h = 20.5$ cm

The final pressure, P_2, is

$P_2 = 76$ cmHg $\times 100$ cm/(100 cm $- 20.5$ cm) $= 95.6$ cmHg $= 956$ Torr

1.15. a. $\dfrac{V_1}{T_1} = C = \dfrac{V_2}{T_2}$ $V_2 = \dfrac{0.250 \text{ dm}^3}{303 \text{ K}}(273.15 \text{ K}) = 0.225 \text{ dm}^3$

b. Molar mass, $M = \dfrac{mRT}{PV}$

$$= \frac{1.08 \times 10^{-3} \text{ kg} \times 8.3145 \text{ J K}^{-1}\text{mol}^{-1} \times 273.15 \text{ K}}{(101.3 \times 10^3 \text{ Pa})(2.25 \times 10^{-4} \text{ m}^3)}$$

$$= 0.1076 \text{ kg mol}^{-1} = 108 \text{ g mol}^{-1}$$

1.16. $M = \dfrac{\rho PT}{P} = \dfrac{1.92 \text{ kg m}^{-3} \times 8.3145 \text{ J mol}^{-1}\text{K}^{-1} \times 298.15 \text{ K}}{150 \times 10^3 \text{ Pa}}$

$$= 0.0317 \text{ kg mol}^{-1} = 31.7 \text{ g mol}^{-1}$$

1.17. $n/V = P/RT = \dfrac{1.01325 \times 10^5 \text{ Pa}}{8.3145 \text{ J K}^{-1}\text{mol}^{-1} \times 298.15 \text{ K}}$

$$= 40.87 \text{ mol m}^{-3} = 0.040 \ 87 \text{ mol dm}^{-3}$$

Thus 0.040   87 mol $= 1.159$ g

1 mol $= 28.36$ g $=$ molar mass

1.18. From Eq. 1.53

$P_t = P_{H_2} + P_{H_2O}$

$P_{H_2} = P_t - P_{H_2O}$

$= 99.99$ kPa $- 3.17$ kPa $= 96.82$ kPa

1.19. The amount of each gas is

$n_{N_2} = 100.00$ g/28.012 g mol^{-1} $= 3.5699$ mol.

$n_{CO_2} = 100.00$ g/44.010 g mol^{-1} $= 2.2722$ mol.

The mole fractions are

$x_{N_2} = \dfrac{3.5699 \text{ mol}}{(3.5699 + 2.2722)\text{mol}} = 0.6111;$ $x_{CO_2} = \dfrac{2.2722 \text{ mol}}{(3.5699 + 2.2722)\text{mol}} = 0.3889.$

The partial pressures are

$$P_{N_2} = 3.5699 \text{ mol} \times \frac{8.3145 \text{ J K}^{-1} \text{ mol}^{-1} \times 298.15 \text{ K}}{2.50 \text{ dm}^3} = 35.399 \text{ bar}$$

$$P_{CO_2} = 2.2722 \text{ mol} \times \frac{8.3145 \text{ J K}^{-1} \text{ mol}^{-1} \times 298.15 \text{ K}}{2.50 \text{ dm}^3} = 22.531 \text{ bar}$$

1.20. P of dry gas at 27.5 °C is

751.4 Torr − 27.5 Torr = 723.9 Torr

$P_1V_1/T_1 = P_2V_2/T_2$ gives $V_2 = V_1P_1T_2/P_2T_1$

$$V_2 = (27.8 \text{ cm}^3) \frac{(723.9 \text{ Torr})(298.15 \text{ K})}{(750.06 \text{ Torr})(300.65 \text{ K})}$$

$$= 26.6 \text{ cm}^3$$

This can also be done using 133.32 Pa/Torr and using 100 000 Pa to eliminate the pascal unit.

1.21. The lifting force comes from the difference between the mass of air displaced and the mass of helium that replaces the air. Assume air to have an average molar mass of 28.8 g mol⁻¹.

(80 mol % N_2 = 22.4 g mol⁻¹ air;   20 mol % O_2 = 6.4 g mol⁻¹ air)

Lifting force = $V(\rho_{air} - \rho_{helium})$ = 1000 kg

$$\rho_{air} = \frac{PM}{RT} = \frac{0.940(\text{atm})101\ 325(\text{Pa atm}^{-1})28.8(\text{g mol}^{-1})}{8.3145(\text{J K}^{-1} \text{ mol}^{-1})\ 290(\text{K})\ 10^3 \text{ g kg}^{-1}}$$

$$= 1.138 \text{ kg m}^{-3}$$

$$\rho_{He} = \frac{PM}{RT} = \frac{0.940(\text{atm})101\ 325(\text{Pa atm}^{-1})4.003(\text{g mol}^{-1})}{8.3145(\text{J K}^{-1} \text{ mol}^{-1})\ 290\ (\text{K})\ 10^3 \text{ g kg}^{-1}}$$

$$= 0.158 \text{ kg m}^{-3}$$

$$V = \frac{1000 \text{ kg}}{(1.138 - 0.158) \text{ kg m}^{-3}} = 1020 \text{ m}^3$$

1.22. The amount of butane present is given by $n = PV/RT$

$$n = \frac{101\ 325 \text{ Pa} \times 40.0 \text{ dm}^3 \times 10^{-3} \text{ m}^3 \text{dm}^{-3}}{8.3145(\text{J K}^{-1} \text{mol}^{-1})298 \text{ K}}$$

$$= 1.64 \text{ mol}$$

The mixture required is 95 parts argon to 5 parts butane, or 19 to 1.

$n_{argon} = 19 \times 1.64 = 31.2 \text{ mol}$

Mass of argon = 31.2 mol × 39.9 g mol⁻¹ = 1240 g

The final pressure, P_f, is proportional to the total amount of gas; thus

P_f = 1 atm × 20/1 = 20 atm

1.23.   a.   The standard gravitational acceleration is 9.807 m s^{-2}. If g decreases by 0.010 m s^{-2} per km of height, this is equivalent to a change of 0.010 m s^{-2}/10^3 m = 10^{-5} s^{-2} z where z is the altitude. Therefore, $g = 9.807$ m s$^{-2} - 10^{-5}$ s^{-2} z and substitution in Eq. 1.74 gives

$$\frac{dP}{P} = -\frac{M}{RT}(9.807 - 10^{-5}\, z)\; dz$$

or

$$\ln(P/P_0) = -\frac{M}{RT}(9.807 \text{ m s}^{-2}\, z - 5 \times 10^{-6} \text{ s}^{-2}z^2)$$

    b. $\ln(P/P_0) = -\dfrac{28.0(\text{g mol}^{-1})(10^{-3} \text{ kg g}^{-1})}{8.3145\,(\text{J K}^{-1}\text{ mol}^{-1})298.15 \text{ K}} \times [9.807(10^5)\text{m}^2 \text{ s}^{-2} - 5 \times 10^{-6}(10^5)^2 \text{ m}^2 \text{ s}^{-2}]$

$$\ln(P/P_0) = -10.51; \quad P/P_0 = 2.73 \times 10^{-5}$$

$$P = 2.73 \times 10^{-5} \text{ atm}$$

1.24. $P = P_0\, e^{-Mgz/RT}$

Assume that the temperature remains constant at 250 K. The value of M is 0.017 kg mol^{-1}.

$$P = 400 \text{ Torr } e^{-0.017(9.807)8000/(8.3145 \times 250)}$$

$$= 400\, e^{-0.642} = 400 \times 0.526 = 210 \text{ Torr}$$

1.25.   Start with the barometric equation and substitute $T_0 - az$ for T.

$$\frac{dP}{P} = \frac{-Mg\, dz}{RT} \longrightarrow \frac{-Mg\, dz}{R(T_0 - az)}$$

The solution requires solving the differential equation. Let

$$(T_0 - az) = x$$

$$dx = -adz \quad \text{or} \quad dz = -dx/a$$

$$\int_{x_0}^{x} \frac{-dx}{ax} = \frac{-1}{a}\ln\frac{x}{x_0} = \frac{1}{a}\ln\left(\frac{T_0}{T_0 - az}\right)$$

Integration of the left-hand side between P_0 and P and the final substitution gives

$$\ln\frac{P}{P_0} = \frac{Mg}{Ra}\ln\left(\frac{T_0 - az}{T_0}\right)$$

1.26.   From Eq. 1.15 we have

$$\theta_{Hg} = \frac{(V_{50} - V_0)_{Hg}}{(V_{100} - V_0)_{Hg}} \times 100$$

Integrating the expression for α, we have

$$V_{50} - V_0 = \int_0^{50} \alpha V_0\, d\theta$$

$$\left[1.817 \times 10^{-4}\, \theta\, \Big|_0^{50} + 2.95 \times 10^{-9}\, \theta^2\, \Big|_0^{50} + 1.15 \times 10^{-10}\, \theta^3\, \Big|_0^{50}\right]V_0$$

$$= 0.009\ 107\ V_0$$

$$V_{100} - V_0 = \left[1.817 \times 10^{-4}\ \theta \Big|_0^{100} + 2.95 \times 10^{-9}\ \theta^2 \Big|_0^{100} + 1.15 \times 10^{-10}\ \theta^3 \Big|_0^{100} \right] V_0$$

$$= 0.018\ 31\ V_0$$

Then

$$\theta_{Hg} = \frac{0.009\ 107\ V_0}{0.018\ 31\ V_0} \times 100 = 49.7\ °C$$

■ Graham's Law, Molecular Collisions, and Kinetic Theory

1.27. Rate of effusion is proportional to $1/T$; therefore, from Graham's law

$$\frac{v_A}{v_B} = \frac{1}{2.3} = \sqrt{\frac{28\ g\ mol^{-1}}{MW\ of\ A}}$$

Molar mass of A $= 28\ (2.3)^2 = 148 \approx 150\ g\ mol^{-1}$

1.28. The time for effusion is inversely proportional to the rate of effusion. (A high rate requires a short time for diffusion.)

$$\frac{rate\ (N_2)}{rate\ (He)} = \frac{t_{He}}{t_{N_2}} = \frac{\sqrt{MW_{He}}}{\sqrt{MW_{N_2}}}$$

$$t_{He} = t_{N_2} \sqrt{\frac{MW_{He}}{MW_{N_2}}} = 5.80 \sqrt{\frac{4}{28}} = 2.19\ min$$

1.29. $PV = \frac{2}{3}\ nE_k;\quad E_k = \frac{3\ PV}{2\ n}$ (Kinetic energy per mole of gas.)

$$E_k = \frac{3}{2} \left(\frac{1}{0.5\ mol} \right) (200\ kPa)\ (8.0\ dm^3) = 4800\ J\ mol^{-1}$$

For 0.5 mol, $nE_k = 2400\ J$

1.30. a. $n = \dfrac{PV}{RT} = \dfrac{152\ 000(Pa)2(dm^3)10^{-3}(m^3\ dm^{-3})}{8.3145(J\ K^{-1}\ mol^{-1})298.15(K)}$

$$= 0.1226\ mol = 0.123\ mol$$

b. The number of molecules = amount of substance × Avogadro's constant

$$= 0.123(mol) \times 6.022 \times 10^{23}(mol^{-1})$$

$$= 7.41 \times 10^{22}$$

c. Rearrangement of Eq. 1.43 gives

$$\overline{u^2} = \frac{3RT}{mL} \quad or \quad \sqrt{\overline{u^2}} = \sqrt{\frac{3RT}{M}}$$

$$\sqrt{\overline{u^2}} = \sqrt{\frac{3(8.3145 \text{ J K}^{-1} \text{ mol}^{-1})(298.15 \text{ K})}{0.028\,0134 \text{ kg mol}^{-1}}}$$

$$= 515.2 \text{ m s}^{-1}$$

d. $\overline{\varepsilon}_k = \frac{1}{2}m\overline{u^2} = \dfrac{0.028\,0134(\text{kg mol}^{-1})(515.2)^2(\text{m}^2\text{s}^{-2})}{2(6.022 \times 10^{23} \text{ mol}^{-1})},$

$$= 6.174 \times 10^{-21} \text{ J}$$

e. $E_k(\text{total}) = 0.123(\text{mol})6.022 \times 10^{23}(\text{mol}^{-1}) \times 6.174 \times 10^{-21}(\text{J molecule}^{-1})$

$$= 457 \text{ J}$$

Alternatively,

$$E_k(\text{total}) = 0.123 \times \frac{3}{2}RT$$

$$= (0.123 \text{ mol}) (3) (8.3145 \text{ J K}^{-1} \text{ mol}^{-1}) (298.15 \text{ K})/2$$

$$= 457 \text{ J}$$

1.31. From Eq. 1.43 we have

$$\frac{\sqrt{\overline{u_2^2}}}{\sqrt{\overline{u_1^2}}} = \sqrt{\frac{T_2}{T_1}} = \sqrt{\frac{400}{300}} = \sqrt{1.333} = 1.15$$

1.32. The value of N_A, the number of molecules formed for use in the mean-free-path equation, must be calculated from the ideal gas law. The number of molecules per unit volume is

$$\frac{N_A}{V} = \frac{LP}{RT}$$

$$= \frac{6.022 \times 10^{23}(\text{mol}^{-1})101\,325(Pa)}{8.3145(\text{J K}^{-1}\text{mol}^{-1})298.15(K)}, = 2.46 \times 10^{25} \text{ m}^{-3}$$

Since $1 \text{ Pa} \equiv \text{kg m}^{-1}\text{ s}^{-2}$ and $J = \text{kg m}^2 \text{ s}^{-2}$

$\lambda = 1/(\sqrt{2}\,\pi d^2 N_A) = 1/(\sqrt{2}(\pi)(3.74 \times 10^{-10}\,m)^2 \times (2.46 \times 10^{25} \text{ m}^{-3})$

$$= 6.54 \times 10^{-8} \text{ m}$$

$Z_A = \sqrt{2}\,\pi d^2\,\overline{u}_A N_A$

$$= \sqrt{2}\,\pi(3.74 \times 10^{-10}\,m)^2\,474.6(\text{m s}^{-1})2.46 \times 10^{25} \text{ m}^{-3}$$

$$= 7.26 \times 10^9 \text{ s}^{-1}$$

$Z_{AA} = \dfrac{1}{\sqrt{2}}\,\pi d^2 \overline{u}_A N_A^2 = \dfrac{1}{2}\,N_A Z_A$

$$= 8.9 \times 10^{34} \text{ m}^{-3}\text{ s}^{-1}$$

1.33. $\lambda = \dfrac{V}{\sqrt{2}\,\pi d^2 N}$ where N = number of molecules.

Since $PV = nRT = \dfrac{N}{L}RT$

$\lambda = \dfrac{RT}{\sqrt{2}\,\pi d^2 LP}$

1.34.   Apply Eq. 1.66 using $N_{Ar} = LPV/RT$ and $\bar{u}_{Ar} = (0.92)\,3RT/LM$

$Z_{Ar} = \sqrt{2}\,\pi d_{Ar}^2\,(0.92)\,(3RT/LM)\,(LP/RT)$

$= \sqrt{2}\,\pi(3.84 \times 10^{-10})^2$

$\dfrac{(0.92)3(8.3145)298.15}{6.022\times10^{23}(39.948\times10^{-3}\,/\,6.022\times10^{23})}\times$

$\dfrac{6.022\times10^{23}\,\text{mol}^{-1}\,\times1\times10^5\,\text{Pa}}{8.3145\times298.15}$

$= 2.73 \times 10^{12}\,\text{s}^{-1}$

$Z_{Ar,Ar} = 0.5(Z_{Ar}) \times N_{Ar}/V = 0.5(2.73 \times 10^{12}) \times 2.24 \times 10^{25}$

$= 3.05 \times 10^{37}\,\text{m}^{-3}\,\text{s}^{-1}$

1.35.   Use the equation from Problem 1.33.

$\lambda = \dfrac{RT}{\sqrt{2}\pi d^2 LP} = \dfrac{8.3145 \times 293.15}{\sqrt{2}\pi(3.84\times10^{-10})^2 \times 6.022\times10^{23} \times 1\times10^5}$

$= 6.18 \times 10^{-8}\,\text{m}$

1.36.   a.   At 133.32 Pa,

$\lambda = \dfrac{(8.3145\text{J K}^{-1}\text{mol}^{-1})(298.15\text{ K})}{\sqrt{2}\pi(0.258\times10^{-9}\text{m})^2(6.022\times10^{23}\text{mol}^{-1})} \times 1/(133.32\text{ Pa})$

$= 1.044 \times 10^{-4}\,\text{m}$ since $\text{J/Pa} = \text{m}^3$

b.   At 101.325 kPa, $\lambda = 1.37 \times 10^{-7}$ m

Alternatively, λ is lowered by a ratio of

$\dfrac{101\,325}{133.32} = 760$

Thus, $1.044 \times 10^{-4}/760 = 1.37 \times 10^{-7}$ m

c.   At 1.0×10^8 Pa,

$\lambda = 1.044 \times 10^{-4}\,\text{m} \times \dfrac{133.32}{1.0\times10^8} = 1.39 \times 10^{-10}\,\text{m}$

1.37.   From Eq. 1.68

$$\lambda = \frac{V}{\sqrt{2}\pi d^2 N} = \frac{1 \text{ m}^3}{\sqrt{2}\pi (2.5 \times 10^{-10})^2 \text{ m}^2}$$

$$= 3.60 \times 10^{18} \text{ m}$$

This is about a hundred times greater than the distance between the earth and the nearest star (Proxima Centauri)!

1.38.   Consider the gamboge particles to be gas molecules in a column of air. Since the number of particles present is proportional to their pressure, we write Eq. 1.75 as

$$dN/N = -(Mg/RT)dz \quad \text{or} \quad \ln\frac{N}{N_0} = -\frac{Mg\Delta z}{RT}$$

$$RT \ln\frac{N}{N_0} = -mLg\Delta z \quad \text{since} \quad M = mL$$

The mass is $\frac{4}{3}\pi r^3 \Delta\rho$, and substitution gives RT $\ln\dfrac{N_0}{N} = \dfrac{4}{3}\pi r^3 Lg\Delta\rho\Delta z$

$$8.3145(\text{J K}^{-1} \text{ mol}^{-1})288.15(\text{K}^{-1})\ln\frac{100}{47}$$

$$= L\frac{4}{3}(3.1416)(2.12 \times 10^{-7} m)^3 \, 9.81 \text{ m s}^{-2} \times$$

$$(35 - 5) \times 10^{-6} \text{ m } (1.206 - 0.999)\text{g cm}^{-3} \times$$

$$\frac{\text{kg } 10^6 \text{ cm}^3}{10^3 \text{ g m}^3}$$

then $L = 7.44 \times 10^{23} \text{ mol}^{-1}$

1.39.   a. $\dfrac{\sqrt{\overline{u^2}}}{\overline{u}} = \sqrt{\dfrac{3\pi}{8}} = 1.085$

b. $\dfrac{u}{u_{mp}}\dfrac{2}{\sqrt{\pi}} = 1.128$

The differences between $\sqrt{\overline{u^2}}$ and $\overline{u}$, and between $\overline{u}$ and u_{mp}, increase with T and decrease with m.

1.40.   From Table 1.3, the average speed is $\overline{u} = \sqrt{\dfrac{8k_B T}{\pi m}}$

and, therefore, $T = \dfrac{\pi m \overline{u}^2}{8k_B}$

a.   The mass of the hydrogen molecule is $\dfrac{2.016 \times 10^3}{6.022 \times 10^{23}} = 3.348 \times 10^{-27}$ kg

$$T = \frac{\pi \times 3.348 \times 10^{-27} \, kg \times (1.07 \times 10^4 \, m \, s^{-1})^2}{8 \times 1.381 \times 10^{-23} \, J \, K^{-1}} = 10\,900 \, K$$

b. $\quad m_{O_2} = \dfrac{32.00 \times 10^3}{6.022 \times 10^{23}} = 5.31 \times 10^{-26} \, kg$

$$T = \frac{\pi \times 5.31 \times 10^{-26} \times (1.07 \times 10^4)^2}{8 \times 1.381 \times 10^{-23}} = 173\,000 \, K$$

1.41. a. From Eq. 1.91, the ratio for two different speeds u_1 and u_2 is

$$\frac{e^{-mu_1^2/2k_BT} \, u_1^2}{e^{-mu_2^2/2k_BT} \, u_2^2} = \left(\frac{u_1}{u_2}\right)^2 e^{-m(u_1^2 - u_2^2)/2k_BT}$$

$$u_1 = 2\bar{u} = 2\sqrt{\frac{8k_BT}{\pi m}} \; ; \quad u_2 = \bar{u} = \sqrt{\frac{8k_BT}{\pi m}}$$

$$\left(\frac{u_1}{u_2}\right)^2 = 4 \quad \text{and} \quad u_1^2 - u_2^2 = \frac{3 \times 8k_BT}{\pi m}$$

$$\frac{m\left(u_1^2 - u_2^2\right)}{2k_BT} = \frac{12}{\pi}$$

The ratio is $4 \, e^{-12/\pi} = 8.77 \times 10^{-2}$

There is no effect of mass or temperature.

b. For two temperatures, T_1 and T_2, the ratio is, from Eq. 1.91,

$$\left(\frac{T_1}{T_2}\right)^{3/2} \frac{e^{-m(\bar{u}_1)^2/k_BT_1}}{e^{-m(\bar{u}_2)^2/k_BT_2}} \left(\frac{u_1}{u_2}\right)^2$$

Insertion of the expression for $\bar{u}_1$ and $\bar{u}_2$ gives

$$\left(\frac{T_2}{T_1}\right)^{3/2} \frac{e^{-4/\pi}}{e^{-4/\pi}} \left(\frac{T_1}{T_2}\right) = \left(\frac{T_2}{T_1}\right)^{1/2}$$

With $T_1 = 100 \, °C = 373.15 \, K$ and $T_2 = 298.15 \, K$, this gives

$$(298.15/373.15)^{1/2} = 0.8939$$

There is no dependence on mass.

1.42. The average molecular speed is

$$\bar{u} = \sqrt{\frac{8k_BT}{\pi m}}$$

The ideal gas law is

$$PV = nRT = nLk_BT$$

and the density ρ is

$$\rho = \frac{nmL}{V}$$

Thus

$$\bar{u} = \sqrt{\frac{8PV}{nL\pi} \times \frac{nL}{\rho V}} = \sqrt{\frac{8P}{\pi\rho}}$$

Since P and ρ are the same in the two gases, the average speeds are the same.

1.43. a. Call 25 °C T_2 and 100 °C T_1. Then the average speed $\bar{u}_{25°}$ is

$$\bar{u}_{25°} = \sqrt{\frac{8k_B T_2}{\pi m}}$$

From Eq. 1.91, the required ratio is

$$\left(\frac{T_2}{T_1}\right)^{3/2} \frac{e^{-8k_B T_2 m / 2k_B T_1 \pi m}}{e^{-k_B m T_2 / 2k_B \pi m T_2}} \left(\frac{T_1}{T_2}\right) = \left(\frac{T_2}{T_1}\right)^{1/2} e^{-\frac{4}{\pi}\left(\frac{T_2}{T_1}-1\right)}$$

$$= \left(\frac{298.15}{373.15}\right)^{1/2} e^{0.2559} = 1.155$$

b. The ratio for a speed $10\,\bar{u}_{25°}$ is

$$= \left(\frac{T_2}{T_1}\right)^{1/2} e^{-\frac{4}{\pi}\left(\frac{T_2}{T_1}-1\right)} = \left(\frac{298.15}{373.15}\right)^{1/2} e^{2.559} = 11.55$$

1.44. The fraction is

$$\frac{dN}{N} = \frac{B e^{-mu_x^2 / 2k_B T} du_x}{B \int\limits_{0}^{\infty} e^{-mu_x^2 / 2k_B T} du_x}$$

From the appendix to this chapter, the integral is $\dfrac{1}{2}\left(\dfrac{2\pi k_B T}{m}\right)^{1/2}$.

Thus

$$\frac{dN}{N} = \left(\frac{2m}{\pi k_B T}\right)^{1/2} e^{-mu_x^2 / (2k_B T)} du_x.$$

A plot of $(dN/N)/du_x$ against u_x is symmetrical about the $(dN/N)\,du_x$ axis; the maximum is ththerefore at $u_x = 0$, which is the most probable speed.

1.45. $\varepsilon_x = \frac{1}{2} mu_x^2$

and therefore

$$\frac{d\varepsilon_x}{du_x} = (2\varepsilon_x m)^{1/2}$$

From the expression obtained in Problem 1.44,

$$\frac{dN_x}{N} = \left(\frac{2m}{\pi k_B T}\right)^{1/2} e^{-\varepsilon_x/k_B T} \frac{1}{(2\varepsilon_x m)^{1/2}} \, d\varepsilon_x$$

$$= (\pi \varepsilon_x k_B T)^{-1/2} e^{-\varepsilon_x/k_B T} \, d\varepsilon_x$$

The average one-dimensional energy is

$$\bar{\varepsilon}_x = \int_0^\infty \varepsilon_x \frac{dN}{N} = (\pi k_B T)^{1/2} \int_0^\infty \varepsilon_x^{1/2} e^{-\varepsilon_x/k_B T} \, d\varepsilon_x$$

$$= (\pi k_B T)^{1/2} \frac{1}{2} k_B T (\pi k_B T)^{-1/2} = \frac{1}{2} k_B T$$

1.46. From Eqs. 1.80 and 1.81, with $\beta = 1/k_B T$ and $u^2 = u_x^2 + u_y^2$,

$$dP_x dP_y = B^2 e^{-mu^2/2k_B T} \, du_x du_y$$

We consider a circular shell of radius u and replace $du_x du_y$ with $2\pi u \, du$. Then

$$dP = 2\pi B^2 e^{-mu^2/2k_B T} u \, du$$

and therefore

$$\frac{dN}{N} = \frac{e^{-mu^2/2k_B T} u \, du}{\int_0^\infty e^{-mu^2/2k_B T} u \, du} = \frac{e^{-mu^2/2k_B T} u \, du}{\frac{1}{2} \times \frac{2k_B T}{m}}$$

$$= \frac{m}{k_B T} e^{-mu^2/2k_B T} u \, du$$

$$\varepsilon = \frac{1}{2} mu^2 \quad \text{and} \quad \frac{d\varepsilon}{du} = (2\varepsilon m)^{1/2}$$

Thus

$$\frac{dN}{N} = \frac{m}{k_B T} e^{-\varepsilon/k_B T} \left(\frac{2\varepsilon}{m}\right)^{1/2} \frac{d\varepsilon}{(2\varepsilon m)^{1/2}}$$

$$= \frac{1}{k_B T} e^{-\varepsilon/k_B T} \, d\varepsilon$$

The fraction with energy in excess of ε^* is

$$\frac{N*}{N} = \int_{\varepsilon*}^{\infty} e^{-\varepsilon/k_{\mathrm{B}}T} d\varepsilon = -e^{-\varepsilon/k_{\mathrm{B}}T}\Big|_{\varepsilon*}^{\infty} = e^{-\varepsilon*/k_{\mathrm{B}}T}.$$

■ Real Gases

1.47. The centers of two spherical molecules in contact with each other cannot approach closer than d, the diameter of a molecule. Therefore, the excluded volume for a pair of molecules is $(4/3)\pi d^3 = (4/3)\pi(2r)^3 = 8 \times (4/3)\pi r^3 = 8V$. Therefore, the excluded volume per molecule, $b = 4V$, where V is the volume occupied by one molecule of radius r.

1.48. The curves are similar to those in Figure 1.21.

1.49. Ideal gas prediction: $P = nRT/V$

$$P = \frac{17.5\ \mathrm{g}\ /\ (2 \times 35.45\ \mathrm{g\ mol^{-1}})8.3145(\mathrm{J\ K^{-1}mol^{-1}})273.15\ \mathrm{K}}{0.8 \times 10^{-3}\ \mathrm{m}^3}$$

$$= 700.7\ \mathrm{kPa}$$

Van der Waals prediction: $P = \dfrac{nRT}{V - nb} - \dfrac{an^2}{V^2}$

$n = 0.2468$;   From Table 1.5, $a = 0.6579$ Pa m^6 mol^{-2}; $b = 0.0562 \times 10^{-3}$ m^3 mol^{-2}.

$$P = \frac{0.2468(\mathrm{mol})8.3145(\mathrm{J\ mol^{-1}K^{-1}})273.15\ \mathrm{K}}{(0.8 \times 10^{-3}\ \mathrm{m}^3) - (0.2468\ \mathrm{mol})(0.0562 \times 10^{-3}\ \mathrm{m^3 mol^{-1}})} - \frac{0.6579(0.2468)^2}{(0.8 \times 10^{-3})^2}\ \mathrm{Pa}$$

$$= (713.0 - 62.6) \times 10^3\ \mathrm{Pa} = 650\ \mathrm{kPa}.$$

1.50. $V_2 = \dfrac{P_1 V_1}{P_2} \times \dfrac{Z_2 T_2}{Z_1 T_1} = \dfrac{800\ \mathrm{bar} \times 1.00\ \mathrm{dm}^3 \times 1.10 \times 373.2\ \mathrm{K}}{200\ \mathrm{bar} \times 1.95 \times 223.2\ \mathrm{K}}$

$$= 3.77\ \mathrm{dm}^3.$$

$$V_2 = \frac{P_1 V_1}{P_2} \times \frac{T_2}{T_1} = \frac{800\ \mathrm{bar} \times 1.00\ \mathrm{dm}^3 \times 373.2\ \mathrm{K}}{200\ \mathrm{bar} \times 223.2\ \mathrm{K}}$$

$$= 6.69\ \mathrm{dm}^3.$$

This represents an error of

$$\frac{6.69 - 3.77}{3.77} \times 100 = 77.5\%.$$

We conclude that using the compression factor gives the more accurate result.

1.51. At the critical point, both the first and second derivatives of P with respect to V at constant T are zero.

$$\left(\frac{\partial P}{\partial V}\right)_T = -\frac{RT}{V - b} + \frac{a}{V^2} = 0.$$

Rearrangement gives

$$\frac{a}{V^2} = \frac{RT}{(V-b)^2}. \tag{1}$$

$$\left(\frac{\partial^2 P}{\partial V^2}\right)_T = \frac{2RT}{(V-b)^2} - \frac{2a}{V^3} = 0,$$

or $\qquad \dfrac{2a}{V^3} = \dfrac{2RT}{(V-b)^3}. \tag{2}$

Division of (2) by (1) gives

$$\frac{2RT/(V-b)^3}{RT/(V-b)^2} = \frac{2a/V^3}{a/V^2} \quad \text{or} \quad \frac{2}{V-b} = \frac{2}{V}.$$

From the last expression, $1/V = 1/(V-b)$. Therefore, $b = 0$. As a result, the two derivatives cannot vanish simultaneously unless $b = 0$. This is contrary to the statement of the problem. Therefore, the gas does not have a critical point.

1.52. Using Figure 1.23 for the compressibility factor, find the reduced temperature and reduced pressure.

$T_r = T/T_c = (273.15 + 97.2)/(308.6) = 1.20$

$P_r = P/P_c = 90.0/61.659 = 1.46$

From Figure 1.23, the value of Z for these two values is 0.7. From $Z = PV/RT$ (Because pressure in the chart is based on atmospheres, we must use 0.0821 atm dm^3 mol^{-1} K^{-1} for R.), we have:

$$V = ZRT/P = \frac{0.7(0.0821)(370.35)}{90.0} = 0.24 \text{ dm}^3 \text{ mol}^{-1}$$

From the ideal gas law:

$$V = RT/P = \frac{0.0821(370.35)}{90.0} = 0.34 \text{ dm}^3 \text{ mol}^{-1}$$

1.53 From Table 1.5, $b = 0.0428 \times 10^{-3}$ m^3 mol^{-1}. For a sphere, $V = (4/3)\pi r^3$. Substituting $b = 4V_m$, the volume per molecule is given by

$$V = \frac{4}{3}\pi r^3 = \frac{b}{4L} = \frac{0.0428 \times 10^{-3} \text{ m}^3 \text{ mol}^{-1}}{4 \times 6.022 \times 10^{23} \text{ mol}^{-1}} = 1.78 \times 10^{-29} \text{ m}^3.$$

Therefore, $r = [1.78 \times 10^{-29}/(4\pi/3)]^{1/3} = 1.62 \times 10^{-10}$ m.
(The actual "radius," i.e., the C-H bond distance, of methane molecule is 1.09×10^{-10} m.)

1.54. This equation reduces to $PV_m/RT = 1$ when $[1 - 4(T_c/T)^2 = 0$. This occurs when $T = T_B$. Therefore, solve for $T = T_B$.

$$4(T_c/T_B)^2 = 1 \quad \text{and} \quad T_B = 2T_c$$

1.55. a. Multiplying both sides of Eq. (1.101) by $V_m/(RT)$, we get

b. Using the series expansion given, we get

$$\frac{PV_m}{RT} = \frac{V_m}{V_m - b} - \frac{a}{RT}\frac{1}{V_m}.$$

$$\frac{PV_m}{RT} = \left(1 - b/V_m\right)^{-1} - \frac{a}{RT}\frac{1}{V_m} = 1 + \frac{b}{V_m} + \left(\frac{b}{V_m}\right)^2 + \dots - \frac{a}{RT}\frac{1}{V_m}$$

$$Z = \frac{PV_m}{RT} = 1 + \left(b - \frac{a}{RT}\right)\frac{1}{V_m} + b^2\frac{1}{V_m^2} + \dots$$

c. Grouping together terms containing the same power of V_m, we get

Comparing this to Eq. (1.117) with $n = 1$, we see that $B(T) = b - a/(RT)$, $C(T) = b^2$.

d. At the Boyle temperature, $B(T_B) = 0$, i.e., $b = a/(R\,T_B)$ or $T_B = a/(bR)$.

1.56. Use the van der Waals law

$$P = \frac{RT}{V - b} - \frac{a}{V^2}$$

and multiply through by V/RT to put it into the form of Z.

$$PV/RT = [V/(V - b)] - a/RT = \frac{1}{1 - (b/V)} - \frac{a}{RTV}$$

Expanding the first term on the right gives

$$Z = 1 + \frac{b}{V} + \left(\frac{b}{V}\right)^2 + \dots - \frac{a}{RTV} = 1 + \left(b - \frac{a}{RT}\right)\frac{1}{V} + \frac{b^2}{V^2} \qquad (1)$$

Boyle's temperature is the temperature at which $Z \to 1$ as P goes to zero, or

$$\lim_{P \to 0} \left(\frac{\partial Z}{\partial P}\right)_T = 0$$

Change the variable V to RT/P in Eq. 1.

$$Z = 1 + \frac{P}{RT}\left(b - \frac{a}{RT}\right) + \left(\frac{b}{RT}\right)^2 P^2$$

Differentiation gives

$$\left(\frac{\partial Z}{\partial P}\right)_T = \frac{1}{RT}\left(b - \frac{a}{RT}\right) + 2\left(\frac{b}{RT}\right)^2 P$$

As P goes to 0, the last term disappears and we have the first term on the right equal to 0. Since $b - a/RT = B(T)$, at this point $T = T_B$. Solving gives

$$T_B = \frac{a}{Rb}$$

1.57. $T_r = \dfrac{T}{T_c} = \dfrac{356}{309.65} = 1.15$

$P_r = \dfrac{P}{P_c} = \dfrac{54.0}{71.7} = 0.753$

From Figure 1.16, $Z = 0.85 = PV/RT$

$$V = \frac{0.85RT}{P} = 0.85(0.0821 \text{ atm dm}^3 \text{ K}^{-1} \text{ mol}^{-1}) \ (356 \text{ K})/54.0 \text{ atm} = 0.46 \text{ dm}^3$$

1.58. For CH_4, $P_r = 2.00/46.0 = 4.35 \times 10^{-2}$; $T_r = 500.0/190.6 = 2.623$. In order for hydrogen to have the same reduced pressure and temperature, we require its pressure and temperature to be

$$P = P_r P_c(H_2) = 4.35 \times 10^{-2} \times 13.0 \text{ bar} = 0.57 \text{ bar}$$

$$T = T_r T_c(H_2) = 2.623 \times 33.2 \text{ K} = 87.1 \text{ K.}$$

1.59. The Dieterici equation is $P = \dfrac{RT}{V_m - b} \ e^{-a/RTV_m}$

$$\left(\frac{\partial P}{\partial V_m} \right)_T = RTe^{-a/RTV_m} \left[\frac{a}{V_m^2 RT(V_m - b)} - \frac{1}{(V_m - b)^2} \right]$$

$$= P \left(\frac{a}{V_m^2 RT} - \frac{1}{V_m - b} \right)$$

$$\left(\frac{\partial^2 P}{\partial V_m^2} \right)_T = P \left\{ -\frac{2a}{V_m^3 RT} + \frac{1}{(V - b)^2} + \left[\frac{a}{V^2 RT} - \frac{1}{(V - b)} \right]^2 \right\}$$

At the critical point $(\partial P/\partial V)_T = 0$, and by substituting the critical constants for P, V, and T, we have

$$P_c \left[\frac{a}{V_c^2 RT_c} - \frac{1}{(V_c - b)} \right] = 0 \tag{1}$$

Similarly for the second derivative, which corresponds to the highest temperature, a horizontal line may exist on the curve:

$$P_c \left\{ -\frac{2a}{V_c^3 RT_c} + \frac{1}{(V_c - b)^2} + \left[\frac{a}{V_c^2 RT_c} - \frac{1}{(V_c - b)} \right]^2 \right\} = 0 \tag{2}$$

After eliminating P_c from both expressions, a is obtained from Eq. 1: $a = V_c^2 RT_c/(V_c - b)$.
Substitution into Eq. 2 yields

$$\frac{-2V_c^2 RT_c}{(V_c - b)(V_c^3 RT_c)} + \frac{1}{(V_c - b)^2} + \left[\frac{V_c^2 RT_c}{(V_c - b)V_c^2 RT_c} - \frac{1}{(V_c - b)}\right]^2 = 0$$

$$\frac{-2}{V_c} + \frac{1}{V_c - b} = 0 \qquad b = V_c/2$$

Substitution of b into Eq. 1 yields

$$a = 2V_c RT_c$$

Therefore at the critical point

$$P_c = \frac{2RT_c}{V_c} e^{-2}$$

1.60.   Starting with

$$Z = 1 + \left(b - \frac{a}{RT}\right)\frac{1}{V_m} + b^2 \frac{1}{V_m^2},$$

and using $V_m = RT/P$, we obtain

$$Z = 1 + \frac{1}{RT}\left(b - \frac{a}{RT}\right)P + \left(\frac{b}{RT}\right)^2 P^2.$$

From Table 1.5, the van der Waals constants for CCl_2F_2 are $a = 1.066$ dm^3 bar mol^{-2}, $b = 0.0973$ dm^3 mol^{-1}. Substituting above, we obtain Z = 1.01.

Therefore, $V_m = Z(RT/P) = 5.05$ dm^3. Using the ideal gas law, we get $V_m = RT/P = 4.99$ dm^3.

1.61.   Write the van der Waals equation (Eq. 1.100) in the form

$$Z = \frac{PV}{RT} = 1 + \frac{Pb}{RT} - \frac{a}{RTV} + \frac{ab}{RTV^2}$$

When the pressure is low, $\underset{P \to 0}{\lim}\frac{ab}{RTV^2} = \underset{V \to \infty}{\lim}\frac{ab}{RTV^2} = 0$.

In the term $\frac{a}{RTV}$, V may be replaced by $\frac{RT}{P}$, and we have

$$Z = \frac{PV}{RT} = 1 + \frac{Pb}{RT} - \frac{a}{RT}\left(\frac{P}{RT}\right) = 1 + \frac{P}{RT}\left(b - \frac{a}{RT}\right)$$

As P goes to zero, $Z = \frac{PV}{RT} = 1$

1.62. For a: 1 atm = 101 325 Pa, $(1 \text{ dm})^6 = (0.1 \text{ m})^6 = 1 \times 10^{-6}$ m

$$a = 5.49 \text{ atm dm}^6 \text{ mol}^{-2} \times \frac{101\ 325 \text{ Pa}}{\text{atm}} \times \frac{1 \times 10^{-6} \text{ m}}{\text{dm}^3} = 0.5563 \text{ Pa m}^6 \text{ mol}^{-2}$$

For b: $(1 \text{ dm})^3 = (0.1 \text{ m})^3 = 10^{-3}$ m^3

$$b = 0.0638 \text{ dm}^3 \text{ mol}^{-1} \times \frac{10^{-3} \text{ m}^3}{\text{dm}^3} = 0.0638 \times 10^{-3} \text{ m}^3 \text{ mol}^{-1}$$

1.63.   Ideal gas equation:

$$P = \frac{nRT}{V} = \frac{3 \text{ mol} \times 8.3145 \text{ J K}^{-1} \text{ mol}^{-1} \times 298.15 \text{ K}}{0.008\ 25 \text{ m}^3} = 901\ 400 \text{ Pa} \times \frac{1 \text{ bar}}{100\ 000 \text{ Pa}} = 9.01 \text{ bar}$$

Van der Waals equation:

$$P = \frac{RT}{V_m - b} - \frac{a}{V_m^2}$$

$$= \frac{8.3145 (\text{J K}^{-1} \text{mol}^{-1}) 298.15 (\text{K})}{2.75 \times 10^{-3} \text{m}^3 \text{mol}^{-1} 4.27 \times 10^{-5} \text{m}^3 \text{mol}^{-1}} - \frac{0.3640 \text{ Pa m}^6 \text{ mol}^{-2}}{(2.75 \times 10^{-3} \text{m}^3 \text{ mol}^{-1})^2}$$

$$= 867\ 473 \text{ Pa} \times \frac{1 \text{ bar}}{100\ 000 \text{ Pa}} = 8.67 \text{ bar}$$

Dieterici equation:

$$P = \frac{RT \exp(-a/V_m RT)}{V_m - b}$$

$$P = \frac{8.3145 \text{ J K}^{-1} \text{mol}^{-1} \times 298.15 \text{ K}}{2.75 \times 10^{-3} \text{m}^3 \text{mol}^{-1} - 4.63 \times 10^{-5} \text{m}^3 \text{mol}^{-1}} \times$$

$$\exp(-0.462/2.75 \times 10^{-3} \times 8.3145 \times 298.15)$$

$$= 856\ 800 \text{ Pa} \times \frac{1 \text{ bar}}{100\ 000 \text{ Pa}} = 8.57 \text{ bar}$$

Beattie-Bridgeman equation:

$$P = \frac{RT}{V_m^2} \left(1 - \frac{c}{V_m T^3} \right) \left(V_m + B_0 - \frac{bB_0}{V_m} \right) - \frac{A_0}{V_m^2} \left(1 - \frac{a}{V_m} \right)$$

$$P = \frac{8.3145(298.15)}{(0.002\ 75)^2} \left(1 - \frac{660}{(298.15)^3 (0.002\ 75)} \right) \times$$

$$\left[0.002\ 75 + 104.76 \times 10^{-6} - \frac{72.35 \times 10^{-6} (104.76 \times 10^{-6})}{0.002\ 75} \right] - \frac{0.507\ 28}{(0.002\ 75)^2}) \times$$

$$\left(1 - \frac{71.32 \times 10^{-6}}{0.002\ 75} \right)$$

$$861\ 027\ \text{Pa} \times \frac{1\ \text{bar}}{100\ 000\ \text{Pa}} = 8.61\ \text{bar}$$

All the calculated values for P are approximately the same under these conditions.

1.64.   From Eq. 1.109, $b = \dfrac{V_c}{3}$ and $V_c = \dfrac{3RT_c}{8P_c}$. So

$$b = \frac{RT_c}{8P_c}$$

$$b = \frac{8.3145(\text{J K}^{-1}\text{mol}^{-1})473\ \text{K}}{8(30\ \text{atm} \times 101\ 325\text{Pa atm}^{-1})}$$

$$= 0.162 \times 10^{-3}\ \text{m}^3\ \text{mol}^{-1}$$

1.65.   The Dieterici equation can be expanded in terms of the infinite series
$e^x = 1 + x + x^2/2! + x^3/3! + \dots$

This gives

$$P = \frac{RT}{(V_m - b)} - \frac{a}{V_m(V_m - b)} + \frac{a^2}{2RTV_m^2(V_m - b)} - \dots$$

Expand $\dfrac{1}{V_m - b}$ and collect terms in powers of V_m to give coefficients that are independent of V_m.

$$\frac{1}{V_m - b} = \frac{1}{V_m}\left(\frac{1}{1 - b/V_m}\right) = \frac{1}{V_m}\left(1 - \frac{b}{V_m} + \frac{b^2}{V_m^2} - \frac{b^3}{V_m^3} + \dots\right)$$

Substitution leads to

$$P = \frac{RT}{V_m} - \frac{a + RTb}{V_m^2} + \frac{1}{V_m^3}\left(\frac{a^2}{2RT} + ab + RTb^2\right) - \dots$$

The second coefficient is $-(a + RTb)$

And the third coefficient is $\left(\dfrac{a^2}{2RT} + ab + RTb^2\right)$.

At low densities the third and higher terms are negligible. Dropping the third and higher terms, and substituting, we obtain

$$P = \frac{RT}{V_m} - \frac{a + RTb}{V_m^2}$$

This is in the same form as the van der Waals equation.

■ Essay Questions

1.68. Starting with a fixed quantity of gas at 150 Torr in the thermometer, the gas is thermally equilibrated with the object whose temperature is being measured while the pressure is maintained at 150 Torr and the volume of the gas is measured. Then the gas thermometer must reach thermal equilibrium with a water triple-point at 273.16 K, keeping the pressure constant at 150 Torr. The volume is read and the ratio is taken. The procedure is repeated, successively lowering the pressure and obtaining the new ratios to $(PV)_T/(PV)_{\text{triple point}}$. The volume ratio is plotted against pressure, and the straight-line data are extrapolated to 0 pressure. The intersection of the line at $V/V_{\text{triple point}}$ axis multiplied by 273.16 K gives the temperature.

2. THE FIRST LAW OF THERMODYNAMICS

■ Energy, Heat, and Work

2.1. Weight of bird $= 1.5 \times 9.81$ kg m s^{-2}

Work required to raise it 75 m $= 1.5 \times 9.81 \times 75$ kg m^2 s$^{-2} = 1103.6$ J = potential energy

Kinetic energy $= \frac{1}{2} mv^2$

$$= \frac{1}{2} \times 1.5 \times 20^2 = 300 \text{ J}$$

Total change of energy $= 1103.6 + 300 = 1403.6$ J

$$= 1.40 \text{ kJ}$$

2.2. 1 mol of ice has a volume of 18.01 g/0.9168 g cm^{-3}

$$= 19.64 \text{ cm}^3$$

1 mol of water has a volume of 18.01 g/0.9998 g cm^{-3}

$$= 18.01 \text{ cm}^3$$

ΔV(ice $\rightarrow$ water) $= -1.63$ cm^3 mol^{-1}

$\Delta(PV) = -0.001\ 63$ atm dm$^3 = -0.165$ J mol^{-1}

$\Delta H = 6025$ J $= \Delta U + \Delta (PV) = \Delta U - (0.165$ J mol$^{-1})$

$\Delta U = 6025$ J mol$^{-1} = 6.025$ kJ mol^{-1}

(Difference between ΔH and ΔU is only 0.165 J mol^{-1}.)

Work done on the system $= 0.165$ J mol^{-1}

2.3. 1 mol of water at 100 °C has a volume of

$$\frac{18.01 \text{ g}}{0.9584 \text{ g cm}^{-3}} = 18.79 \text{ cm}^3$$

Volume of steam $= \dfrac{18.01 \text{ g}}{0.000\ 596 \text{ g cm}^{-3}} = 30\ 218.1 \text{ cm}^3$

Volume increase $= 30\ 199$ cm^3 mol$^{-1} = 30.20$ dm^3 mol^{-1}

$$\Delta(PV) = 30.20 \text{ atm dm}^3 \text{ mol}^{-1} = 3059.3 \text{ J mol}^{-1}$$

$$= 3.06 \text{ kJ mol}^{-1}$$

$$\Delta H = 4063 \text{ J mol}^{-1} = \Delta U + (3059.3 \text{ J mol}^{-1})$$

$$\Delta U = 37.6 \text{ kJ mol}^{-1}$$

Work done by the system $= -w = 3.06 \text{ kJ}$

2.4. Heat the water from -10 °C to 0 °C: (1)

$$q_1 = C_P \, dT = C_P(T_2 - T_1) = 753 \text{ J mol}^{-1}$$

Freeze the water at 0 °C: (2)

$$q_2 = 6025 \text{ J mol}^{-1}$$

Cool the ice from 0 °C to -10 °C: (3)

$$q_3 = -377 \text{ J mol}^{-1}$$

$$\text{Net } q = 753 - 6025 - 377 = -5649 \text{ J mol}^{-1}$$

$$= \Delta H = -5.65 \text{ kJ mol}^{-1}$$

2.5. Heat evolved $= 1.69 \times 6937 = 11\,724 \text{ J}$

Molar mass of $CH_3COCH_3 = 3 \times 12.01 + 6 \times 1.008 + 16.00 = 58.08 \text{ g mol}^{-1}$

Heat evolved in the combustion of 1 mol

$$= \frac{11\,724 \times 58.08}{0.700} \text{ J} = 972.7 \text{ kJ}$$

a. $\Delta U = -972.7 \text{ kJ mol}^{-1}$

b. $CH_3COCH_3(l) + 4O_2(g) \rightarrow 3CO_2(g) + 3H_2O(l)$

$\Delta n(\text{gases}) = -1$

$$\Delta H = -972\,700 - 8.3145 \times 299.8 \text{ J mol}^{-1}$$

$$= -975.2 \text{ kJ mol}^{-1}$$

2.6. a. Heat capacity of man $= 292\,600 \text{ J K}^{-1}$

Temperature rise $= \dfrac{10\,460\,000 \text{ J}}{292\,600 \text{ J K}^{-1}} = 35.7 \text{ °C}$

Final temperature $= 37 + 35.7 = 72.7 \text{ °C}$

b. $\Delta H = 43\,400 \text{ J mol}^{-1}/18.0 \text{ g mol}^{-1} = 2411 \text{ J g}^{-1}$

Mass of water required $= \dfrac{10\,460\,000 \text{ J}}{2411 \text{ J g}^{-1}} = 4340 \text{ g} = 4.34 \text{ kg}.$

2.7.   The reaction is

$$Zn + H_2SO_4 \rightarrow ZnSO_4 + H_2(g)$$

Thus, 1 mol of gas is liberated by each mole of Zn, i.e., by 65.37 g. One hundred grams therefore liberates $(100/65.37)$mol $= 1.53$ mol of H_2. The work done by the system is $-P\Delta V$:

$$-w = P\Delta V = n_{H_2}RT$$

$$= 1.53 \text{ mol} \times 8.3145 \text{ J K}^{-1} \text{ mol}^{-1} \times 298.15 \text{ K}$$

$$= 3790 \text{ J} = 3.79 \text{ kJ}$$

The work in a sealed vessel ($\Delta V = 0$) is zero.

2.8.   a.   Since $P \propto$ diameter, D, we write $P_1 = k\,D_1$ where $k = \dfrac{1 \text{ bar}}{0.50 \text{ m}}$. At 4 bar,

$$D_2 = \frac{P_2}{k} = 5 \text{ bar} \times \frac{0.5 \text{ m}}{1 \text{ bar}} = 2.5 \text{ m}$$

From geometry,

$$V = \frac{4}{3}\,\pi\,(D/2)^3 = \frac{1}{6}\,\pi D^3$$

and from the calculus

$$dV = \frac{3}{6}\,\pi D^2\,dD = \frac{1}{2}\,\pi D^2\,dD$$

        b.   Substituting work into the equation for PV gives

$$-w_{rev} = \int_{D_1}^{D_2} P\,dV = \int_{D_1}^{D_2} k\,D \times \frac{1}{2}\,\pi D^2\,dD$$

$$= \frac{1}{2}\,\pi k \int_{D_1}^{D_2} D^3\,dD = \frac{1}{8}\,\pi k(D_2{}^4 - D_1{}^4)$$

$$= \frac{1}{8}\,(3.141\,59)\,(1 \text{ bar}/0.50 \text{ m})\,(2.50^4 - 0.50^4)\ \text{m}^4$$

$$= 0.7854 \text{ bar}\,(39.06 - 0.06)\ \text{m}^3 = 30.6 \text{ bar m}^3 = 30.6 \text{ kJ}$$

2.9.   One gram of water is 1 g/18.02 g mol^{-1} = 0.0555 mol.

One calorie is 4.184 J. The heat capacity is thus

$$\frac{1 \text{ cal}}{1 \text{ g} \times 1 \text{ K}} = \frac{4.184 \text{ J}}{0.0555 \text{ mol} \times 1 \text{ K}} = 75.4 \text{ J K}^{-1} \text{ mol}^{-1}$$

2.10.   One kilogram of water is 55.5 mol. The temperature range is 75 K, and the heat required therefore

$$= 55.5 \text{ mol} \times 75 \text{ K} \times 75.4 \text{ J K}^{-1} \text{ mol}^{-1}$$

$$= 313\,900 \text{ J} = 314 \text{ kJ}$$

A 1-kW heater supplies this energy in 314 s [J = W s].

2.11. The apparent mass of the metal decreases by the mass of liquid displaced.

M_2 (mass of liquid displaced) $= V \, m^3 (d \, kg \, m^{-3})$

The apparent mass of metal $= (M - V d) \, kg$

The work done $= [(M - V d) \, kg] \, (g \, m \, s^{-2}) \times (h \, m)$

$$= (M - V d) \, g \, h \, kg \, m^2 \, s^{-2}$$

$$= (M - V d) \, g \, h \, J$$

where g is the gravitational constant.

The change in potential energy $= (M - Vd) \, g \, h \, J$.

2.12. For one mole of an ideal gas, $PV = RT$. Then, from Appendix C, the total derivative of P is a function of T and V. Thus, $P = RT/V$, and

$$dP = -\frac{RT}{V^2} \, dV + \frac{R}{V} \, dT.$$

Applying Euler's theorem gives

$$\frac{\partial}{\partial T}\left(-\frac{RT}{V^2}\right) = -\frac{R}{V^2} = \frac{\partial}{\partial V}\left(\frac{R}{V}\right).$$

The mixed partial derivatives are equal. Therefore, from Euler's test for exactness, P is exact and is a function of the state of the system.

2.13. From Appendix C and the Euler criteria for exactness,

$$\frac{\partial(xy^2)}{\partial y} = 2xy \text{ and } \frac{\partial(x^2y)}{\partial x} = 2xy.$$

Since these are equal, the differential is exact. The integral is

$$\int\limits_{}^{x,y} dU = \int_A xy^2 dx + \int_B x^2 y \, dy,$$

where A and B are two segments to the final position (x,y). The idea is to use a path that will simplify the integration. Thus, if we choose 0 to x on the A segment, $y = 0$, so that the first integral is zero. In the second segment, x has a specific value and y varies from 0 to y. A sketch of this path is shown below.

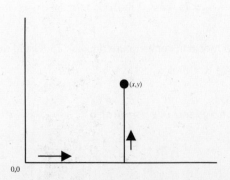

The integral is

$$\int dU = x^2 \int_0^y y\,dy = \frac{1}{2}x^2y^2.$$

$$U = \frac{1}{2}x^2y^2 + \text{constant}.$$

■ Thermochemistry

2.14. The enthalpy for this reaction at 298 K is the enthalpy of formation of two moles of water, i.e.,

$$\Delta H°(298\ \text{K}) = 2\ \text{mol} \times (-241.826\ \text{kJ mol}^{-1}) = -483.652\ \text{kJ}.$$

At 800 K, using Eq. (2.52), we obtain

$$\Delta H°(800\ \text{K})/\text{J} = \Delta H°(298\ \text{K}) + \Delta d(800{-}298) + \frac{1}{2}\Delta e(800^2{-}298^2) - \Delta f(800^{-1}{-}298^{-1}),$$

where $\Delta d = 2d_{H_2O} - (2d_{H_2} + d_{O_2})$, and Δe and Δf are defined similarly.

$$\Delta d/(\text{J K}^{-1}) = 2 \times 30.54 - (2 \times 27.28 + 29.96) = -23.44$$

$$\Delta e/(\text{J K}^{-2}) = 2 \times 10.29 \times 10^{-3} - (2 \times 3.26 \times 10^{-3} + 4.18 \times 10^{-3}) = 9.88 \times 10^{-3}$$

$$\Delta f/(\text{J K}) = 2 \times 0 - (2 \times 5.0 \times 10^4 - 1.67 \times 10^5) = 6.7 \times 10^4.$$

Therefore,

$$\Delta H°(800\ \text{K})/\text{J} = -483652 - 23.44(800{-}298) + \frac{1}{2} \times 9.88 \times 10^{-3} \times (800^2{-}298^2)$$

$$- 6.7 \times 10^4(800^{-1}{-}298^{-1}) = -4.9254 \times 10^5 \text{ or } -492.55\ \text{kJ}.$$

2.15. Molar mass of benzene $\quad = 6 \times 12.01 + 6 \times 1.008$

$$= 78.11\ \text{g mol}^{-1}$$

Heat evolved in the combustion of 1 mol

$$= \frac{26.54\ \text{kJ} \times 78.11\ \text{g mol}^{-1}}{0.633\ \text{g}} = 3274.9\ \text{kJ}$$

a. $\Delta U = -3274.9\ \text{kJ mol}^{-1}$

b. $C_6H_6(l) + \frac{15}{2}O_2(g) \rightarrow 6CO_2(g) + 3H_2O(l)$

$\Delta v(\text{gases}) = -1.5$

$$\Delta H = -3\ 274\ 900 - 1.5 \times 8.3145 \times 298.15\ \text{J mol}^{-1}$$

$$= -3278.6\ \text{kJ mol}^{-1}$$

2.16. $\Delta H° = \Delta H°(C_2H_6,g) + \Delta H°(H_2,g) - 2[\Delta H°(CH_4,g)]$

$$\Delta H° = -84.0 - 2(-74.6) = 65.2\ \text{kJ mol}^{-1}$$

2.17. Molar mass of CH_3OH $= 12.01 + 4 \times 1.008 + 16.00$

$= 32.04$ g mol^{-1}

Amount of methanol $= 5.27$ g$/32.04$ g mol$^{-1} = 0.164$ mol

Heat evolved $= \dfrac{119.50 \text{ kJ} \times 32.04 \text{ g mol}^{-1}}{5.27 \text{ g}} = 726.5$ kJ mol$^{-1} = -\Delta_c U°$

a. $\Delta_c U° = -726.5$ kJ mol^{-1}

$CH_3OH(l) + \frac{3}{2}O_2(g) \rightarrow CO_2(g) + 2H_2O(l)$

$\Delta v(\text{gases}) = -0.5$

$\Delta_c H° = -726\ 500 - 0.5 \times 8.3145 \times 298.15$ J mol^{-1}

$= -727.7$ kJ mol^{-1}

b. $CH_3OH(l) + \frac{3}{2}O_2(g) \rightarrow CO_2(g) + 2H_2O(l)$　　(1)

$\Delta H° = -727.7$ kJ mol^{-1}

$H_2(g) + \frac{1}{2}O_2(g) \rightarrow H_2O(l)$　　(2)

$\Delta_f H° = -285.83$ kJ mol^{-1}

$C(s) + O_2(g) \rightarrow CO_2(g)$　　(3)

$\Delta_f H° = -393.51$ kJ mol^{-1}

$2 \times (2) + (3) - (1)$ gives

$C(s) + 2H_2(g) + \frac{1}{2}O_2(g) \rightarrow CH_3OH(l)$　　(4)

$\Delta_f H° = 2(-285.83) - 393.51 + 727.7$

$= -237.5$ kJ mol^{-1}

This value is slightly different from the value listed in Appendix D. Experimental data is constantly being evaluated causing changes in listed values. This problem may have used older data.

c. $CH_3OH(l) \rightarrow CH_3OH(g)$ $\Delta_v H° = 35.27$ kJ mol^{-1}　　(5)

$(4) + (5)$ gives

$C(s) + 2H_2(g) + \frac{1}{2}O_2(g) \rightarrow CH_3OH(g)$

$\Delta_f H° = -202.2$ kJ mol^{-1}

2.18. $2C(\text{graphite}) + 3H_2(g) \rightarrow C_2H_6(g)$　　(1)

$\Delta_f H° = -84.0$ kJ mol^{-1}

$C(\text{graphite}) + O_2(g) \rightarrow CO_2(g)$　　(2)

$$\Delta_f H° = -393.51 \text{ kJ mol}^{-1}$$

$$H_2(g) + \frac{1}{2}O_2(g) \rightarrow H_2O(l) \tag{3}$$

$$\Delta_f H° = -285.83 \text{ kJ mol}^{-1}$$

$3 \times (3) + 2 \times (2) - (1)$ gives

$$C_2H_6(g) + \frac{7}{2}O_2(g) \rightarrow 2CO_2(g) + 3H_2O(l)$$

$$\Delta_c H° = -1560.0 \text{ kJ mol}^{-1}$$

2.19.   First, perform a multiple regression on $z = a + bx + cy$ using the definitions $z = C_{P,m}$; $x = T$ and $y = 1/T^2$. The result is $z = 1.7267 + 9.3424 \times 10^{-2}x - 871.4y$. In other words, we find that

$d = 1.7267$ J K^{-1} mol^{-1}; $e = 9.3424 \times 10^{-2}$ J K^{-2} mol^{-1}; $f = -8.714 \times 10^2$ J K mol^{-1}.

Below, we present two plots of this function, one in the range $15 \leq T \leq 275$, and another, in the range $10 \leq T \leq 25$. It can be seen that the function becomes negative at $T \leq 16.1$ K. A negative heat capacity is obviously unphysical. This is an indication that the temperature dependence of heat capacities of solids at low temperature cannot be expressed using the model used here (see Chapter 16, Section 5).

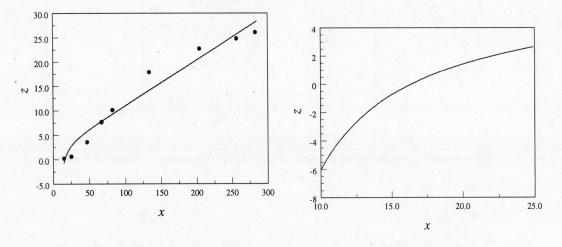

2.20.   Measure the heats of combustion of graphite, CO(g), and of CO$_2$(g) and use Hess's law.

2.21. $$C_3H_6(g) + \frac{9}{2}O_2(g) \rightarrow 3CO_2(g) + 3H_2O(l) \tag{1}$$

$$\Delta_c H° = -2091.2 \text{ kJ mol}^{-1}$$

$$C(\text{graphite}) + O_2(g) \rightarrow CO_2(g) \tag{2}$$

$$\Delta_f H° = -393.51 \text{ kJ mol}^{-1}$$

$$H_2(g) + \frac{1}{2}O_2(g) \rightarrow H_2O(l) \tag{3}$$

$$\Delta_f H° = -285.83 \text{ kJ mol}^{-1}$$

$3 \times (2) + 3 \times (3) - (1)$ gives

$$3C(\text{graphite}) + 3H_2(g) \rightarrow C_3H_6(g)$$

$$\Delta_f H° = 53.2 \text{ kJ mol}^{-1}$$

2.22. Perform a multiple regression on $z = a + bx + cy$ using the definitions $z = C_{P,m}$; $x = T$ and $y = 1/T^2$. The result is $z = 0.80053 + 1.303 \times 10^{-3}x - 21991.0y$. In other words, we find that

$$d = 0.801 \text{ J K}^{-1} \text{ mol}^{-1}; \ e = 1.303 \times 10^{-3} \text{ J K}^{-2} \text{ mol}^{-1}; \ f = -2.199 \times 10^4 \text{ J K mol}^{-1}.$$

A plot of the fit is shown below.

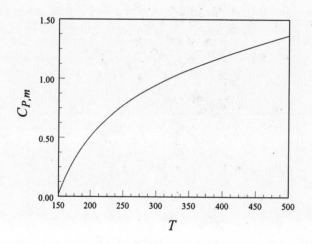

2.23. $\Delta H° = -277.6 - 52.4 + 285.83$

$\quad\quad = -44.2 \text{ kJ mol}^{-1}$

2.24. $C_2H_5OH + 3O_2 \rightarrow 2CO_2 + 3H_2O \tag{1}$

$$\Delta_c H° = -1370.7 \text{ kJ mol}^{-1}$$

$CH_3CHO + \frac{5}{2}O_2 \rightarrow 2CO_2 + 2H_2O \tag{2}$

$$\Delta_c H° = -1167.3 \text{ kJ mol}^{-1}$$

$CH_3COOH + 2O_2 \rightarrow 2CO_2 + 2H_2O \tag{3}$

$$\Delta_c H° = -876.1 \text{ kJ mol}^{-1}$$

Reaction (a) is (1) – (2); $\Delta H°$ $= -1370.7 + 1167.3$

$= -203.4$ kJ mol^{-1}

Reaction (b) is (2) – (3); $\Delta H°$ $= -1167.3 + 876.1$

$= -291.2$ kJ mol^{-1}

2.25.   Let us list all the reactions involved:

1. $CH_2CHCN + O_2(g) \rightarrow 3\ CO_2(g) + \frac{3}{2} H_2O(g) + \frac{1}{2} N_2(g);$ $\Delta_c H° = -1760.90$ kJ mol^{-1}
2. $C(graphite) + O_2(g) \rightarrow CO_2(g);$ $\Delta_f H° = -393.50$ kJ mol^{-1}
3. $H_2(g) + \frac{1}{2} O_2(g) \rightarrow H_2O(g);$ $\Delta_f H° = -241.83$ kJ mol^{-1}
4. $2C(graphite) + H_2(g) \rightarrow C_2H_2(g);$ $\Delta_f H° = 226.73$ kJ mol^{-1}
5. $C(graphite) + \frac{1}{2} H_2(g) + \frac{1}{2} N_2(g) \rightarrow HCN(g);$ $\Delta_f H° = 135.10$ kJ mol^{-1}

The desired reaction is $HCN(g) + C_2H_2(g) \rightarrow CH_2CHCN$. To generate this equation from the five given above, we need to manipulate them to get 3(Eq. 2) + (3/2)(Eq. 3) – (Eq. 1) – (Eq. 4) – (Eq. 5). Performing the same manipulations with the enthalpies, we obtain

$$\Delta H° = 3(-393.50) + (3/2)(-241.83) - (-1760.90) - 226.73 - 135.10 = -144.2 \text{ kJ mol}^{-1}.$$

2.26.   From Appendix D

$$2C + 3H_2 + \frac{1}{2} O_2 \rightarrow C_2H_5OH \tag{1}$$

$\Delta_f H° = -277.6$ kJ mol^{-1}

$$2C + 2H_2 + O_2 \rightarrow CH_3COOH \tag{2}$$

$\Delta_f H° = -484.3$ kJ mol^{-1}

$$H_2 + \frac{1}{2} O_2 \rightarrow H_2O \tag{3}$$

$\Delta_f H° = -285.83$ kJ mol^{-1}

Reaction is (2) + (3) – (1); $\Delta H° = -492.5$ kJ mol^{-1}

This differs slightly from the result of Problem 2.24, namely

$-203.4 - 291.2 = -494.6$ kJ mol^{-1}

2.27. $\Delta H°$ $= 2(-1263.1) - (-2238.3) - (-285.83)$

$= -2.1$ kJ mol^{-1}

2.28. -131.4 kJ mol$^{-1} = \Delta_f H°$ (succinate) $- (-777.4$ kJ mol$^{-1})$

$\Delta_f H° = -908.8$ kJ mol^{-1}

2.29.   α–D–glucose(aq) → β–D–glucose(aq)

$\Delta H° = -1.16$ kJ mol^{-1}

α–D–glucose(s) $\rightarrow$ α–D–glucose(aq)

$\quad \Delta H^\circ = 10.72$ kJ mol^{-1}

β–D–glucose(aq) $\rightarrow$ β–D–glucose(s)

$\quad \Delta H^\circ = -4.68$ kJ mol^{-1}

Adding:

α–D–glucose(s) $\rightarrow$ β–D–glucose(s)

$\quad \Delta H^\circ = 4.88$ kJ mol^{-1}

2.30. The equation for the reaction is

$$H_2NCONH_2 + H_2O \rightarrow CO_2 + 2NH_3$$

$$\Delta H^\circ = -413.26 + (2 \times -80.29) + 333.1 + 285.83$$

$$= 45.09 \text{ kJ mol}^{-1}$$

2.31. From Appendix D, $\Delta_f H^\circ$ for ethanol and acetic acid are –277.6 and –484.3 kJ mol^{-1}, respectively. Therefore, the enthalpy change for the reaction $C_2H_5OH(l) + O_2(g) \rightarrow CH_3COOH(l) + H_2O$ is –484.3 + (-285.83) (–277.6) = –494.53 kJ mol^{-1}. Since ethanol is fed in at the rate of 45.00 kg h^{-1}, and only 42 mole % of ethanol is converted, the actual heat released in the reaction per hour is (MW of ethanol = 46.069 g mol^{-1})

$$\frac{45.00 \text{ kg h}^{-1}}{0.046069 \text{ kg mol}^{-1}} \times \left(-492.53 \text{ kJ mol}^{-1} \right) \times 0.42 = -20\,2060 \text{ kJ h}^{-1}$$

Therefore, heat will have to be removed at the rate of 202 MJ h^{-1}.

2.32. a. Assume that all the ice melts. The process would absorb

$$100 \text{ g} \times 6.025 \text{ kJ mol}^{-1}/18.02 \text{ g mol}^{-1} = 33.44 \text{ kJ}$$

Suppose that the final temperature is t °C; then

$$33\,440 \text{ J} = 75.3 \text{ J K}^{-1} \text{ mol}^{-1} \times 1000 \text{ g} (20 - t) \text{ K}/18.02 \text{ g mol}^{-1}$$

$$20 - t = 8.00$$

$$t = 12 \text{ °C}$$

Since this is not below 0 °C, it follows that all of the ice in fact melts. The final temperature is 12 °C.

b. It is obvious that now not all of the ice will melt. (If we assumed it all melted, we would find the final temperature to be below 0 °C.) The final temperature of the water is now 0 °C, and if we suppose that x g of the ice melts, the heat balance equation is

$$x \text{ g} \times 6025 \text{ J mol}^{-1} = 75.3 \text{ J K}^{-1} \text{ mol}^{-1} \times 1000 \text{ g} \times 20 \text{ K}$$

$$x = 250 \text{ g}, 750 \text{ g remains}$$

2.33. From Appendix D, $\Delta_f H°$ at 25°C = –393.51 kJ mol^{-1}.

From the values in Table 2.1, we obtain

Δd = (44.22 – 29.96 – 16.86) = –2.60 J K^{-1} mol^{-1}

Δe = (8.79 – 4.77 – 4.18) × 10^{-3}

     = –0.16 × 10^{-3} J K^{-2} mol^{-1}

Δf = –(8.62 – 8.54 – 1.67) × 10^{5}

     = 1.59 × 10^{5} J K mol^{-1}

Then, from Eq. 2.52,

$$\Delta H_{1000 \text{ K}}/\text{J mol}^{-1} = -393\,510 - 2.60\,(1000 - 298) + \frac{1}{2}\,(-0.16 \times 10^{-3})(1000^2 - 298^2) - $$

$$1.59 \times 10^5 \left(\frac{1}{1000} - \frac{1}{298} \right)$$

$$= -393\,510 - 1825 - 73 + 375$$

$$= -395\,033$$

$$\Delta H_{1000 \text{ K}} = -395.03 \text{ kJ mol}^{-1}$$

2.34. In propane there are 2 C—C bonds and 8 C—H bonds; heat of atomization is thus

(2 × 348) + (8 × 413) = 4000 kJ mol^{-1}

Then

3C(g) + 8H(g) → C$_3$H$_8$(g)	$\Delta H°$ = –4000 kJ mol^{-1}
3C(graphite) → 3C(g)	$\Delta H°$ = 2150.1 kJ mol^{-1}
4H$_2$(g) → 8H(g)	$\Delta H°$ = 1744 kJ mol^{-1}

Adding

3C(graphite) + 4H$_2$(g) → C$_3$H$_8$(g) $\Delta H°$ = –105.9 kJ mol^{-1}

The agreement with experiment is relatively good, –103.8 kJ mol^{-1}.

2.35. a. Since a bomb calorimeter is a constant-volume instrument, the heat evolved is ΔU. The heat evolved per gram is thus

2186.0 J/0.1328 g = 16 461 J g^{-1}

The molecular weight of sucrose is 342.30 g mol^{-1}, and therefore

$\Delta_c U_m$ = –16 461 J g^{-1} × 342.3 g mol^{-1} = –5 634 600 J mol^{-1}

     = –5635 kJ mol^{-1}

The equation for the combustion reaction is

C$_{12}$H$_{22}$O$_{11}$(s) + 12O$_2$(g) → 12CO$_2$(g) + 11H$_2$O(l)

The change Δn is therefore zero, and by Eq. 2.41, $\Delta H = \Delta U$; thus

$$\Delta_c H_m = -5635 \text{ kJ mol}^{-1}$$

b. From the equation, with the use of Hess's law, it can be deduced that

$$\Delta_f H_m = 12\Delta_f H_m(CO_2, g) + 11\Delta_f H_m(H_2O, l) - \Delta_c H_m(\text{sucrose})$$

$$= (12 \times -393.51) + (11 \times -285.83) + 5635 \text{ kJ mol}^{-1}$$

$$= -4722.12 - 3144.13 + 5635 = -2231 \text{ kJ mol}^{-1}$$

2.36. $\Delta n = 1CO_2 - 1CO - \frac{1}{2}O_2 = -\frac{1}{2}$.

$$\Delta H = \Delta U + \Delta n RT$$

$$\Delta U = \Delta H - \Delta n RT$$

$$= -282.98 \text{ kJ mol}^{-1} + \frac{1}{2}\text{mol} \times 8.3145 \text{ J mol}^{-1} \text{ K}^{-1} \times 10^{-3} \text{ kJ J}^{-1} \times 298.15 \text{ K}$$

$$= -280.00 \text{ kJ mol}^{-1}$$

■ Ideal Gases

2.37. Note that we need to find the intersection of the isotherm that passes through the initial state and the adiabat that passes through the final state. Let this point be (P_0, V_0) at the temperature of the isotherm T_i. For adiabatic processes, (Eq. 2.90), $T_f / T_i = (V_f / V_0)^{\gamma - 1}$, where

$$\gamma = \left(\frac{3}{2}R + R\right)\bigg/\frac{3}{2}R = 5/3. \quad \text{Therefore, } \gamma - 1 = 2/3.$$

The final volume is

$$V_f = RT_f/P_f = (0.08314 \text{ dm}^3 \text{ bar K}^{-1} \text{ mol}^{-1} \times 253.2 \text{ K})/2.0 \text{ bar} = 10.526 \text{ dm}^3.$$

Therefore, $V_0 = V_f (T_f/T_i)^{3/2} = 10.526 \times (253.2/298.0)^{3/2} = 8.244 \text{ dm}^3$.

2.38 a. 1 mol in 22.7 dm^3 at 273 K exerts a pressure of 1 bar; $\therefore$ 2 mol in 11.35 dm^3 exert a pressure of 4 bar.

b. $PV = 4 \times 11.35 = 45.40 \text{ bar dm}^3$

1 bar dm^3 = 100 J

45.40 bar dm^3 = 4.540 kJ

c. $C_{V,m} = C_{P,m} - R = 29.4 - 8.3 = 21.1 \text{ J K}^{-1} \text{ mol}^{-1}$

2.39. a. Zero

b. $2 \times 21.1 \times 100 = 4220 \text{ J} = 4.22 \text{ kJ}$

c. $4220 \text{ J} = 4.22 \text{ kJ}$

d. $P = 4 \times 373/273 = 5.47 \text{ bar} = 547 \text{ kPa}$

e. $P_2V_2 = 5.47 \times 11.35 = 62.08$ bar dm^3

$\qquad\qquad\qquad = 6208$ J $= 6.208$ kJ

f. $\Delta H = 4220 + 6208 - 4540 = 5888$ J $= 5.89$ kJ

2.40. a. $V = 11.35 \times 373/273 = 15.5$ dm^3

b. $w = P\Delta V = 4 \times (15.5 - 11.35) = 16.6$ bar dm^3

$\qquad\qquad\qquad\qquad\qquad = 1660$ J $= 1.66$ kJ

c. $q = 2 \times 29.4 \times 100 = 5880$ J $= 5.88$ kJ

d. $\Delta H = 5880$ J $= 5.88$ kJ

e. $\Delta U = \Delta H - P\Delta V = 5880 - 1660 = 4220$ J $= 4.22$ kJ

$\qquad (= 2C_{V,m}\Delta T = 21.1 \times 2 \times 100)$

2.41. a. Zero

b. $P_2 = 8$ bar $= 800$ kPa

c. $w = 2 \times 8.3145 \times 273 \ln 2 = 3150$ J $= 3.15$ kJ

d. $-q = 3147$ J $= 3.15$ kJ

e. Zero

2.42. Initial volume of gas $= \dfrac{nRT}{P} = \dfrac{2 \times 0.083\ 09 \times 273.15}{10} = 4.539$ dm^3

Final volume $\quad = 22.70$ dm^3

$\qquad \Delta V \quad = 18.16$ dm^3

a. Work done by the gas $= 2.0 \times 18.16$ dm^3

$\qquad\qquad = 36.32$ bar dm$^3 = 3632$ J $=$ heat transferred to surroundings.

b. $\Delta U = \Delta H = 0$

c. $q = 3632$ J

2.43. a. Work done by gas $= nRT \ln \dfrac{V_{final}}{V_{initial}}$

$\qquad\qquad\qquad = 2 \times 8.3145 \times 273.15 \ln 5$

$\qquad\qquad\qquad = 7310$ J

b. $\Delta U = \Delta H = 0$

c. $q = 7310$ J $= 7.31$ kJ

2.44. $C_{V,m} = 28.80 - 8.3145 = 20.49$ J K^{-1} mol^{-1}

a. $P_2 = P_1 \left(\dfrac{V_1}{V_2} \right)^\gamma = 3.0 \left(\dfrac{1.5}{5.0} \right)^{1.406}$

$\qquad = 0.552$ bar

$T_2 = T_1 \left(\dfrac{V_1}{V_2} \right)^{\gamma - 1} = 298.15 \left(\dfrac{1.5}{5.0} \right)^{0.406}$

$\qquad = 182.9$ K

b. $\Delta U_m = C_{V,m}(T_2 - T_1)$

$\qquad = 20.49(182.9 - 298.15) = -2361 \text{ J mol}^{-1}$

Amount of H_2,

$$n = \frac{3.0 \times 1.5}{0.083\ 14 \times 298.15} = 0.182 \text{ mol}$$

$\Delta U = -430$ J for 0.182 mol

$\Delta H_m = 28.80(182.9 - 298.15) = -3319 \text{ J mol}^{-1}$

$\Delta H = -604$ J for 0.182 mol

2.45. a. Initial volume, $V_1 = \dfrac{nRT}{P} = 0.1 \times 0.083\ 09 \times 353.15 = 2.934 \text{ dm}^3$

$V_2 = V_1 \left(\dfrac{P_1}{P_2} \right)^{1/\gamma}$

$\qquad = 2.934\ (10)^{0.763} = 17.00 \text{ dm}^3$

b. $T_2 = \dfrac{T_1 P_2 V_2}{P_1 V_1} = \dfrac{353.15 \times 17.00}{10 \times 2.934} = 204.6$ K

c. $C_{P,m} - C_{V,m} = 8.3145 \text{ J K}^{-1} \text{ mol}^{-1}$

$\dfrac{C_{P,m}}{C_{V,m}} = 1.31; \quad \dfrac{C_{P,m}}{C_{V,m}} - 1 = 0.31; \quad C_{P,m} - C_{V,m} = 0.31\ C_{V,m}$

$\therefore 0.31\ C_{V,m} = 8.3145 \text{ J K}^{-1} \text{ mol}^{-1}$

$\qquad C_{V,m} = 26.82 \text{ J K}^{-1} \text{ mol}^{-1}$

$\qquad C_{P,m} = 35.13 \text{ J K}^{-1} \text{ mol}^{-1}$

$\Delta U_m = 26.82(204.6 - 353.15) = 3984 \text{ J K}^{-1} \text{ mol}^{-1}$

$\Delta U = -398.4 \text{ J K}^{-1}$ for 0.1 mol

$\Delta H_m = 35.13\ (204.6 - 353.15) = -5219 \text{ J K}^{-1} \text{ mol}^{-1}$

$\Delta H = -522 \text{ J K}^{-1}$ for 0.1 mol

2.46. a. $C_{P,m} - C_{V,m} = R = 8.3145 \text{ J K}^{-1} \text{ mol}^{-1}$

$$\Delta H = -522 \text{ J K}^{-1} \text{ for } 0.1 \text{ mol}$$

2.46. a. $C_{P,m} - C_{V,m} = R = 8.3145 \text{ J K}^{-1} \text{ mol}^{-1}$

$$C_{P,m} = 29.83 + 8.2 \times 10^{-3} (T/K)$$

b. $\Delta T = 0$, $\Delta U = \Delta H = 0$. It could be made to occur adiabatically by allowing free expansion, with $w = q = 0$.

2.47. The accompanying diagram shows two adiabatics intersected by two isotherms corresponding to temperatures T_h and T_c (compare Figure 3.2):

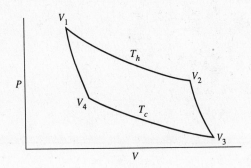

From Eq. 2.90,

$$T_h/T_c = (V_4/V_1)^\gamma \quad \text{and} \quad T_h/T_c = (V_3/V_2)^\gamma$$

Thus

$$V_4/V_1 = V_3/V_2 \quad \text{or} \quad V_2/V_1 = V_3/V_4$$

Thus, if any isotherm is drawn to intersect the two adiabatics, the ratio of the volume at the intersection points is always the same. The adiabatics therefore cannot intersect.

2.48. By definition

$$C_V = \left(\frac{\partial U}{\partial T} \right)_V$$

$$\left(\frac{\partial C_V}{\partial V} \right)_T = \frac{\partial}{\partial V} \left(\frac{\partial U}{\partial T} \right)_V = \frac{\partial}{\partial T} \left(\frac{\partial U}{\partial V} \right)_T$$

Since $(\partial U/\partial V)_T = 0$ for an ideal gas, $(\partial C_V/\partial V)_T = 0$, and C_V is therefore independent of V and of P. Similarly,

$$C_P = \left(\frac{\partial H}{\partial T} \right)_P \quad \text{and} \quad \left(\frac{\partial C_P}{\partial P} \right)_T = \frac{\partial}{\partial P} \left(\frac{\partial H}{\partial T} \right)_P = \frac{\partial}{\partial T} \left(\frac{\partial H}{\partial P} \right)_T$$

Thus, $(\partial C_P / \partial P)_T = 0$; C_P is therefore independent of P and of V.

2.49. The reversible work done by a gas in a reversible isothermal expansion is

$$-w = nRT \ln \frac{V_2}{V_1} \qquad \text{(Eq. 2.74)}$$

In this case $n = 1$ mol and $V_2/V_1 = 2$. Thus,

$$1000 \text{ J mol}^{-1} = (8.3145 \text{ J K}^{-1} \text{ mol}^{-1})T \ln 2$$

$$T = \frac{1000}{8.3145 \ln 2} \text{ K} = 173.5 \text{ K}$$

2.50. $C_{V,m} = \frac{5}{2} R$ and $C_{P,m} = \frac{7}{2} R$

$$\gamma = \frac{7}{5}$$

$$\frac{T_2}{T_1} = \left(\frac{V_1}{V_2}\right)^{\gamma-1} \qquad \text{(Eq. 2.90)}$$

Therefore,

$$T_2 = T_1 \left(\frac{V_1}{V_2}\right)^{\gamma-1} = 298.15 \text{ K} \times \left(\frac{1}{2}\right)^{2/5} = 226.0 \text{ K}$$

$$\Delta U_m = C_{V,m}(T_2 - T_1) = \frac{5}{2} R(226.0 - 298.15) \text{ K}$$

$$= -1501 \text{ J mol}^{-1} = -1.5 \text{ kJ mol}^{-1}$$

$$\Delta H_m = C_{P,m}(T_2 - T_1) = \frac{7}{2} R(226.0 - 298.15) \text{ K}$$

$$= -2100 \text{ J mol}^{-1} = -2.1 \text{ kJ mol}^{-1}$$

2.51. a. $\Delta H = C_P \Delta T$

$$C_{V,m} = \frac{3}{5} R = 12.47 \text{ J K}^{-1} \text{ mol}^{-1}$$

$$C_{P,m} = \frac{5}{2} R = 20.79 \text{ J K}^{-1} \text{ mol}^{-1}$$

$$1500 \text{ J mol}^{-1} = 20.79 \, (T_2 - 300 \text{ K}) \text{ J K}^{-1} \text{ mol}^{-1}$$

$$T_2 = 372.15 \text{ K}$$

$$\frac{P_2}{P_1} = \frac{T_2}{T_1} \cdot \frac{V_1}{V_2} = \frac{372.15}{300} \cdot \frac{1}{2} = 0.62 \text{ bar}$$

$$P_2 = 0.62 \text{ bar}$$

b. $\Delta U_m = C_{V,m} \Delta T$

$$= 12.47 \times 72.15 \text{ J mol}^{-1} = 900 \text{ J mol}^{-1}$$

$$w = \Delta U - q = (900 - 1000) \text{ J mol}^{-1} = -100 \text{ J mol}^{-1}$$

2.52. The definition of C_V is

$$C_V = \left(\frac{\partial U}{\partial T}\right)_V$$

From Euler's chain rule (Appendix C),

$$\left(\frac{\partial U}{\partial T}\right)_V = -\left(\frac{\partial U}{\partial V}\right)_T \left(\frac{\partial V}{\partial T}\right)_U$$

Therefore,

$$C_V = -\left(\frac{\partial U}{\partial V}\right)_T \left(\frac{\partial V}{\partial T}\right)_U$$

2.53. We have the general relationship (Appendix C),

$$dP = \left(\frac{\partial P}{\partial T}\right)_V dT + \left(\frac{\partial P}{\partial V}\right)_T dV$$

For an ideal gas,

$$P = \frac{nRT}{V}$$

$$\left(\frac{\partial P}{\partial T}\right)_V = \frac{nR}{V} = \frac{P}{T} \; ; \quad \left(\frac{\partial P}{\partial V}\right)_T = -\frac{nRT}{V^2} = -\frac{P}{V}$$

Therefore,

$$dP = \frac{P}{T} \, dT - \frac{P}{V} \, dV \quad \text{and} \quad \frac{1}{P}\frac{dP}{dt} = \frac{1}{T}\frac{dT}{dt} - \frac{1}{V}\frac{dV}{dt}$$

2.54. The initial temperature is T_1 (= 298.15 K) and the final temperature is T_2. Then

$$dU = nC_{V,m} \, dT$$

Also

$$dU = dq + dw = -P_2 dV \quad \text{since} \quad dq = 0$$

Thus

$$nC_{V,m} \, dT = -P_2 dV$$

$$nC_{V,m}(T_2 - T_1) = -P_2(V_2 - V_1)$$

$$= -P_2\left(\frac{nRT_2}{P_2} - \frac{nRT_1}{P_1}\right)$$

$$= -nRT_2 + \frac{nRT_1P_2}{P_1}$$

$$T_2 = \frac{C_{V,m}T_1 + \frac{RT_1P_2}{P_1}}{C_{V,m} + R}$$

$$= \frac{(20.8 \times 298.15) + (8.3145 \times 298.15 \times 0.1)}{20.8 + 8.3145} \text{ K}$$

$$= 221 \text{ K}$$

$$\Delta U = 5 \times 20.8 \ (221 - 298.15) = -8020 \text{ J}$$

$$\Delta H = 5 \times 29.1 \ (221 - 298.15) = -11\ 200 \text{ J}$$

2.55. If there are n mol initially,

$$P_1V_1 = nRT$$

Finally,

$$P_2V_2 = (n + 0.27 \text{ mol}) RT$$

$$\Delta(PV) = 0.27 \ RT$$

$$\Delta H = \Delta U + \Delta(PV)$$

$$= 9400 + (0.27 \times 8.3145 \times 300) \text{ J}$$

$$= 9400 + 673 = 10\ 073 \text{ J}$$

$$= 10.07 \text{ kJ}$$

2.56. The work is

$$w = P(V_2 - V_1)$$

$$= 10^5 \text{ Pa } (0.0100 - 0.1000) \text{ m}^3$$

$$= -9000 \text{ J}$$

The final temperature is

$$T = PV/nR = 1 \text{ bar} \times 10.00 \text{ dm}^3/0.0831 \text{ dm}^3 \text{ bar K}^{-1}$$

$$= 120.3 \text{ K}$$

C_V for an ideal gas is $(3/2)R = 12.47$ J K^{-1} mol^{-1}.

$$\Delta U = 12.47 \ (120.3 - 298.15) = -2218 \text{ J}$$

$$q = \Delta U - w = -2218 - (-9000) = +6782 \text{ J} = +6.78 \text{ kJ}$$

$$\Delta H = (5/2)R \ (120.3 - 298.15) = -20.79 \times 177.9$$

$$= -3\ 697 \text{ J} = -3.70 \text{ kJ}$$

2.57. Isobaric, so $P_2 = P_1 = P$

Adiabatic, so $q = 0$

$$w = P(V_2 - V_1)$$

$$\Delta U = w + q = w + 0 = w$$

$$T_2 = \frac{PV_2}{nR}$$

$$\Delta H = C_P \Delta T$$

For ideal monatomic gas, $C_P = \frac{5}{2} R$

$$\Delta H = \frac{5}{2} R \left(\frac{PV_2}{nR} - T_1 \right)$$

2.58. $T_2 = \dfrac{PV_2}{nR} = \dfrac{(1\ \text{bar})\,(10.00\ \text{dm}^3)}{(1\ \text{mol})(0.0831\ \text{dm}^3\ \text{bar}\ \text{K}^{-1})} = 120.3\ \text{K}$

$w = nRT \ln \dfrac{V_1}{V_2} = (1\ \text{mol})(8.3145\ \text{J}\ \text{K}^{-1}\ \text{mol}^{-1})(120.3\ \text{K}) \ln(\frac{1}{10}) = -2304\ \text{J}$

$\Delta H = C_P \Delta T = \frac{5}{2}(8.3145\ \text{J}\ \text{K}^{-1}\ \text{mol}^{-1})(120.3 - 298.15) \times \text{K} = -3697\ \text{J} = -3.70\ \text{kJ}$

$\Delta U = C_V \Delta T = \frac{3}{2}(8.3145\ \text{J}\ \text{K}^{-1}\ \text{mol}^{-1})(120.3 - 298.15) \times \text{K} = -2218\ \text{J} = -2.22\ \text{kJ}$

$q = \Delta U - w = -2.22\ \text{kJ} + 2.30\ \text{kJ} = 0.08\ \text{kJ}$

2.59. The volume of the inflated balloon is

$(4/3)\pi(7.5\ \text{m})^3 = 1767\ \text{m}^3$

a. The amount of helium can be calculated from $PV = nRT$:

$$n = \frac{PV}{RT} = \frac{1.013\ 25 \times 10^5\ \text{Pa} \times 1767\ \text{m}^3}{8.3145 \times 293.15\ \text{J}\ \text{mol}^{-1}}$$

$$= 73\ 460\ \text{mol}^{-1}$$

The atomic mass of helium is $4.026\ \text{g}\ \text{mol}^{-1}$, so that the mass of helium is

$73\ 460 \times 4.026\ \text{g} = 295\ 751\ \text{g} = 295.8\ \text{kg}$

b. The work done by the balloon during an expansion at constant pressure P is (Eq. 2.12)

$-w = P(V_2 - V_1)$ and $V_1 = 0$

$= 1.013\ 15 \times 10^5\ \text{Pa} \times 1767\ \text{m}^3$

$= 1.7902 \times 10^8\ \text{J} = 1.790 \times 10^5\ \text{kJ}$

2.60. a. Calculate the original volume from the ideal gas law:

$$V_1 = \frac{nRT}{P} = (1 \text{ mol}) \frac{(0.083 \text{ dm}^3 \text{ bar K}^{-1} \text{ mol}^{-1})(300 \text{ K})}{2 \text{ bar}} = 12.45 \text{ dm}^3$$

From the combined gas laws:

$$V_2 = \frac{P_1 V_1}{P_2} = \frac{2.0 \text{ bar } (12.45 \text{ dm}^3)}{1.5 \text{ bar}} = 16.6 \text{ dm}^3$$

$$w = -P_{external} \, \Delta V = -1.5 \text{ bar} \times (16.6 - 12.45) \text{ dm}^3 = -6.225 \text{ dm}^3 \text{ bar}$$

Since 1 bar = 100 kPa and 1 Pa = kg m^{-1} s^{-2},

$$w = -622 \text{ kPa dm}^3 = -622 \text{ kg m}^2 \text{ s}^{-2} = -622 \text{ J}$$

The work is negative; work is done by the gas.

b. $$w_{rev} = -\int_{V_1}^{V_2} P \, dV = -nRT \ln \frac{V_2}{V_1}$$

$$V_1 = 12.45 \text{ dm}^3$$

$$w_{rev} = -1 \text{ mol } (8.3145 \text{ J K}^{-1} \text{ mol}^{-1})(300 \text{ K}) \ln \frac{16.6}{12.45} = -718 \text{ J}$$

Thus, 718 J of work is done in the reversible process, but only 622 J is done in the nonreversible process in (a).

2.61. For an ideal gas, $PV = nRT$, or $V = nRT/P$. So, we evaluate the values of the two differentials for 1 mol of gas.

$$\left(\frac{\partial V}{\partial T} \right)_P = \frac{R}{P} \quad \text{and} \quad \left(\frac{\partial V}{\partial P} \right)_T = -\frac{RT}{V^2},$$

and $$\beta = -\frac{1}{V} \left(-\frac{RT}{P^2} \right) = \frac{RT}{VP^2} \quad \text{and} \quad \alpha = \frac{1}{V} \frac{R}{P}.$$

Substitution in $C_P - C_V = TV\alpha^2/\beta$ gives

$$C_P - C_V = \frac{TV(1/V^2)(R^2/P^2)}{RT/(VP^2)} = R.$$

■ Real Gases

2.62. For 1 mol of a van der Waals gas

$$\left(P + \frac{a}{V_m^2} \right)(V_m - b) = RT \qquad \text{(Eq. 1.100)}$$

Therefore

$$T = \frac{P(V_m - b)}{R} + \frac{a}{V_m^2 R}(V_m - b) \quad \text{and} \quad \left(\frac{\partial T}{\partial P}\right)_V = \left(\frac{V_m - b}{R}\right).$$

2.63. a. Work done on the system $= -\int_{V_1}^{V_2} P\,dV = RT \ln \frac{V_1}{V_2}$

$$= 8.3145 \times 300 \ln \frac{10}{0.2} = 9757 \text{ J} = 9.76 \text{ kJ mol}^{-1}$$

b. $w = -RT \int_{V_1}^{V_2} \frac{dV}{V - b}$

Put $V - b = x; \quad dV = dx$

$$= -RT \int_{V_1}^{V_2} \frac{dx}{x} = -RT \ln(V - b) \Big|_{V_1}^{V_2}$$

$$= RT \ln \frac{V_1 - b}{V_2 - b} = 8.3145 \times 300 \ln \frac{9.97}{0.17}$$

$$= 10\,155 \text{ J mol}^{-1} = 10.16 \text{ kJ mol}^{-1}$$

More work is done in (b) because of the greater ratio of free volumes.

2.64. a. $w = RT \ln\frac{V_1}{V_2} = 8.3145 \times 100 \ln\frac{20}{5}$

$$= 1153 \text{ J mol}^{-1} = 1.15 \text{ kJ mol}^{-1}$$

b. $P = \frac{RT}{V} - \frac{a}{V^2}$

$$w = -\int_{V_1}^{V_2} P\,dV = -RT \int_{V_1}^{V_2} \frac{dV}{V} + a \int_{V_1}^{V_2} \frac{dV}{V^2}$$

$$= +RT \ln\frac{V_1}{V_2} - a\left(\frac{1}{V_1} - \frac{1}{V_2}\right)$$

$$= 1153 - 3.84 \times 101.3 \left(\frac{1}{20} - \frac{1}{5}\right)$$

$$= 1211 \text{ J} = 1.21 \text{ kJ mol}^{-1}$$

In (b), additional work has to be done against the molecular repulsions.

2.65. We have the general relationship (Appendix C),

$$dP = \left(\frac{\partial P}{\partial V_m}\right)_T dV_m + \left(\frac{\partial P}{\partial T}\right)_{V_m} dT$$

For 1 mol of a van der Waals gas

$$P = \frac{RT}{V_m - b} - \frac{a}{V_m^2}$$

Then

$$\left(\frac{\partial P}{\partial T}\right)_{V_m} = \frac{R}{V_m - b}; \quad \left(\frac{\partial P}{\partial V_m}\right)_T = -\frac{RT}{(V_m - b)^2} + \frac{2a}{V_m^3}$$

Therefore,

$$dP = -\frac{RT\, dV_m}{(V_m - b)^2} + \frac{2a\, dV_m}{V_m^3} + \frac{R\, dT}{V_m - b}$$

$$= -\frac{P + \dfrac{a}{V_m^2}}{V_m - b}\, dV_m + \frac{2a\, dV_m}{V_m^3} + \frac{P\, dT}{T} + \frac{a}{V_m^2}\frac{dT}{T}$$

$$= -\frac{P}{V_m - b}\, dV_m - \frac{a(3V_m - 2b)}{V_m^3(V_m - b)}\, dV_m + \frac{P\, dT}{T} + \frac{a\, dT}{V_m^2 T}$$

2.66. The reaction is

$$CH_4(g) + 2O_2(g) \rightarrow CO_2(g) + 2H_2O(g)$$

$$\Delta H^\circ = \Delta_f H^\circ(CO_2, g) + 2\Delta_f H^\circ(H_2O, g) - \Delta_f H^\circ(CH_4, g)$$

$$\Delta H^\circ/\text{kJ mol}^{-1} = -393.51 + (2 \times -241.826) + 74.6 = -802.56$$

For the product gases, $CO_2 + 2H_2O$,

$$C_P(CO_2)/\text{J K}^{-1}\,\text{mol}^{-1} = 44.22 + 8.79 \times 10^{-3}\ (T/\text{K})$$

$$2C_P(H_2O)/\text{J K}^{-1}\,\text{mol}^{-1} = 61.08 + 2.06 \times 10^{-2}\ (T/\text{K})$$

$$C_P(\text{products})/\text{J K}^{-1} = 105.30 + 2.939 \times 10^{-2}\ (T/\text{K})$$

Since $C_{P,m} - C_{V,m} = R$, $C_{V,m} = C_{P,m} - R$.

$$802\,560 = \int_{298.15}^{T_2/\text{K}} (105.30 - 8.3145 + 2.939 \times 10^{-2}(T/\text{K})\, d(T/\text{K})$$

$$= 96.99[T_2/\text{K} - 298.15] + \frac{1}{2}(2.939 \times 10^{-2})[(T_2/\text{K})^2 - 298.15^2]$$

$$1.4695 \times 10^{-2}\ (T_2/\text{K})^2 + 96.99\ (T_2/\text{K}) - 835\,261.48 = 0$$

$$T_2/\text{K} = \frac{-96.99 \pm \sqrt{9406.19 + 49096.67}}{2.939 \times 10^{-2}}$$

Using the positive value of the square root yields

$$T_2/\text{K} = 4929.69\ \text{K} \text{ or } 4930\ \text{K} \text{ as the maximum temperature.}$$

This value will be reduced to 4763 K if the gas were allowed to expand under constant pressure conditions.

2.67. The reversible work done on the system is

$$w_{rev} = -\int_{V_1}^{V_2} P\,dV = -nRT \int_{V_1}^{V_2} \frac{dV}{V - nb}$$

$$= -nRT \ln \frac{V_2 - nb}{V_1 - nb}$$

$$= -2 \times 8.3145 \times 300 \text{ (J) } \ln\frac{1.00 - 0.08}{10.00 - 0.08}$$

$$= -4\,989 \text{ J } \ln(0.92/9.92)$$

$$= 4\,989 \times 2.378 = 11\,860 \text{ J} = 11.9 \text{ kJ}$$

From Eq. 2.125, ΔU is zero since a is zero. To obtain ΔH we must calculate $\Delta(PV)$:

$$P_1 = 2RT/(V_1 - 0.08) = 2 \times 0.0831 \times 300/9.92$$

$$= 5.03 \text{ bar}$$

$$P_2 = 2RT/(V_2 - 0.08) = 2 \times 0.0831 \times 300/0.92$$

$$= 54.2 \text{ bar}$$

$$P_1 V_1 = 5.03 \times 10.0 = 50.3 \text{ bar dm}^3 = 5030.0 \text{ Pa m}^3$$

$$= 5030.0 \text{ J}$$

$$P_2 V_2 = 54.2 \times 1.00 = 54.2 \text{ bar dm}^3$$

$$= 5420.0 \text{ J}$$

$$\Delta(PV) = P_2 V_2 - P_1 V_1 = 390 \text{ J}$$

$$\Delta H = \Delta U + \Delta(PV) = 0 + 390 = 390 \text{ J}$$

2.68. The reversible work is

$$w_{rev} = -\int_{V_1}^{V_2} P\,dV = -\int_{V_1}^{V_2} \left(\frac{nRT}{V} - \frac{n^2 a}{V^2}\right) dV$$

$$= -nRT \ln \frac{V_2}{V_1} - n^2 a \left(\frac{1}{V_2} - \frac{1}{V_1}\right)$$

$$= -3 \times 8.3145 \times 300 \text{ (J) } \ln (1/20) - (3.00)^2 \times 0.55 \text{ (Pa m}^6) \left(\frac{1}{10^{-3} \text{ m}^3} - \frac{1}{20 \times 10^{-3} \text{ m}^3}\right)$$

$$= 22\,417 - 4703 = 17\,714 \text{ J} = 17.7 \text{ kJ}$$

The change in internal energy is obtained from the relationship:

$$dU = \left(\frac{\partial U}{\partial V}\right)_T dV = \frac{n^2 a}{V^2} dV$$

$$\Delta U = \int_{V_1}^{V_2} \frac{n^2 a}{V^2} dV = -n^2 a \left(\frac{1}{V}\right) \Big|_{V_1}^{V_2}$$

$$= -n^2a\left(\frac{1}{V_2} - \frac{1}{V_1}\right)$$

$$= -(3.00)^2 \times 0.55 \text{ (Pa m}^6\text{)}\left(\frac{1}{10^{-3} \text{ m}^3} - \frac{1}{20 \times 10^{-3} \text{ m}^3}\right)$$

$$= -4703 \text{ J} = -4.70 \text{ kJ}$$

To obtain ΔH we calculate P_1V_1 and P_2V_2:

$V_1 = 20 \text{ dm}^3$

$P_1 = (nRT/V_1) - n^2a/V_1^2$

$\quad = [3 \times 0.0831 \times 300 \text{ (bar)}/20] \text{ (}10^5 \text{ Pa/bar)} [-9 \times 0.55 \text{ (Pa m}^6\text{)}]/(20 \times 10^{-3} \text{ m}^3)^2$

$\quad = 374\ 000 \text{ Pa} - 12\ 375 \text{ Pa} = 361.6 \text{ kPa}$

$V_2 = 1 \text{ dm}^3$

$P_2 = [3 \times 0.0831 \times 300 \text{ (bar)}/1] \text{ (}10^5 \text{ Pa/bar)} [-9 \times 0.55 \text{ (Pa m}^6\text{)}]/(10^{-3} \text{ m}^3)^2$

$\quad = 7\ 479\ 000 \text{ Pa} - 4\ 950\ 000 \text{ Pa} = 2\ 529\ 000 \text{ Pa} = 2529 \text{ kPa}$

$P_1V_1 = (20 \times 10^{-3} \text{ m}^3) \times 361\ 600 \text{ Pa} = 7232 \text{ J}$

$P_2V_2 = (10^{-3} \text{ m}^3) \times 2\ 529\ 000 \text{ Pa} = 2\ 529 \text{ J}$

$\Delta(PV) = 2\ 529 - 7232 = -4703 \text{ J}$

$\Delta H = -4703 - 4703 = -9406 \text{ J} = -9.41 \text{ kJ}$

2.69. From Eq. 2.123 with $n = 1$,

$$w_{\text{rev}} = -RT \ln\frac{20 - 0.064}{60 - 0.064} - 0.556 \text{ Pa m}^6\left(\frac{1}{20 \text{ dm}^3} - \frac{1}{60 \text{ dm}^3}\right) \cdot \frac{1000 \text{ dm}^3}{\text{m}^3}$$

$$= -2494.3 \text{ (J) } \ln\frac{19.94}{59.94} - 0.556 (50.00 - 16.67) \text{ J}$$

$$= 2745.6 - 18.53 \text{ J} = 2727 \text{ J} = 2.73 \text{ kJ}$$

From Eq. 2.126,

$$\Delta U = 0.556 \left(\frac{1}{60 \times 10^{-3} \text{ m}^3} - \frac{1}{20 \times 10^{-3} \text{ m}^3}\right) \text{J}$$

$$= -0.566 \times 33.33 \text{ J} = -18.9 \text{ J}$$

To calculate ΔH we first calculate P_1V_1 and P_2V_2:

$$P_1 = \frac{RT}{V_1 - b} - \frac{a}{V_1^2} = \frac{2494.3 \text{ J}}{59.94 \times 10^{-3} \text{ m}^3} - \frac{0.556 \text{ Pa m}^6}{(60 \times 10^{-3} \text{ m}^3)^2}$$

$$= 41\ 613 - 154.4 \text{ Pa} = 41\ 459 \text{ Pa}$$

$$P_2 = \frac{2494.3 \text{ J}}{19.94 \times 10^{-3} \text{ m}^3} - \frac{0.556 \text{ Pa m}^6}{(20 \times 10^{-3} \text{ m}^3)^2}$$

$$= 125\ 090 - 1390\ \text{Pa} = 123\ 700\ \text{Pa}$$

$$P_1 V_1 = 41\ 459\ \text{Pa} \times 60 \times 10^{-3}\ \text{m}^3 = 2487.5\ \text{J}$$

$$P_2 V_2 = 123\ 700 \times 20 \times 10^{-3}\ \text{J} = 2474.0\ \text{J}$$

$$\Delta(PV) = -13.5\ \text{J}$$

$$\Delta H = \Delta U + \Delta(PV) = -18.9\ \text{J} - 13.5\ \text{J} = -32.4\ \text{J}$$

2.70. From Eqs. 2.108 and 2.110 upon expanding the partial derivative,

$$\mu = (\partial T/\partial P)_H = -(\partial T/\partial H)_P\ (\partial H/\partial P)_T = \frac{-1}{C_P}\ (\partial H/\partial P)_T.$$

Then from the statement of the problem,

$$\mu = \frac{-1}{C_P}(\partial H/\partial P)_T = \frac{2a/RT - b}{C_P}$$

$$\int dH = -\int_{P_1}^{P_2} (2a/RT - b)\ dP$$

$$\Delta H = (b - 2a/RT)\ (P_2 - P_1)$$

$$\Delta H = \left[0.0391 \times 10^{-3}\ \text{m}^3\ \frac{2(0.1408\ \text{Pa m}^6)}{(8.3145\text{J K}^{-1})(300\ \text{K})} \right](100 - 1)\ 100\ \text{kPa}$$

$$\Delta H = [0.0391 \times 10^{-3} - (1.129 \times 10^{-4})]\ (99)\ (100)\ \text{kPa}$$

$$= -0.731\ \text{m}^3\ \text{kPa}$$

Since $1\ \text{Pa} = 1\ \text{kg m}^{-1}\ \text{s}^{-2}$ and $1\ \text{J} = 1\ \text{kg m}^2\ \text{s}^{-2}$

$$\Delta H = -0.731\ \text{kJ} = -731\ \text{J mol}^{-1}.$$

3. THE SECOND AND THIRD LAWS OF THERMODYNAMICS

■ The Carnot Cycle

3.1. a. Efficiency $= \dfrac{T_h - T_c}{T_h} = \dfrac{1000 - 200}{1000} = 0.8 = 80\%$

   b. Heat rejected $= 150 \times \dfrac{1000}{200} = 30$ kJ

   c. Entropy increase $= \dfrac{150\ 000}{1000} = 150$ J K^{-1}

   d. Entropy decrease $= \dfrac{30\ 000}{200} = 150$ J K^{-1}

   e. $\Delta S = 0$

   f. $\Delta S = 150$ J K^{-1}; $\Delta H = 0$

   $\Delta G = \Delta H - T\Delta S = 0 - 1000 \times 150 = -150\ 000$ J

   $= -150$ kJ

3.2. Efficiency $= \dfrac{w}{q_h} = \dfrac{398 - 313}{398} = 0.214$

   $\dfrac{1500}{0.214} = 7010$ J must be withdrawn.

3.3. a.

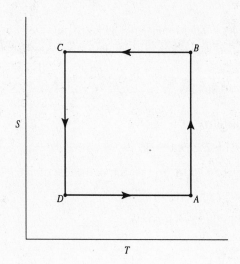

b. $\dfrac{q_h}{T_h} + \dfrac{q_c}{T_c} = 0 = \dfrac{q_h}{T_h} + \dfrac{q_c}{300 \text{ K}} = 0$

Work performed by system, $-w = q_h + q_c = 10$ kJ

$\Delta S_h = \dfrac{q_h}{T_h} = 100$ J K^{-1}

100 J K$^{-1} + \dfrac{q_c}{300 \text{ K}} = 0$

$q_c = -30\,000$ J $= -30$ kJ

$q_h = 40$ kJ

$T_h = \dfrac{40}{30} \times 300 \text{ K} = 400$ K

3.4. a. Efficiency $= (T_h - T_c)/T_h = 0.25 = 25\%$

b. Heat absorbed at 400 K $= 800/0.25 = 3200$ J $= 3.2$ kJ

c. Heat rejected at 300 K $= 3200 - 800 = 2400$ J $= 2.4$ kJ

d. Entropy change for A $\rightarrow$ B $= 3200$ J/400 K $= 8.0$ J K^{-1}

e. Zero

f. $\Delta H = 0$ and thus $\Delta G = -T\Delta S = -400 \times 8.0 = -3200$ J $= -3.2$ kJ

g. The efficiency is 25%, so to produce 2000 J of work,

$\dfrac{2000 \text{ J}}{0.25} = 8000$ J $= 8.0$ kJ

must be absorbed.

3.5. The heat transferred from the water to the ice in melting it is

$(10^{12} \text{ g}/18 \text{ g mol}^{-1}) \times 6.025 \text{ kJ mol}^{-1} = 3.35 \times 10^{11} \text{ kJ}$

The fraction of this that can be converted into work is (Eq. 3.21)

$$\frac{T_h - T_c}{T_h} = \frac{20}{293.15} = 0.0682$$

The available work is therefore

$0.0682 \times 3.35 \times 10^{11} = 2.29 \times 10^{10} \text{ kJ}$

One day is $24 \times 60 \times 60 = 86\,400$ s. The power, or the rate of doing work, is thus

$2.29 \times 10^{10}/86\,400 = 2.65 \times 10^5 \text{ kJ s}^{-1}$

$= 2.65 \times 10^8 \text{ J s}^{-1} = 2.65 \times 10^8 \text{ W}$

$= 265 \text{ MW}$

3.6. A thermodynamic equation of state (Eq. 3.128) is

$$\left(\frac{\partial U}{\partial V}\right)_T = -P + T\left(\frac{\partial P}{\partial T}\right)_V.$$

From the ideal gas law, we have $P = RT/V_m$; therefore, $(\partial P/\partial T)_V = R/V_m$. Substituting above, and recognizing that $RT/V_m = P$, we obtain

$$\left(\frac{\partial U}{\partial V}\right)_T = -P + \left(\frac{RT}{V_m}\right) = 0.$$

■ Entropy Changes

3.7. C_6H_6: $30\,800/353 = 87.3 \text{ J K}^{-1} \text{ mol}^{-1}$

$CHCl_3$: $29\,400/334 = 88.0 \text{ J K}^{-1} \text{ mol}^{-1}$

H_2O: $40\,600/373 = 108.8 \text{ J K}^{-1} \text{ mol}^{-1}$

C_2H_5OH: $38\,500/351 = 109.7 \text{ J K}^{-1} \text{ mol}^{-1}$ $\left.\right\}$ $\left[\begin{array}{l}\text{Hydrogen-bonded} \\ \text{structure in liquids} \\ H_2O \text{ and } C_2H_5OH.\end{array}\right.$

3.8. a. The reaction is

$C(\text{graphite}) + 2H_2(g) + \frac{1}{2}O_2(g) \rightarrow CH_3OH(l)$

$\Delta_f H° = 126.8 - [5.74 + (2 \times 130.68) + \frac{1}{2}(205.14)\,]$

$= -242.9 \text{ J K}^{-1} \text{ mol}^{-1}$

b. The reaction is $C(\text{graphite}) + 2H_2(g) + \frac{1}{2}O_2(g) + N_2(g) \rightarrow H_2NCONH_2(s)$

$\Delta_f H° = 104.60 - [5.74 + (2 \times 130.68) + \frac{1}{2}(205.14) + 191.61]$

$= -456.8 \text{ J K}^{-1} \text{ mol}^{-1}$

3.9. a. $\Delta_r S^\circ = [2(192.77) - 191.61 - 3(130.68)]$ J K^{-1} mol^{-1}

$\qquad = -198.11$ J K^{-1} mol^{-1}.

b. $\Delta_r S^\circ = 2(240.1) - 304.2$ J K^{-1} mol^{-1} = -176.0 J K^{-1} mol^{-1}.

3.10. First, calculate the value of $\Delta_r S^\circ$ at 298.15 K. Then, using this value, calculate the increase in entropy for the increase in temperature. Assume C_P is constant throughout the range.

For the dissociation H$_2$(g) $\rightarrow$ 2H(g),

$\qquad \Delta_r S^\circ(298.15$ K$) = [2(114.717) - 130.680]$ J K^{-1} mol^{-1} = 98.754 J K^{-1} mol^{-1}.

The value of $\Delta_r S^\circ$ at different temperatures may be calculated from

$$\Delta_r S^\circ(T) = \Delta_r S^\circ(198.15 \text{ K}) + \int_{298.15}^{1273.15} \frac{\Delta_r C_P^\circ}{T} dT,$$

where $\Delta_r C_P^\circ$ is $\sum_i \nu_i C_{P,i}^\circ$ for the reaction. If $\Delta_r C_P^\circ$ is independent of temperature, the quantity $\Delta_r C_P^\circ$ can be taken out of the integral. Using the C_P° values at 298.15 K, we get

$$\Delta_r S^\circ(1273.15 \text{ K}) = 98.754 + [2(20.784) - 28.824] \int_{298.15}^{1273.15} \frac{dT}{T}$$

$$= 98.754 + 12.744 \ln(1273.15/298.15)$$

$$= 186.46 \text{ J K}^{-1} \text{ mol}^{-1}.$$

A more accurate value could be obtained if values of S° or C_P° were available at 1273.15 K.

3.11. a. $\Delta S_m = \int_{298}^{353} \frac{C_{P,m} dT}{T} \qquad = \frac{5}{2} R \ln \frac{353}{298}$

$\qquad\qquad\qquad\qquad\qquad = 3.52$ J K^{-1} mol^{-1}

b. $\Delta S_m = \int_{298}^{353} \frac{C_{V,m} dT}{T} \qquad = \frac{3}{2} R \ln \frac{353}{298}$

$\qquad\qquad\qquad\qquad\qquad = 2.11$ J K^{-1} mol^{-1}

3.12. For N$_2$ and O$_2$, the volume increases by a factor of 2.5, increase of entropy for each

$\qquad \Delta S = R \ln 2.5 = 7.62$ J K^{-1}

For H$_2$, volume increases by a factor of 5:

$\qquad \Delta S = 0.5 \times 8.3145 \ln 5 = 6.69$ J K^{-1}

$\qquad$ Total $\Delta S = 2 \times 7.62 + 6.69 = 21.9$ J K^{-1}

3.13. By Eq. 3.63,

$$\Delta S = 8.3145 \ln\frac{3}{1} + 5 \times 8.3145 \ln\frac{3}{2}$$

$$= 9.13 + 16.86 = 25.99 \text{ J K}^{-1}$$

3.14. The mole fractions are

$$x_{N_2} = 0.79 \quad x_{O_2} = 0.20 \quad x_{Ar} = 0.01$$

The entropy change per mole of mixture is thus, by Eq. 3.65,

$$\Delta S = -8.3145(0.79 \ln 0.79 + 0.20 \ln 0.20 + 0.01 \ln 0.01)$$

$$= -8.3145(-0.186 - 0.322 - 0.046) = 4.61 \text{ J K}^{-1} \text{ mol}^{-1}$$

3.15. The equation for the formation of ethanol from its elements is

$$2C(\text{graphite}) + 3H_2(g) + \tfrac{1}{2}O_2(g) \rightarrow C_2H_5OH(l)$$

The standard molar entropy of formation is thus

$$\Delta_f S^\circ{}_m = S^\circ{}_{\text{product}} - \Sigma S^\circ{}_{\text{reactants}}$$

$$= 160.7 - [(2 \times 5.74) + (3 \times 130.68) + \tfrac{1}{2}(205.14)\,]$$

$$= -345.4 \text{ J K}^{-1} \text{ mol}^{-1}$$

3.16. a. $\Delta S_{\text{gas}} = R \ln\dfrac{V_2}{V_1} = 8.3145 \ln 10.0 = 19.1 \text{ J K}^{-1} \text{ mol}^{-1}$

$\Delta S_{\text{surr}} = -19.1 \text{ J K}^{-1} \text{ mol}^{-1}$

b. For an adiabatic process, $g = 0$, but with no work done, $w = 0$

Therefore, $\Delta U = 0$, and since $\Delta U = m \, mC \, \Delta T$

there is no change in temperature.

$$T_c = 25^\circ \text{ C}$$

Because ΔS_{gas} must be calculated from a reversible process,

$$\Delta S_{\text{gas}} = 19.1 \text{ J K}^{-1} \text{ mol}^{-1}; \quad \Delta S_{\text{surr}} = 0 \text{ since } q = 0$$

Net $\Delta S = 19.1 \text{ J K}^{-1} \text{ mol}^{-1}$

3.17. a. Taking into account the entropy change for heating liquid water and the entropy change during vaporization, we obtain

$$\Delta S^\circ(373.0 \text{ K}) / (\text{J K}^{-1}\text{mol}^{-1}) = \int_{273}^{373} \frac{C_{P,m(l)}}{T} dT + \frac{40\,670}{373.0}$$

$$= 75.48 \ln\left(\frac{373}{272}\right) + \frac{40\,670}{373.0} = 132.59.$$

b. We recognize that changing the pressure on the surface of a liquid dies not affect the entropy. Therefore, we take into account the entropy changes for vaporization, heating the vapor at constant pressure and, finally, compressing the vapor to the final pressure at constant temperature (see Eq. 3.51). This yields

$$\Delta S°(373.0 \text{ K})/(\text{J K}^{-1}\text{mol}^{-1}) = \frac{44\ 920}{273.0} + \int_{273}^{373} \frac{C_{P,m(g)}}{T} dT + 8.3145 \ln\left(\frac{V_2}{V_1}\right)$$

Substituting $C_{P,m} = 30.54 + 10.29 \times 10^{-3}T$ and replacing V_2/V_1 in the last term by $P_1/P_2 = 0.00602/1.00$, we get

$$\Delta S°(373.0 \text{ K})/(\text{J K}^{-1}\text{mol}^{-1}) = \frac{44\ 920}{273.0} + \int_{273}^{373}\left(\frac{30.54}{T} + 10.29 \times 10^{-3}\right) dT + 8.3145 \ln\left(\frac{P_1}{P_2}\right)$$

$$= \frac{44\ 920}{273.0} + 30.54 \ln\left(\frac{373}{273}\right) + 10.29 \times 10^{-3}(373 - 272)$$

$$+ 8.3145 \ln\left(\frac{0.00602}{1.00}\right)$$

$$= 132.59.$$

Thus the entropy change for the system depends only on the initial and final states and not the path taken to go from one to the another.

3.18.   a.   Positive (increase in number of molecules)

   b.   Positive (decrease in electrostriction)

   c.   Negative (increase in electrostriction)

   d.   Positive (decrease in electrostriction)

3.19.   Heat absorbed when temperature rises $= dq = C_P dT$

Corresponding entropy change:

$$\Delta S = \int \frac{C_P dT}{T}$$

Entropy increase when the temperature rises from T_1 to T_2:

$$\int_{T_1}^{T_2} \frac{C_P}{T}\ dT = n\int_{T_1}^{T_2} \frac{C_{P,m}}{T}\ dT$$

If the gas is ideal, C_P is constant and

$$\Delta S = C_P \ln\frac{T_2}{T_1} = nC_{P,m} \ln\frac{T_2}{T_1}$$

3.20.   By Eq. 3.57 the entropy change is

$$\Delta S = 5 \times 12.5 \ln\frac{373}{300} + 5 \times 8.3145 \ln\frac{10}{5}$$

$$= 13.6 + 28.8 = 42.4 \text{ J K}^{-1}$$

3.21.   Let the value of the final Celsius temperature be T^u. Then

$$200 \times 0.140 \, (100 - T^u) = 100 \times 4.18 \, (T^u - 20)$$

$$6.70 - 0.0670 \, T^u = T^u - 20$$

$$T^u = \frac{26.70}{1.0670} = 25.02; \quad T = 25.02 \,°\text{C}$$

a. $\Delta S_{\text{mercury}}$ $\quad = 200 \times 0.140 \displaystyle\int_{373.15}^{298.17} \frac{dT}{T}$

$$= 200 \times 0.140 \ln \frac{298.17}{373.15}$$

$$= -6.28 \text{ J K}^{-1}$$

b. $\Delta S_{\text{water + vessel}}$ $\quad = 100 \times 4.18 \displaystyle\int_{293.15}^{298.17} \frac{dT}{T}$

$$= 100 \times 4.18 \ln \frac{298.17}{293.15}$$

$$= 7.10 \text{ J K}^{-1}$$

c. Net ΔS $\quad = -6.28 + 7.10 = 0.82 \text{ J K}^{-1}$

3.22. Heat required to melt 20 g of ice is

$$\frac{20}{18} \times 6020 = 6689 \text{ J}$$

Heat required to heat 20 g of water from 0 °C to T °C is

$$(20 \times 4.184 \, T) \text{ J}$$

Heat required to cool 70 g of water from 30 °C to T °C is

$$70 \times 4.184 \, (30 - T) \text{ J}$$

Heat balance equation:

$$6689 + 20 \times 4.184 \, T = 70 \times 4.184 \, (30 - T)$$

$$T = 5.57 \,°\text{C}$$

Reversible processes:

a. Cool 70 g of water to 0 °C:

$$\Delta S_{\text{system}} = 70 \times 4.184 \ln \frac{273.15}{303.15} = -30.52 \text{ J K}^{-1}$$

$$\Delta S_{\text{surr}} = 30.52 \text{ J K}^{-1}$$

b. Melt 20 g of ice at 0 °C:

$$\Delta S_{\text{system}} = \frac{20 \times 6020}{18 \times 273.15} = 24.49 \text{ J K}^{-1}$$

$$\Delta S_{\text{surr}} = -24.49 \text{ J K}^{-1}$$

c. Heat 90 g of water to 5.57 °C:

$$\Delta S_{system} = 90 \times 4.184 \ln \frac{278.72}{273.15} = 7.60 \text{ J K}^{-1}$$

$$\Delta S_{surr} = -7.60 \text{ J K}^{-1}$$

$$\text{Net } \Delta S_{system} = 1.57 \text{ J K}^{-1}; \quad \Delta S_{surr} = -1.57 \text{ J K}^{-1}$$

3.23.
$$\Delta S_m = \int_{300}^{1000} \frac{C_{P,m}}{T} \, dT$$

$$\Delta S_m / \text{J K}^{-1} \text{ mol}^{-1} = \int_{300}^{1000} \frac{28.58}{T/\text{K}} + 0.00376 - \frac{50\,000}{(T/\text{K})^3} \, d(T/\text{K})$$

$$= 28.58 \ln \frac{1000}{300} + 0.00376 \,(1000 - 300) + 25\,000 \left(\frac{1}{1000^2} - \frac{1}{300^2} \right)$$

$$= 34.41 + 2.63 - 0.25 = 36.79$$

$$\Delta S_m = 36.8 \text{ J K}^{-1} \text{ mol}^{-1}$$

3.24. In an isothermal reversible expansion
$$\Delta E = 0 = q + w \quad \text{and} \quad q_{rev} = \frac{\Delta S}{T}$$
If the process were reversible, the work done by the system would be

$$-w_{rev} = T\Delta S = 300 \times 50 = 15\,000 \text{ J} = 15 \text{ kJ}$$

Since the actual work was less than this, the process is irreversible.

$$\text{Degree of irreversibility} = \frac{6}{15} = 0.4$$

3.25. It is necessary to devise a process in which the freezing occurs reversibly:

Step 1: Heat the supercooled water reversibly from –3 °C to 0 °C:

$$\Delta S_1 = 75.3 \ln \frac{273.15}{270.15} = 0.83 \text{ J K}^{-1} \text{ mol}^{-1}$$

Step 2: Freeze the water at 0 °C:

$$\Delta S_2 = \frac{-6020}{273.15} = -22.04 \text{ J K}^{-1} \text{ mol}^{-1}$$

Step 3: Cool the ice reversibly from 0 °C to –3 °C:

$$\Delta S_3 = 37.7 \ln \frac{270.15}{273.15} = -0.42 \text{ J K}^{-1} \text{ mol}^{-1}$$

The net entropy change in the system is therefore

$$\Delta S = 0.83 - 22.04 - 0.42 = -21.63 \text{ J K}^{-1} \text{ mol}^{-1}$$

To calculate the entropy change in the environment, we calculate the heat that has been gained by the environment in the three steps:

Step 1: $-3 \times 75.3 = -225.9$ J mol^{-1}

Step 2: 6020 J mol^{-1}

Step 3: $3 \times 37.7 = 113.1 \text{ J mol}^{-1}$

The net heat gained by the environment is thus

$-225.9 + 6020 + 113.1 = 5907.2 \text{ J mol}^{-1}$

This heat was gained by the environment at -3 °C, and the entropy change is therefore

$$\frac{5907.2}{270.15} = 21.87 \text{ J K}^{-1} \text{ mol}^{-1}$$

The net entropy change in the system and environment is thus

$-21.63 + 21.87 = 0.24 \text{ J K}^{-1} \text{ mol}^{-1}$

3.26. The solution contains 0.1 mol, and the volume ratio is $1000/200 = 5$. Then, by Eq. 3.51,

$n = 0.5 \times 0.2 = 0.1$

$\Delta S = 0.1 \times 8.3145 \times \ln 5$

$= 1.338 \text{ J K}^{-1} = 1.34 \text{ J K}^{-1}$

3.27. 0.1 mol of substance A is present and the volume increases by a factor of 4:

$\Delta S(A) = 0.1 \times 8.3145 \times \ln 4$

$= 1.153 \text{ J K}^{-1}$

0.15 mol of B is present and the volume increases by a factor of 4/3:

$\Delta S(B) = 0.15 \times 8.3145 \times \ln (4/3)$

$= 0.359 \text{ J K}^{-1}$

Net $\Delta S = 1.51 \text{ J K}^{-1}$

3.28. The final temperature is 40 °C. The water at 60 °C can be cooled reversibly to 40 °C, and the water at 20°C can be heated reversibly to 40 °C.

$$\Delta S_1 = 10 \times 75.3 \ln \frac{313.15}{333.15} = -46.62 \text{ J K}^{-1}$$

$$\Delta S_2 = 10 \times 75.3 \ln \frac{313.15}{293.15} = 49.70 \text{ J K}^{-1}$$

The net entropy change is thus $49.70 - 46.62 = 3.08 \text{ J K}^{-1}$.

3.29. For O_2: $\Delta S = 5 \times 8.3145 \times \ln 2$

$= 28.8 \text{ J K}^{-1}$

The entropy change for the expansion of the N_2 is the same, and the net entropy change is thus

57.6 J K^{-1}

3.30. The process can be imagined as occurring by the following reversible processes:

Step 1: The water freezes to ice at 0 °C:

$$\Delta S_2 = -\frac{6020}{273.15} = -22.04 \text{ J K}^{-1}$$

Step 2: The ice is cooled reversibly to -12 °C:

$$\Delta S_3 = 37.7 \ln \frac{261.15}{273.15} = -1.69 \text{ J K}^{-1}$$

Net ΔS(system) $= -22.04 - 1.69 = -23.73$ J K^{-1}

The heat gained by the freezer $= 6020 + 12 \times 37.7 = 6472.4$ J

This was gained at -12 °C, and therefore

$$\Delta S(\text{surroundings}) = \frac{6472.4}{261.15} = 24.78 \text{ J K}^{-1}$$

The net entropy change is ΔS(system) $+ \Delta S$(surroundings) $= 1.05$ J K^{-1}.

3.31. For the freezing of 1 mol of water at 0 °C: $\Delta S = -\frac{6020}{273.15} = -22.04$ J K^{-1}

For the reversible cooling of the ice from 0 °C to -10 °C:

$$\Delta S = 37.7 \ln \frac{263.15}{273.15} = -1.41 \text{ J K}^{-1}$$

Net ΔS(system) $= -22.04 - 1.41 = -23.45$ J K^{-1}

The heat gained by the freezer $= 6020 + (10 \times 37.7) = 6397$ J

This was gained at -10 °C, so that the entropy change in the freezer is

$$\frac{6397}{263.15} = 24.31 \text{ J K}^{-1}$$

The net entropy change is 0.86 J K^{-1}.

3.32. The final temperature T is calculated in terms of the heat balance:

$$2 \times 75.3 \times (60 - T) = 4 \times 75.3 \times (T - 20)$$
$$T = 33.33 \text{ °C} = 306.48 \text{ K}$$
$$\Delta S = 2 \times 75.3 \ln \frac{306.48}{333.15} + 4 \times 75.3 \ln \frac{306.48}{293.15}$$
$$= -12.6 + 13.4 = 0.8 \text{ J K}^{-1}$$

3.33. There are obviously several reversible paths that can be constructed between the initial and final states in this case. Let us consider four of them.

1. *Isothermal expansion to the final volume followed by constant volume cooling to the final temperature.*

$$\Delta U = 0 + C_{V,m}(T_f - T_i) = \frac{3}{2} R(253.2 - 298.0) = -558.7 \text{ J mol}^{-1}$$

$$\Delta H = 0 + C_{P,m}(T_f - T_i) = \frac{5}{2} R(253.2 - 298.0) = -931.2 \text{ J mol}^{-1}$$

$$\Delta S = R \ln \left(\frac{V_f}{V_i} \right) + C_{V,m} \ln \left(\frac{T_f}{T_i} \right) = R \ln \left(\frac{10.526}{2.478} \right) + \frac{3}{2} R \ln \left(\frac{253.2}{298.0} \right) = 9.994 \text{ J K}^{-1}\text{mol}^{-1}$$

2. *Isothermal expansion to the final pressure followed by constant pressure cooling to the final temperature.*

$$\Delta U = 0 + C_{V,m}(T_f - T_i) = \tfrac{3}{2}R(253.2 - 298.0) = -558.7 \text{ J mol}^{-1}$$

$$\Delta H = 0 + C_{P,m}(T_f - T_i) = \tfrac{5}{2}R(253.2 - 298.0) = -931.2 \text{ J mol}^{-1}$$

$$\Delta S = R\ln\left(\frac{P_i}{P_f}\right) + C_{P,m}\ln\left(\frac{T_f}{T_i}\right) = R\ln\left(\frac{10.0}{2.0}\right) + \tfrac{3}{2}R\ln\left(\frac{253.2}{298.0}\right) = 9.995 \text{ J K}^{-1}\text{mol}^{-1}$$

3. *Isothermal expansion to (P_0, V_0) followed by adiabatic expansion to the final state.*

Note that we need to find the intersection of the isotherm that passes through the initial state and the adiabat that passes through the final state. This intersection is (P_0, V_0), at T_i. Using the relationships for adiabatic processs (Eq. 2.90),

$$V_0 T_i^{(3/2)} = V_f T_f^{(3/2)}; \text{ Therefore,}$$

$$V_0 = 10.526 \text{ dm}^3 \times \left(\frac{253.2}{298.0}\right)^{3/2} = 8.244 \text{ dm}^3.$$

$$\Delta U = 0 + C_{V,m}(T_f - T_i) = \tfrac{3}{2}R(253.2 - 298.0) = -558.7 \text{ J mol}^{-1}$$

$$\Delta H = 0 + C_{P,m}(T_f - T_i) = \tfrac{5}{2}R(253.2 - 298.0) = -931.2 \text{ J mol}^{-1}$$

$$\Delta S = R\ln\left(\frac{V_0}{V_i}\right) + 0 = R\ln\left(\frac{8.244}{2.478}\right) = 9.994 \text{ J K}^{-1}\text{mol}^{-1}$$

4. *Constant Pressure heating to the final volume followed by constant volume cooling to the final pressure.*

The gas will have to be heated to $T_0 = 1266.0$ K in order for it to reach the volume of 10.526 dm^3 at 10.0 bar pressure. Therefore,

$$\Delta U = C_{V,m}(T_0 - T_i) + C_{V,m}(T_f - T_0) = \tfrac{3}{2}R(1266.0 - 298.0 + 253.2 - 1266.0) = -558.7 \text{ J mol}^{-1}$$

$$\Delta H = C_{P,m}(T_0 - T_i) + C_{P,m}(T_f - T_0) = \tfrac{5}{2}R(1266.0 - 298.0 + 253.2 - 1266.0) = -931.2 \text{ J mol}^{-1}$$

$$\Delta S = C_{P,m}\ln\left(\frac{T_0}{T_i}\right) + C_{V,m}\ln\left(\frac{T_f}{T_0}\right) = \tfrac{5}{2}R\ln\left(\frac{1266.0}{298.0}\right) + \tfrac{3}{2}R\ln\left(\frac{253.2}{1266.0}\right) = 9.995 \text{ J K}^{-1}\text{mol}^{-1}$$

Yet another path we can try is constant volume cooling to the final pressure followed by constant pressure heating to the final temperature.

In each of these cases, we have verified that ΔU, ΔH and ΔS are the same, thus proving that they are independent of the path taken, as any state property should be. We now have to find the entropy change of the surroundings.

Entropy change of the surroundings.

The actual process is the expansion of the gas against a constant external pressure of 2 bar. For this process, according to the first law,

$$\Delta U = q_{\text{act}} - P_{\text{ext}}(V_f - V_i); \text{ Therefore,}$$

$$q_{\text{act}} = \Delta U + P_{\text{ext}}(V_f - V_i) = -558.7 + 2.0 \times (10.526 - 2.478) \times (8.3145/0.083145)$$

$$= 1050.9 \text{ J mol}^{-1}.$$

$$\Delta S_{surr} = -q_{act}/T_{surr} = -1050.9 \text{ J mol}^{-1}/298 \text{ K} = -3.526 \text{ J K}^{-1}\text{mol}^{-1}$$

$$\Delta S_{univ} = \Delta S + \Delta S_{surr} = 6.468 \text{ J K}^{-1}\text{mol}^{-1}.$$

This is, therefore, a spontaneous process.

3.34. $\Delta S(\text{system}) = 5 \times 75.3 \ln \dfrac{276.15}{323.15} = -59.18 \text{ J K}^{-1}$

The heat accepted by the refrigerator is

$$5 \times 75.3 \times (50 - 3) = 17\ 696 \text{ J}$$

$$\Delta S(\text{refrigerator}) = 17\ 696/276.15 = 64.08 \text{ J K}^{-1}$$

$$\text{Net } \Delta S = 64.08 - 59.18 = 4.90 \text{ J K}^{-1}$$

3.35. a. In this case all of the ice melts and the final temperature is 12 °C. (See the solution to Problem 2.32.) The entropy changes are

1. Reversible melting of the ice at 0 °C:

$$\Delta S_1 = 6025 \text{ J mol}^{-1} \times (100/18) \text{ mol}/273.15 \text{ K}$$

$$= 122.5 \text{ J K}^{-1}$$

2. Reversible heating of 100 g of water from 0 °C to 12 °C:

$$\Delta S_2 = 75.3 \text{ J K}^{-1} \text{ mol}^{-1} \times (100/18) \text{ mol } \ln (285.15/273.15)$$

$$= 18.0 \text{ J K}^{-1}$$

3. Reversible cooling of 1 kg of water from 20 °C to 12 °C:

$$\Delta S_3 = (1000/18) \text{ mol} \times 75.3 \text{ J K}^{-1} \text{ mol}^{-1} \ln (285.15/293.15)$$

$$= -115.7 \text{ J K}^{-1}$$

The net entropy change is therefore

$$\Delta S = 122.5 + 18.0 - 115.7 = 24.8 \text{ J K}^{-1}$$

b. In this case only 250 g of the ice melts, and the final temperature of the water is 0 °C. (See the solution to Problem 2.32.) The entropy changes are now

1. For the reversible melting of 250 g of ice at 0 °C:

$$\Delta S_1 = (250/18) \text{ mol} \times 6025 \text{ J mol}^{-1}/273.15 \text{ K}$$

$$= 306.4 \text{ J K}^{-1}$$

2. For the cooling of 1 kg of water from 20 °C to 0 °C:

$$\Delta S_2 = (1000/18) \text{ mol} \times 75.3 \text{ J K}^{-1} \text{ mol}^{-1} \ln (273.15/293.15)$$

$$= -295.6 \text{ J K}^{-1}$$

The net entropy change is

$$\Delta S = 306.4 - 295.6 = 10.8 \text{ J K}^{-1}.$$

3.36. Using the expression for $C_{P,m}$ given in Table 2.1,

$$S°(800.0 \text{ K})/(\text{J K}^{-1}\text{mol}^{-1}) = 151.94 + \int_{77.32}^{800.0} \left(\frac{28.58 + 3.76\times10^{-3}T - 5.0\times10^4/T^2}{T} \right) dT$$

$$= 217.3 \text{ J K}^{-1}\text{mol}^{-1}$$

■ Gibbs and Helmholtz Energies

3.37. From Eq. 3.91,

$$\Delta G° = 2(-181.64) + 2(-394.36) - (-914.54)$$

$$= -237.46 \text{ kJ mol}^{-1}$$

$$\Delta H° = 2(-288.3) + 2(-393.51) - (-1263.07)$$

$$= -100.55 \text{ kJ mol}^{-1}$$

$$\Delta S° = \frac{\Delta H° - \Delta G°}{T} = \frac{-100\ 550 + 237\ 460}{298.15}$$

$$= 459.2 \text{ J K}^{-1}\text{mol}^{-1}$$

3.38. Work done by system $= P\Delta V = 30.19 \text{ dm}^3 \text{ atm mol}^{-1}$

$$= 3059 \text{ J mol}^{-1}$$

$$\Delta H = 40\ 600 \text{ J mol}^{-1}$$

$$\Delta U = \Delta H - \Delta (PV)$$

$$= 40\ 600 - 3\ 059 = 37\ 540 \text{ J mol}^{-1}$$

$$\Delta G = 0 \qquad \Delta S = \frac{40\ 600}{373.15} = 108.8 \text{ J K}^{-1}\text{mol}^{-1}$$

3.39. From Example 3.6, ΔH and ΔS are available.

a. At 0°C, $\Delta H = -6020 \text{ J mol}^{-1}$

$$\Delta S = -22.04 \text{ J K}^{-1}\text{mol}^{-1}$$

$$\Delta G = -6020 + 22.04 \times 273.15 = 0$$

b. At -10°C, $\Delta H = -5644 \text{ J mol}^{-1}$

$$\Delta S = -20.64 \text{ J K}^{-1}\text{mol}^{-1}$$

$$\Delta G = -5644 + 20.64 \times 263.15 = -213 \text{ J mol}^{-1}$$

3.40. $\Delta U = 0$

$$\Delta H = \Delta U + \Delta (PV) = 0$$

$$\Delta S = \frac{q_{rev}}{T} = \frac{1}{T} \int_{V_1}^{V_2} P dV \quad = R \ln \frac{V_2}{V_1} = 8.3145 \ln 10$$

$$= 19.14 \text{ J K}^{-1} \text{ mol}^{-1}$$

$$\Delta A = \Delta U - T\Delta S = -298.15 \times 19.14 = -5.706 \text{ kJ mol}^{-1}$$

$$\Delta G = -5.706 \text{ kJ mol}^{-1}$$

The quantities are all state functions, and the preceding values therefore do not depend on how the process is carried out.

3.41. a. (a) $\Delta G \quad = -85\ 200 + (300 \times 170.2) \text{ J mol}^{-1}$

$$= -34.14 \text{ kJ mol}^{-1}$$

   (b) $\Delta G \quad = -85\ 200 + (600 \times 170.2) \text{ J mol}^{-1}$

$$= 16.92 \text{ kJ mol}^{-1}$$

   (c) $\Delta G \quad = -85\ 200 + (1000 \times 170.2) \text{ J mol}^{-1}$

$$= 85.00 \text{ kJ mol}^{-1}$$

   b. At $T = 85\ 200/170.2 = 500.6$ K

3.42. From Eq. (3.169), we write

$$\left[\frac{\partial}{\partial T} \left(\frac{\Delta_c G^\circ}{T} \right) \right]_P \approx \frac{1}{T_2 - T_1} \left(\frac{\Delta_c G_2^\circ}{T_2} - \frac{\Delta_c G_1^\circ}{T_1} \right) = -\frac{\Delta_c H^\circ}{T^2},$$

Where T is the mid-point of the temperature range (T_1, T_2). In the limit $\Delta T \to 0$, this will yield Eq. (3.169). Substituting, we get

$$\frac{1}{348.0 - 298.0} \left(\frac{-802.57}{348.0} - \frac{-815.04}{298.0} \right) = 8.576 \times 10^{-3} = -\frac{\Delta_c H^\circ}{323.0^2}$$

Therefore, $\Delta_c H^\circ = -(323.0)^2 \times 8.567 \times 10^{-3} = -894.73 \text{ kJ mol}^{-1}$.

3.43. Conversion of water to vapor at 0.0313 atm:

$$\Delta H \quad = 44.01 \text{ kJ mol}^{-1}$$

$$\Delta S \quad = \frac{44\ 010}{298.15} = 147.6 \text{ J K}^{-1} \text{ mol}^{-1}$$

Reversible isothermal expansion from 0.0313 atm to 10^{-5} atm:

$$\Delta H = 0$$

$$\Delta S \quad = 8.3145 \ln \frac{0.0313}{10^{-5}} = 66.9 \text{ J K}^{-1} \text{ mol}^{-1}$$

Net $\Delta H \quad = 44.01 \text{ kJ mol}^{-1}$

$$\Delta S \;\; = 147.6 + 66.9 = 214.5 \text{ J K}^{-1} \text{ mol}^{-1}$$

$$\Delta G \;\; = 44\,010 - 298.15 \times 214.5 = -19\,940 \text{ J mol}^{-1}$$

$$= -19.94 \text{ kJ mol}^{-1}$$

3.44.
a. $\Delta U, \Delta H$
b. ΔS
c. ΔH
d. ΔH
e. None
f. ΔG
g. ΔU
h. None

3.45. $\left(\dfrac{\partial G}{\partial P}\right)_T = V$

Therefore

$$\Delta G = \int V dP$$

The molar volume of mercury is

$$V_m = \frac{200.6 \text{ g mol}^{-1}}{13.5 \text{ g cm}^{-3}} = 1.486 \times 10^{-5} \text{ m}^3 \text{ mol}^{-1}$$

Then $\Delta G_m \;\; = 1.486 \times 10^{-5} \text{ m}^3 \times 999 \text{ bar} \times 10^5 \text{ Pa bar}^{-1}$

$$= 1485 \text{ J mol}^{-1} = 1.485 \text{ kJ mol}^{-1}$$

3.46. $\left(\dfrac{\partial G_m}{\partial T}\right)_P = -S_m$

$$= -(A + B \ln T)$$

where $A = 36.36 \text{ J K}^{-1} \text{ mol}^{-1}$

$B = 20.79 \text{ J K}^{-1} \text{ mol}^{-1}$

$$\Delta G_m \;\; = -\int_{298.15}^{323.15} (A + B \ln T)\; dT$$

$$= -[AT + B(T \ln T - T)] \;\Big|_{298.15}^{323.15}$$

$$= -(A - B)\,50 - B\,(323.15 \ln 323.15) - 298.15 \ln 298.15)$$

$$= -778.5 - 3502.3 \text{ J mol}^{-1} = -4.28 \text{ kJ mol}^{-1}.$$

3.47. The entropy at 300.0 K is calculated as

$$S°(300.0 \text{ K})/(\text{J K}^{-1}\text{mol}^{-1}) = 1.26 + \int_{15.0}^{197.64} \frac{32.65}{T} dT + \frac{7402}{197.64} + \int_{197.64}^{263.08} \frac{87.20}{T} dT$$

$$+ \frac{24\,937}{263.08} + \int_{263.08}^{300.0} \frac{39.88}{T} dT = 247.86 \text{ J K}^{-1}\text{mol}^{-1}$$

3.48. ΔU and ΔH are zero for the isothermal expansion of an ideal gas.

$$q = -w = 500 \text{ J mol}^{-1}$$

$$\Delta S = R \ln 2 = 5.76 \text{ J K}^{-1} \text{ mol}^{-1}$$

$$\Delta G = -T\Delta S = -1728 \text{ J mol}^{-1} = -1.73 \text{ kJ mol}^{-1}$$

$$w_{rev} = +1.73 \text{ kJ mol}^{-1}; \quad q_{rev} = -1.73 \text{ kJ mol}^{-1}$$

3.49. Since $P_{ext} = 0$ (evacuated vessel), no work is done; $w = 0$.

$$\Delta U = q + w = q = 30 \text{ kJ mol}^{-1}.$$

For $H_2O(l) \rightarrow H_2O(g)$, $\Delta n = 1$ (see Example 2.5)

$$\Delta H = \Delta U + \Delta(PV) = \Delta U + \Delta n RT$$

$$= 30\,000 + 3102 = 33\,102 \text{ J mol}^{-1} = 33.1 \text{ kJ mol}^{-1}$$

To obtain the entropy change, consider the reversible processes:

(1) H_2O (l, 100 °C) $\rightarrow H_2O$ (g, 100 °C, 1 atm)

$$\Delta S_1 = \frac{40\,600}{373.15} = 108.8 \text{ J K}^{-1}$$

(2) H_2O (g, 100 °C, 1 atm) $\rightarrow H_2O$ (g, 100 °C, 0.5 atm)

$$\Delta S_2 = R \ln \frac{V_2}{V_1} = R \ln 2 = 5.76 \text{ J K}^{-1} \text{ mol}^{-1}$$

Net $\Delta S = 114.6 \text{ J K}^{-1} \text{ mol}^{-1}$

Net $\Delta G = \Delta H - T\Delta S$

$$= 33\,102 - 42\,760 = -9658 \text{ J mol}^{-1}$$

$$= -9.66 \text{ kJ mol}^{-1}$$

3.50. Allow the process to occur by the following reversible steps:

(1) H_2O (g, 100 °C, 2 atm) $\rightarrow H_2O$ (g, 100 °C, 1 atm)

$$\Delta H_1 = 0$$

$$\Delta S_1 = R \ln V_2/V_1 = R \ln 2 = 5.76 \text{ J K}^{-1} \text{ mol}^{-1}$$

$$\Delta G_1 = -T\Delta S_1 = -2150 \text{ J mol}^{-1}$$

(2) H_2O (g, 100 °C, 1 atm) $\rightarrow H_2O$ (l, 100 °C, 1 atm)

$$\Delta H_2 = -40\,600 \text{ J mol}^{-1}$$

$$\Delta S_2 = -\frac{40\,600}{373.15} = -108.8 \text{ J K}^{-1} \text{ mol}^{-1}$$

$$\Delta G_2 = 0 \text{ (reversible process at constants } P \text{ and } T)$$

(3) H_2O (l, 100 °C, 1 atm) → H_2O (l, 100 °C, 2 atm)

The ΔH, ΔS, and ΔG changes are negligible for this process.

The overall changes are thus

$$\Delta H = -40\ 600 \text{ J mol}^{-1} = -40.6 \text{ kJ mol}^{-1}$$

$$\Delta S = 5.76 - 108.8 = -103 \text{ J K}^{-1} \text{ mol}^{-1}$$

$$\Delta G = -2.15 \text{ kJ mol}^{-1}$$

3.51. $\Delta U = \displaystyle\int_{T_1}^{T_2} C_{V,m} dT$

$$C_{V,m} = C_{P,m} - R$$

$$= (20.27 + 1.76 \times 10^{-2}\ T/\text{K}) \text{ J K}^{-1} \text{ mol}^{-1}$$

$$q = 0; \quad C_{V,m} dT = dw = -P_2 dV$$

$$\Delta U_m = 300 \int_{T_1}^{T_2} (20.27 + 1.76 \times 10^{-2}\ T\ /\ K)\ dT$$

$$= -P_2 \int_{V_1}^{V_2} dV = -P_2 \left(\frac{RT_2}{P_2} - \frac{300R}{P_1} \right)$$

$$= -4 \times 8.3145 \left(\frac{T_2}{4} - \frac{300}{10} \right)$$

$$20.27\ (T_2 - 300) + 0.0088\ (T_2^2 - 300^2) = -4 \times 8.3145 \left(\frac{T_2}{4} - 30 \right)$$

$$0.0088\ T_2^2 + 28.58\ T_2 - 7871 = 0$$

$$T_2 = \frac{-28.58 \pm (816.816 + 277.06)^{1/2}}{0.0176}$$

$$= \frac{-28.58 + 33.07}{0.0176} = 255.3 \text{ K}$$

$$\Delta U_m = 20.27\ (255.3 - 300) + 0.0088\ (255.3^2 - 300^2) = -1124.5 \text{ J mol}^{-1}$$

$$\Delta H_m = 28.58\ (255.3 - 300) + 0.0088\ (255.3^2 - 300^2) = -1495.9 \text{ J mol}^{-1}$$

$$dS = \frac{dq}{T} = \frac{dU + PdV}{T}$$

For 1 mol of ideal gas, $PV_m = RT$

$$d(PV_m) = RdT = PdV_m + V_m dP$$

$$PdV_m = RdT - V_m dP = RdT - \frac{RT}{P}\ dP$$

$$dS_m = \frac{dU_m + RdT - \dfrac{RTdP}{P}}{T} = \frac{C_{P,m}dT}{T} - R\frac{dP}{P}$$

$$\Delta S_m / \text{J K}^{-1} \text{ mol}^{-1} = \int_{300}^{255.3} \frac{(28.58 + 0.0176\ T)\ dT}{T} - 8.314 \ln \frac{4}{10}$$

$$= 28.58 \ln \frac{255.3}{300} + 0.0176\ (255.3 - 300) + 7.618 = 2.22$$

$$\Delta S_m = 2.22 \text{ J K}^{-1} \text{ mol}^{-1}$$

3.52. $\Delta H° = 2(-285.85) - 393.51 + 74.81 = -890.4 \text{ kJ mol}^{-1}$

$\Delta G° = 2(-237.13) - 394.36 + 50.72 = -817.9 \text{ kJ mol}^{-1}$

$$\Delta S° = \frac{\Delta H° - \Delta G°}{T} = \frac{(890.4 + 817.9) \times 1000}{298.15} = -243.2 \text{ J K}^{-1} \text{ mol}^{-1}$$

3.53. a. and b. True only for an ideal gas.

c. True only if the process is reversible.

d. True only for the total entropy.

e. True only for an isothermal process occurring at constant pressure.

3.54. First vaporize the water at 1 atm pressure:

$$\Delta S_1 = \frac{40\ 600}{373.15} = 108.8 \text{ J K}^{-1}$$

Then expand from 1 bar to 0.1 bar:

$$\Delta S_2 = R \ln 10 = 19.1 \text{ J K}^{-1}$$

The net entropy change is

$$\Delta S = \Delta S_1 + \Delta S_2 = 108.8 + 19.1$$

$$= 127.9 \text{ J K}^{-1}$$

The Gibbs energy change is

$$\Delta G = \Delta H - T\Delta S = 40\ 600 - (373.15 \times 127.9)$$

$$= -7125.9 \text{ J} = -7.13 \text{ kJ}$$

3.55. $\Delta H° = -205.0 + 104.6 = -100.4 \text{ kJ mol}^{-1}$

$\Delta G° = -108.74 + 32.2 = -76.5 \text{ kJ mol}^{-1}$

$$\Delta S° = \frac{\Delta H° - \Delta G°}{T} = \frac{-100\ 400 + 76\ 540}{298.15}$$

$$= -80.0 \text{ J K}^{-1} \text{ mol}^{-1}$$

■ Energy Conversion

3.56. 100-atm engine: $T_h = 585$ K; $T_c = 303$ K

Efficiency = $\dfrac{585 - 303}{585} = 48.2\%$

5-atm engine: $T_h = 425$; $T_c = 303$ K

Efficiency = $\dfrac{425 - 303}{425} = 28.7\%$

3.57. $\dfrac{w}{q_c} = \dfrac{293 - 269}{269} = 0.089$

$w = 0.089 \times 10^4 = 890$ J min^{-1} = 14.8 J s^{-1}

At 40% efficiency, power = 14.8/0.4 = 37.0 J s^{-1}

$= 37.0$ W

3.58. Performance factor = $\dfrac{T_h}{T_h - T_c}$

a. 59.6%;   b. 11.9%;   c. 6.6%

3.59. Efficiency = $\dfrac{1073}{2273} = 0.472$

1 liter = 8.0 kg = $\dfrac{8000 \text{ g}}{114.2 \text{ g mol}^{-1}} = 70.05$ mol

Energy produced = 70.05 mol $\times$ 5500 kJ mol^{-1}

$= 3.85 \times 10^5$ kJ

Work = $0.472 \times 3.85 \times 10^5 = 1.82 \times 10^5$ kJ

3.60. Let q_h be the heat supplied to the building at 20 °C and q_c be the heat taken in by the heat pump at 10 °C:

$\dfrac{q_h}{q_c} = \dfrac{293.15}{283.15}$

Work supplied to heat pump:

$w = q_h - q_c = q_h \left(1 - \dfrac{283.15}{293.15} \right) = 0.034\, q_h$

Let q_h' be the heat produced by the fuel at 1000 °C and q_c' be the heat rejected at 20 °C. Work performed by the heat engine and supplied to the heat pump is

$w = q_h' - q_c' = q_h' \left(1 - \dfrac{293.15}{1273.15} \right) = 0.770\, q_h'$

Thus

$$0.034\, q_h = 0.770\, q_h'$$

and the performance factor is

$$\frac{q_h}{q_h'} = \frac{0.770}{0.034} = 22.6$$

3.61. The heat that must be removed from 1 kg (= 55.5 mol) of water in order to freeze it is

$$6.02 \times 55.5 = 334 \text{ kJ} = q_c$$

This is the heat gained by the refrigerator.

a. If the efficiency were 100%,

$$\frac{T_h}{T_c} = \frac{298.15}{273.15} = -\frac{q_h}{q_c} = \frac{-q_h}{334 \text{ kJ}}$$

Thus, the heat discharged at 25 °C is

$$-q_h = 365 \text{ kJ}$$

Work required to be supplied to the refrigerator is

$$365 - 334 = 31 \text{ kJ}$$

With an efficiency of 40%, the actual work will be

$$\frac{100}{40} \times 31 = 78 \text{ kJ}$$

b. The heat discharged at 25 °C will be

$$334 + 78 = 412 \text{ kJ}.$$

■ Thermodynamic Relationships

3.62. a. Using the given relationship and the definitions of α and κ, we have

$$\left(\frac{\partial P}{\partial T}\right)_T = -\frac{(\partial V / \partial T)_P}{(\partial V / \partial P)_T} = -\frac{\alpha}{\kappa}.$$

Substituting into Eq. (3.128), we obtain

$$\left(\frac{\partial U}{\partial V}\right)_T = -P + T\left(\frac{\alpha}{\kappa}\right) = \frac{\alpha T - \kappa P}{\kappa}.$$

b. Using the chain rule of partial differentiation, we obtain

$$\left(\frac{\partial U}{\partial P}\right)_T = \left(\frac{\partial U}{\partial V}\right)_T \left(\frac{\partial V}{\partial P}\right)_T = \left(\frac{\alpha T - \kappa P}{\kappa}\right)(-\kappa V) = V(\kappa P - \alpha T).$$

3.63.
$$\left(\frac{\partial H}{\partial P}\right)_T = T\left(\frac{\partial S}{\partial P}\right)_T + V.$$

Since $-\left(\frac{\partial S}{\partial P}\right)_T = \left(\frac{\partial V}{\partial T}\right)_P$, we get

$$\left(\frac{\partial H}{\partial P}\right)_T = V - T\left(\frac{\partial V}{\partial T}\right)_P.$$

Since $\alpha = \frac{1}{V}\left(\frac{\partial V}{\partial T}\right)_P$, we get

$$\left(\frac{\partial H}{\partial P}\right)_T = V\left(1 - \alpha T\right).$$

3.64.   a. $V_m = RT/P$. Therefore, $(\partial V_m/\partial T)_P = R/P$, and the cubic expansion coefficient is

$$\alpha = \frac{1}{V_m}\left(\frac{\partial V_m}{\partial T}\right)_P = R/(PV_m) = 1/T.$$

   b.   Since $V_m = RT/P$, $(\partial V_m/\partial P)_T = -RT/P^2$. Therefore,

$$\kappa = -\frac{1}{V_m}\left(\frac{\partial V_m}{\partial P}\right)_T = RT^2/(P^2 V_m) = 1/P,$$

since $(RT/P\,V_m) = 1$.

3.65.   From the first and second laws

$$dU = TdS - PdV$$

Therefore

$$\left(\frac{\partial U}{\partial V}\right)_T = T\left(\frac{\partial S}{\partial V}\right)_T - P$$

$$= T\left(\frac{\partial P}{\partial T}\right)_V - P$$

using the Maxwell equation, Eq. 3.124. From the van der Waals equation

$$\left(\frac{\partial P}{\partial T}\right)_V = \frac{P}{V_m - b} = \frac{1}{T}\left(P + \frac{a}{V_m^2}\right)$$

$$\left(\frac{\partial U}{\partial V}\right)_T = \frac{a}{V_m^2}$$

3.66. $\mu \equiv \left(\frac{\partial T}{\partial P}\right)_H = \frac{-(\partial H/\partial P)_T}{(\partial H/\partial T)_P} = -\frac{1}{C_P}\left(\frac{\partial H}{\partial P}\right)_T$

Since $dH = TdS + VdP$,

$$\left(\frac{\partial H}{\partial P}\right)_T = T\left(\frac{\partial S}{\partial P}\right)_T + V$$

$$= -T\left(\frac{\partial V}{\partial T}\right)_P + V$$

from the Maxwell equation, Eq. 3.125. Thus

$$\mu = \frac{T\left(\dfrac{\partial V}{\partial T}\right)_P - V}{C_P}$$

$$= \frac{T\left(\dfrac{\partial V_m}{\partial T}\right)_P - V_m}{C_{P,m}} \text{ for 1 mol}$$

The equation $P(V_m - b) = RT$ applies to 1 mol of gas, and it follows that

$$\left(\frac{\partial V}{\partial T}\right)_P = \frac{R}{P}$$

Therefore

$$\mu = \frac{\dfrac{RT}{P} - V_m}{C_{P,m}}$$

3.67.   a. $\left(\dfrac{\partial G}{\partial T}\right)_P = -S$ (Eq. 3.119)

Therefore

$$\left(\frac{\partial^2 G}{\partial T^2}\right)_P = -\left(\frac{\partial S}{\partial T}\right)_P$$

$$dS = \frac{q_{\text{rev}}}{T} = \frac{C_P dT}{T}$$

$$\left(\frac{\partial S}{\partial T}\right)_P = \frac{C_P}{T}$$

$$C_P = -T\left(\frac{\partial^2 G}{\partial T^2}\right)_P$$

b. $\left(\dfrac{\partial C_P}{\partial P}\right)_T = \left[\dfrac{\partial}{\partial P}\left(\dfrac{\partial H}{\partial T}\right)_P\right]_T = \left[\dfrac{\partial}{\partial T}\left(\dfrac{\partial H}{\partial P}\right)_T\right]_P$

$\left(\dfrac{\partial H}{\partial P}\right)_T = -T\left(\dfrac{\partial V}{\partial T}\right)_P + V$ (see Problem 3.66)

$$\left[\frac{\partial}{\partial T} \left(\frac{\partial H}{\partial P} \right)_T \right]_P = -T \left(\frac{\partial^2 V}{\partial T^2} \right)_P - \left(\frac{\partial V}{\partial T} \right)_P + \left(\frac{\partial V}{\partial T} \right)_P$$

$$\left(\frac{\partial C_P}{\partial P} \right)_T = -T \left(\frac{\partial^2 V}{\partial T^2} \right)_P$$

3.68. $A = U - TS$

$dA = dU - TdS - SdT$

At constant temperature

$dA = dU - TdS = dq + dw - TdS$

But $dq = TdS$, and therefore $dA = dw$.

3.69. $dU = TdS - PdV$ (Eq. 3.105)

Dividing by dV at constant T:

$$\left(\frac{\partial U}{\partial V} \right)_T = T \left(\frac{\partial S}{\partial V} \right)_T - P = 0$$

But

$$\left(\frac{\partial S}{\partial V} \right)_T = \left(\frac{\partial P}{\partial T} \right)_V$$ (Eq. 3.124)

and therefore

$$\left(\frac{\partial P}{\partial T} \right)_V = \frac{P}{T}$$

Integrating, $\ln P = \ln T + \text{const}$, or $P \propto T$.

Thus

$PV = \text{const} \times T$

3.70. By Euler's reciprocity theorem (Appendix C),

$$\left(\frac{\partial S}{\partial V} \right)_U = - \frac{(\partial U / \partial V)_S}{(\partial U / \partial S)_V}$$

$$\left(\frac{\partial U}{\partial V} \right)_S = -P \quad \text{(Eq. 3.116)} \qquad \left(\frac{\partial U}{\partial S} \right)_V = T \quad \text{(Eq. 3.116)}$$

$$\therefore \left(\frac{\partial S}{\partial V} \right)_U = \frac{P}{T}$$

For an ideal gas, U depends only on T so that

$$\left(\frac{\partial S}{\partial V} \right)_U = \left(\frac{\partial S}{\partial V} \right)_T$$

For an isothermal process involving n mol of an ideal gas

$$dS = nRd \ln V = \frac{nRdV}{V} = \frac{PdV}{T}$$

Thus, $(\partial S/\partial V)_T = P/T$ and therefore $(\partial S/\partial V)_U = P/T$.

3.71. a. Eq. (3.160) gives the following relationship for the fugacity of a gas:

$$RT \ln\left(\frac{f_2}{P_2}\right) = \int_{P_1}^{P_2}\left(V_m - \frac{RT}{P}\right)dP = \int_{P_1}^{P_2}\left(\frac{PV_m - RT}{P}\right)dP.$$

Dividing both sides by RT and using the definition of the compression factor Z, we get

$$\ln\left(\frac{f_2}{P_2}\right) = \int_{P_1}^{P_2}\left(\frac{Z-1}{P}\right)dP,$$

Which is the desired result.

b. Dividing both sides of the given equation of state by RT, we get

$Z = 1 + [b - A/(RT^{2/3})](P/RT) = 1 + [b/(RT) - A/(RT^{5/3})]P$, which means

$(Z-1)/P = b/(RT) - A/(RT^{5/3})$.

Now, setting $P_1 = 0$ and integrating, we obtain

$$\ln\left(\frac{f}{P_2}\right) = \int_0^{P_2}\left(\frac{Z-1}{P}\right)dP = \left(\frac{b}{RT} - \frac{A}{RT^{5/3}}\right)P_2.$$

3.72. Using the expression for fugacity derived in Problem (3.71), we get

$$\ln\left(\frac{f}{P}\right) = \int_0^P\left(\frac{Z-1}{P}\right)dP = \frac{1}{RT}\left(b - \frac{a}{RT}\right)\int_0^P dP + \left(\frac{b}{RT}\right)^2\int_0^P PdP$$

which yields

$$\ln\left(\frac{f}{P}\right) = \frac{1}{RT}\left(b - \frac{a}{RT}\right)P + \frac{1}{2}\left(\frac{b}{RT}\right)^2 P^2 = -0.622\,84.$$

Therefore, the fugacity of the gas $f = P \times \exp(-0.622\,84) = 268$ bar.

4. CHEMICAL EQUILIBRIUM

■ Equilibrium Constants

4.1.

	2A	$\rightleftarrows$	Y	+	2Z	
Initial amounts:	4		0		0	mol
Amounts at equilibrium:	1		1.5		3.0	mol
Concentrations at equilibrium:	$\dfrac{1}{5}$		$\dfrac{1.5}{5}$		$\dfrac{3.0}{5}$	mol dm^{-3}

$$K_c = \frac{(1.5/5)\,(3.0/5)^2}{(1/5)^2} \text{ mol dm}^{-3} = \frac{(1.5)\,(3)^2}{(5)} \text{ mol dm}^{-3}$$

$$= 2.7 \text{ mol dm}^{-3}$$

4.2.

	A	+	B	$\rightleftarrows$	Y	+	Z	
Initial amounts:	x		3		0		0	mol
Amounts at equilibrium:	$x-2$		1		2		2	mol

$$\frac{2 \times 2}{(x-2) \times 1} = 0.1; \quad x - 2 = 40$$

$$x = 42$$

Thus, initially there must be 42 mol of A.

4.3.

	A	+	2B	$\rightleftarrows$	Z	
Initial amounts:	x		4		0	mol
Amounts at equilibrium:	$x-1$		2		1	mol
Concentrations at equilibrium:	$\dfrac{x-1}{5}$		$\dfrac{2}{5}$		$\dfrac{1}{5}$	mol dm^{-3}

$$\frac{1/5}{[(x-1)/5]\,(2/5)^2} = 0.25 = \frac{25}{4(x-1)}$$

$$25 = x - 1 \quad x = 26$$

Thus, initially there must be 26 mol of A.

4.4. Two moles of $SO_3(g)$ produce 3 mol of product; thus $\Sigma v = +1$ mol. Then, from Eq. 4.26,

$$K_P = K_c(RT)^{+1}$$

$$= (0.0271 \text{ mol dm}^{-3})\,(8.3145 \text{ J K}^{-1} \text{ mol}^{-1} \times 1100 \text{ K})$$

$$= 247.8 \text{ J dm}^{-3} = 2.478 \times 10^5 \text{ J m}^{-3}$$

$$= 2.478 \times 10^5 \text{ Pa} = 2.48 \text{ bar}$$

4.5

	I_2	$\rightleftarrows$	$2I$
Initial amounts:	0.0061		0 mol
Equilibrium amounts:	0.0061(1 − 0.0274)		0.0061 × 2 × 0.0274 mol
	= 5.93 × 10⁻³		= 3.3428 × 10⁻⁴ mol

$$K_c = \frac{(3.3428 \times 10^4 / 0.5)^2}{5.93 \times 10^3 / 0.5} \text{ mol dm}^{-3} = 3.77 \times 10^{-5} \text{ mol dm}^{-3}$$

$$K_P = 3.77 \times 10^{-5} \text{ mol dm}^{-3}\,(8.3145 \text{ J K}^{-1} \text{ mol}^{-1} \times 900 \text{ K})$$

$$= 0.282 \text{ Pa} = 2.82 \times 10^{-6} \text{ bar}$$

4.6. Addition of N_2 at constant *volume* and temperature necessarily requires the equilibrium to shift to the *right*.

$$K_c = \frac{[NH_3]^2}{[N_2][H_2]^3} = \frac{n^2_{NH_3} V^2}{n_{N_2} n^3_{H_2}}$$

If n_{N_2} is increased at constant V, the equilibrium must shift so as to produce more ammonia. If the *pressure* (as well as the temperature) is held constant, however, addition of N_2 requires that V is increased. If the proportional increase in V^2 is greater than the increase in n_{N_2} the equilibrium will shift to the *left* when N_2 is added.

The volume V is proportional to $n_{N_2} + n_{H_2} + n_{NH_3}$, and V^2 is proportional to

$$(n_{N_2} + n_{H_2} + n_{NH_3})^2$$

If n_{N_2} is very much larger than $n_{H_2} + n_{NH_3}$, V^2 will increase approximately with $n^2_{N_2}$ and therefore increases more strongly than n_{N_2}. If n_{N_2} is not much larger than $n_{H_2} + n_{NH_3}$, an increase in n_{N_2} will have a relatively smaller effect on V^2. The increase in ammonia dissociation when N_2 is added is therefore expected when N_2 is in excess, but not otherwise.

On the other hand, $n^3_{H_2}$ appears in the equilibrium expression; this varies more strongly than V^2, and added H_2 therefore cannot lead to the dissociation of ammonia.

4.7. Suppose that in 1 dm^3 there are

$$x \text{ mol of } N_2O_4$$

$$y \text{ mol of } NO_2$$

$$\text{Pressure} = 0.597 \text{ bar} = (x+y)RT/V$$

$$= \frac{(x+y)\text{mol} \times 0.0831 \text{ bar dm}^3\text{K}^{-1}\text{mol}^{-1} \times 298.15\text{K}}{1 \text{ dm}^3}$$

Therefore, $x + y = 0.024\ 08$ (1)

$$\text{Density} = 1.477 \text{ g dm}^{-3} = \frac{(92.02x + 46.01y) \text{ g}}{1 \text{ dm}^3} = \frac{(2x+y) \times 46.01 \text{ g}}{1 \text{ dm}^3}$$

Therefore, $2x + y = 0.032\ 10$ (2)

From Eqs. (1) and (2), we get $x = 0.00802$; $y = 0.0161$

$$N_2O_4(g) \quad \textbf{Err} \quad 2NO_2 \text{ (g)}$$

$$P(1-\alpha) \qquad\qquad 2\,P\alpha$$

Since partial pressures are proportional to the number of moles of each species present, $x \propto P(1 - \alpha)$ and $y \propto 2P\alpha$, which means that

$$\frac{y}{x} = \frac{2P\alpha}{P(1-\alpha)},$$

from which we get

$$\alpha = \frac{y}{(2x+y)} = 0.500.$$

$$K_c = \frac{(0.0161)^2}{0.00802} \text{ mol dm}^{-3} = 0.0322 \text{ mol dm}^{-3}$$

$$K_P = K_c RT \quad \text{(from Eq. 4.26, with } \Sigma v = 1)$$

$$= 0.0322 \text{ mol dm}^{-3} \times 0.083145 \times 298.15 \text{ dm}^3 \text{ bar mol}^{-1}$$

$$= 0.798 \text{ bar}$$

$$K_x = K_P P^{-1} \text{ (from Eq. 4.32)}$$

$$= K_P (1 \text{ bar})^{-1} = 0.798$$

Addition of He produces no effect, since concentrations, partial pressures, and mole fractions remain unchanged.

4.8. $1.10 \text{ g NOBr} = \dfrac{1.10 \text{ g NOBr}}{14.01 \text{ g mol}^{-1} \text{ N} + 16.00 \text{ g mol}^{-1} \text{ O} + 79.91 \text{ g mol}^{-1} \text{ Br}} = 0.010 \text{ mol NOBr}$

If α is the degree of dissociation,

$$n_{\text{NOBr}} = 0.01 \, (1 - \alpha) \text{ mol}; \quad n_{\text{NO}} = 0.01 \, \alpha \text{ mol}$$

$$n_{\text{Br}_2} = 0.005 \, \alpha \text{ mol}$$

Total amount, $n = 0.01 + 0.005 \, \alpha \text{ mol}$

$$= \frac{PV}{RT} = \frac{0.355 \text{ bar} \times 1 \text{ dm}^3}{0.0831 \times 298.15 \text{ bar dm}^3 \text{mol}^{-1}}$$

$$= 0.0143 \text{ mol}$$

$\alpha = 0.861; \quad n = 0.0143$

$n_{\text{NOBr}} = 1.39 \times 10^{-3} \text{ mol}; \quad n_{\text{NO}} = 8.61 \times 10^{-3} \text{ mol}$

$n_{\text{Br}_2} = 4.305 \times 10^{-3} \text{ mol}$

$$K_c = \frac{(8.61 \times 10^{-3} \text{ mol dm}^{-3})^2 \, (4.305 \times 10^{-3} \text{ mol dm}^{-3})}{(1.39 \times 10^{-3} \text{ mol dm}^{-3})^2}$$

$$= 0.165 \text{ mol dm}^{-3}$$

$K_P = K_c RT \text{ (from Eq. 4.26)}$

$$= 0.165 \text{ mol dm}^{-3} \times 0.0831 \times 298.15 \text{ dm}^3 \text{ atm mol}^{-1}$$

$$= 4.09 \text{ bar}$$

$K_x = K_P \, (0.355 \text{ bar})^{-1} \text{ (from Eq. 4.32)} = 11.5$

4.9.   Suppose that, if there were no dissociation, the partial pressure of $COCl_2$ was P; then the actual partial pressures are

$$COCl_2(g) \quad \rightleftarrows \quad CO(g) + Cl_2(g)$$

$$P(1 - \alpha) \qquad\qquad P\alpha \qquad P\alpha$$

$P + P\alpha = 2 \text{ bar}$

With $\alpha = 6.30 \times 10^{-5}$

$$P = \frac{2 \text{ bar}}{1 + (6.3 \times 10^{-5})} \approx 2 \text{ bar}$$

$$K_P = \frac{(2 \times 6.3 \times 10^{-5})^2}{2[1 - (6.3 \times 10^{-5})]} \approx 2 \times (6.3 \times 10^{-5})^2 \text{ bar}$$

$$= 7.94 \times 10^{-9} \text{ bar}$$

$K_c = K_P (RT)^{-1} \text{ (from Eq. 4.26)}$

$$= 7.94 \times 10^{-9} \text{ bar} \times (0.0831 \times 373.15 \text{ dm}^3 \text{ bar mol}^{-1})^{-1}$$

$$= 2.56 \times 10^{-10} \text{ mol dm}^{-3}$$

$$K_x = K_P P^{-1} \text{ (from Eq. 4.32)} = 7.94 \times 10^{-9} \text{ bar } (2 \text{ bar})^{-1} = 3.97 \times 10^{-9}$$

4.10.

	H_2	+	I_2	$\rightleftarrows$	2HI
Initially:	1		3		0 mol
At equilibrium:	$1 - \frac{x}{2}$		$3 - \frac{x}{2}$		x mol
After addition of 2 mol H_2:	$3 - x$		$3 - x$		$2x$ mol

$$K = \frac{x^2}{\left(1 - \frac{x}{2}\right)\left(3 - \frac{x}{2}\right)} = \frac{4x^2}{(3-x)^2}$$

$$x = 3/2$$

$$K = \frac{4(3/2)^2}{(3/2)^2} = 4$$

4.11. 12.7 g iodine = 0.05 mol I_2

When all of the solid iodine has just gone, the iodine pressure is 0.10 atm. The consumption of 0.05 mol I_2 leads to the formation of 0.10 mol HI, which exerts a pressure of

$$P_{HI} = \frac{0.1 \text{ mol} \times 0.0831 \text{ dm}^3 \text{ bar K}^{-1} \text{ mol}^{-1} \times 313.15 \text{ K}}{10 \text{ dm}^3}$$

$$= 0.260 \text{ bar}$$

Then, if P_{H_2} is the partial pressure of H_2 after equilibrium is established,

$$\frac{(0.260 \text{ bar})^2}{P_{H_2} \times 0.10 \text{ bar}} = 20$$

$$P_{H_2} = 0.0338 \text{ bar}$$

$$n_{H_2} = \frac{0.0338 \text{ bar} \times 10 \text{ dm}^3}{0.0831 \times 313.15 \text{ bar dm}^3 \text{ mol}^{-1}}$$

$$= 0.0130 \text{ mol}$$

Thus, 0.0130 mol of H_2 is produced in the equilibrium mixture, and 0.05 mol of H_2 is required to remove the 0.05 mol of I_2. Therefore, 0.065 mol of H_2 must be added.

4.12. Assuming that we start with one mole of $N_2O_4(g)$,

$$N_2O_4(g) \rightleftarrows 2NO_2 (g)$$

$$1 - \alpha \qquad\qquad 2\alpha$$

We obtain $(1+\alpha)$ moles of gas at equilibrium. Therefore, equilibrium partial pressures are

$$P_{N_2O_4} = \frac{1-\alpha}{1+\alpha}P; \quad P_{NO_2} = \frac{2\alpha}{1+\alpha}P,$$

and

$$K_P = \frac{4\alpha^2}{1-\alpha^2}P; \quad K_c = K_P(RT)^{-1}; \quad K_x = K_P/P.$$

Therefore, at 0.597 bar,

$$K_P = 1.08 \text{ bar}, K_c = 4.35 \times 10^{-2} \text{ mol dm}^{-3}, K_x = 1.81,$$

and at 6.18 bar,

$$K_P = 1.45 \text{ bar}, K_c = 5.84 \times 10^{-2} \text{ mol dm}^{-3}, K_x = 0.234.$$

4.13. Rewriting the reaction in terms of one mole of HCl, we get

$$HCl(g) \quad + \quad \tfrac{1}{4}O_2(g) \quad \rightleftarrows \quad \tfrac{1}{2}Cl_2(g) \quad + \quad \tfrac{1}{2}H_2O(g)$$

$$(1-y) \qquad\qquad P_{O_2} \qquad\qquad\qquad y/2 \qquad\qquad y/2$$

From examining the equation above, it is possible to establish the following relationships:

$$\frac{x_{Cl_2}}{x_{HCl}} = \frac{y}{2(1-y)}; \quad \frac{x_{H_2O}}{x_{Cl_2}} = 1.$$

It is important to remember that the same ratios hold for partial pressures also. Now,

$$K_P = \frac{P_{Cl_2}^{1/2} \times P_{H_2O}^{1/2}}{P_{HCl} \times P_{O_2}^{1/4}} = \frac{P_{Cl_2}}{P_{HCl} \times P_{O_2}^{1/4}} = \frac{y}{2(1-y)} \times \frac{1}{P_{O_2}^{1/4}}.$$

4.14. Assuming ideal behavior, the partial pressure of oxygen is 0.51 bar. Therefore,

$$K_P = \frac{0.76}{2(1-0.76)} \times \frac{1}{0.51^{0.25}} = 1.87 \text{ bar}^{-1/4}.$$

4.15. Since no reactants are initially present, we write

$$H_2(g) \quad + \quad I_2(g) \quad \rightleftarrows \quad HI(g)$$

$$x \qquad\qquad x \qquad\qquad\qquad n-2x$$

where n is the initial amount of HI: $n = 10.0 \text{ g}/(127.912 \text{ g mol}^{-1}) = 7.8179 \times 10^{-2}$ mol.

Since $K_P = 65.0 = \frac{(n-2x)^2}{x^2}$, or $\sqrt{65} = \frac{n-2x}{x}$, we can solve for x. The solution is $x = 7.7695 \times 10^{-3}$ mol. Therefore, the equilibrium mole fractions are $x_{H_2} = x_{I_2} = x/n = 9.94 \times 10^{-2}$; $x_{HI} = (n-2x)/n = 0.801$.

■ Equilibrium Constants and Gibbs Energy Changes

4.16. a. $\Delta G^\circ/\text{J mol}^{-1} = -8.3145 \times 283.15 \ln (2.19 \times 10^{-3})$

$\Delta G^\circ = 14\ 417 \text{ J mol}^{-1} = 14.42 \text{ kJ mol}^{-1}$

b.

	$(C_6H_5COOH)_2$	$\rightleftarrows$	$2C_6H_5COOH$
Initially:	0		0.1 mol dm^{-3}
At equilibrium:	x		$0.1 - 2x \text{ mol dm}^{-3}$

$\dfrac{(0.1 - 2x)^2}{x} = 2.19 \times 10^{-3}$

$4x^2 - 0.402\ 19x + 0.01 = 0$

$x = \dfrac{0.402\ 19 \pm \sqrt{0.001\ 76}}{8} = \dfrac{0.402\ 19 \pm 0.041\ 95}{8}$

$= 0.0555 \text{ or } 0.04503 \text{ mol}$

Only the second answer is possible, and this leads to

Dimer: $0.045 \text{ mol dm}^{-3}$;   Monomer: 0.01 mol dm^{-3}

4.17. K_P at 3000 K $= \dfrac{(0.4)^2 \times 0.2}{(0.6)^2} = 0.0889 \text{ atm}$

$\Delta G^\circ/\text{J mol}^{-1} = -8.3145 \times 3000 \ln 0.0889$

$\Delta G^\circ = 60\ 370 \text{ J mol}^{-1} = 60.4 \text{ kJ mol}^{-1}$

4.18. a. $\ln K_c = -\dfrac{2930}{8.3145 \times 310.15} = -1.136$; $K_c = 0.321$

b. $\ln K_c = +\dfrac{15\ 500}{8.3145 \times 310.15} = 6.01$; $K_c = 408$

c. $K_c = 0.321 \times 408 = 130.9$

$\Delta G^\circ = 2.93 - 15.5 = -12.6 \text{ kJ mol}^{-1}$

4.19. a. $\Delta G^\circ = \Delta G^\circ_{\text{Products}} - \Delta G^\circ_{\text{Reactants}} = 2(-16.63) - 0 - (3 \times 0) = -33.26 \text{ kJ mol}^{-1}$

$K_P = \exp(-\Delta G^\circ/RT) = 6.71 \times 10^5 \text{ bar}^{-2}$

b. $\Delta G^\circ = -32.82 - 209.20 = -242.02 \text{ kJ mol}^{-1}$

$K_P = 2.5 \times 10^{42} \text{ bar}^{-2}$

c. $\Delta G^\circ = -32.82 - 68.15 = -100.97 \text{ kJ mol}^{-1}$

$K_P = 4.9 \times 10^{17} \text{ bar}^{-1}$

d. $\Delta G°$ $= -32.82 + (2 \times 50.72) = 68.62$ kJ mol^{-1}

   K_P $= 9.5 \times 10^{-13}$

4.20.

	K_P	Σv	$K_c = K_P \, (RT)^{-\Sigma v}$	$K_x = K_P \, P^{-\Sigma v}$
a.	6.71×10^5 bar^{-2}	-2	1091.9 (mol dm^{-3})$^{-2}$	6.71×10^5
b.	2.5×10^{42} bar^{-2}	-2	4.0682×10^{39} (mol dm^{-3})$^{-2}$	2.5×10^{42}
c.	4.9×10^{17} bar^{-1}	-1	1.9766×10^{16} (mol dm^{-3})$^{-1}$	4.9×10^{17}
d.	9.5×10^{-13}	0	9.5×10^{-13}	9.5×10^{-13}

4.21. a. $\Delta G°$ $= -RT \ln K$

$= -8.3145 \times 298.15 \ln 10^{-5}$

$= 28\,540$ J mol^{-1} $= 28.5$ kJ mol^{-1}

$\Delta H°$ $= \Delta G° + T\Delta S° = 28\,540 - 298.15 \times 41.8$

$= 16\,077$ J mol^{-1} $= 16.1$ kJ mol^{-1}

b.

	CO	+	H$_2$O	$\rightleftarrows$	CO$_2$	+	H$_2$
Initially:	2		2		0		0 mol
At equilibrium:	$2 - x$		$2 - x$		x		x mol

$$K_P = K_c = \frac{x^2}{(2-x)^2} = 1.00 \times 10^{-5}$$

(Note that the total volume cancels out and so need not be considered.)

$$\frac{x}{2-x} = 3.16 \times 10^{-3}; \quad x = 6.30 \times 10^{-3}$$

The amounts are therefore

1.994 mol (CO);   1.994 mol (H$_2$O)

6.30×10^{-3} mol (CO$_2$); 6.30×10^{-3} mol (H$_2$)

4.22. For reaction (1)

K_1 $= \exp(-23\,800/8.3145 \times 310.15)$

$= 9.81 \times 10^{-5}$ dm^3 mol^{-1}

For reaction (2)

K_2 $= \exp(31\,000/8.3145 \times 310.15) = 1.66 \times 10^5$ mol dm^{-3}

For the coupled reaction (3)

$$K_3 = K_1K_2 = 16.3$$

4.23. $K_1 = \dfrac{2.9}{90.9} = 0.0319 \quad K_2 = \dfrac{6.2}{2.9} = 2.138$

$K(\text{overall}) = K_1K_2 = 0.0682$

$\Delta G° = -RT \ln K = 8.3145 \times 298.15 \ln 0.0682$

$= -6656.8 \text{ J mol}^{-1} = -6.66 \text{ kJ mol}^{-1}$

4.24. Let the solubility be c moles per dm^3. Then

$$Cr(OH)_3(s) \rightleftarrows Cr^{3+} + 3OH^-$$
$$ c \quad\; 3c$$

Then the value of the solubility product is $K_{sp} = c(3c)^3 = 27c^4$. Therefore,

$c = (K_{sp}/27)^{1/4} = 3.2 \times 10^{-8} \text{ mol dm}^{-3}$.

■ Temperature Dependence of Equilibrium Constants

4.25. a. Zero

b. $\Delta G° = 0; \quad \Delta H° > 0; \quad \therefore \Delta S° > 0$

c. $K_c = K_P(RT)^{-\Sigma v} = (0.0831 \times 298.15)^{-1}$

$= 0.0404 \text{ mol dm}^{-3}$

$\Delta G° = -8.3145 \times 298.15 \ln 0.0404 = 7955 \text{ J mol}^{-1}$

$= 7.96 \text{ kJ mol}^{-1}$

d. $K_P > 1$ bar

e. $\Delta G° < 0$

4.26. a. Yes

b. Yes

c. $\Delta S° > 0$

4.27. a. $\Delta G° = -32.82 - 68.15 = -100.97 \text{ kJ mol}^{-1}$

$\Delta H° = -84.68 - 52.26 = -136.94 \text{ kJ mol}^{-1}$

$\Delta S° = \dfrac{-136\,940 + 100\,970}{298.15} = -120.6 \text{ J K}^{-1} \text{ mol}^{-1}$

Standard state is 1 bar

b. $\ln (K_P/\text{bar}^{-1}) = -\dfrac{\Delta G^\circ}{RT} = \dfrac{+100\ 970}{8.3145 \times 298.15} = 40.73$

$K_P = 4.89 \times 10^{17}\ \text{bar}^{-1}$

c. $K_c = K_P (RT) = 4.89 \times 10^{17} \times 0.0831 \times 298.15$

$= 1.21 \times 10^{19}\ \text{dm}^3\ \text{mol}^{-1}$

d. $\Delta G^\circ = -RT \ln K_c = -8.3145 \times 298.15 \ln (1.21 \times 10^{19})$

$= -108\ 925\ \text{J mol}^{-1} = -108.9\ \text{kJ mol}^{-1}$

e. $\Delta S^\circ = \dfrac{\Delta H^\circ - \Delta G^\circ}{T} = \dfrac{-136\ 940 + 108\ 925}{298.15} = -93.96\ \text{J K}^{-1}\ \text{mol}^{-1}$

f. $\Delta G^\circ(100\ ^\circ\text{C}) = \Delta H^\circ - T\Delta S^\circ = -136\ 940 + (373.15 \times 120.6)$

$= -91\ 938\ \text{J mol}^{-1}$

$\ln (K_P/\text{bar}^{-1}) = \dfrac{-\Delta G^\circ}{RT} = \dfrac{91\ 938}{8.3145 \times 373.15} = 29.63$

$K_P(100\ ^\circ\text{C}) = 7.38 \times 10^{12}\ \text{bar}^{-1}$

4.28. a. $2\text{H}_2(g) + \text{O}_2(g) \rightleftarrows 2\text{H}_2\text{O}(g)$

$\Delta_f G^\circ = 2\Delta_f G^\circ_{\text{H}_2\text{O}} - 2\Delta_f G^\circ_{\text{H}_2} - \Delta_f G^\circ_{\text{O}_2}$

$= 2(-228.57) - 2(0) - 0 = -457.14\ \text{kJ mol}^{-1}$

$\Delta_f H^\circ = 2\Delta_f H^\circ_{\text{H}_2\text{O}} - 2\Delta_f H^\circ_{\text{H}_2} - \Delta_f H^\circ_{\text{O}_2}$

$= 2(-241.82) - 2(0) - 0 = -483.64\ \text{kJ mol}^{-1}$

$\Delta G^\circ = \Delta H^\circ - T\Delta S^\circ$

$\Delta_f S^\circ = -\dfrac{-457.14 - (-483.64)}{298.15} = -0.08888\ \text{kJ mol}^{-1} = -88.88\ \text{J K}^{-1}\ \text{mol}^{-1}$

b. $\Delta G^\circ = -RT \ln K_P^\circ$

$\ln (K_P^\circ/\text{bar}^{-1}) = \dfrac{457\ 140\ \text{J mol}^{-1}}{8.3145\ \text{J K}^{-1}\ \text{mol}^{-1}\ 298.15\ \text{K}} = 184.4$

$K_P^\circ = 1.222 \times 10^{80}\ \text{bar}^{-1}$

c. $\Delta G(2000\ ^\circ\text{C})/\text{kJ mol}^{-1} = \Delta_f H^\circ - 2273.15\ \Delta_f S^\circ$

$= -483.64 - 2273.15\ (-0.08888)$

$= -281.6\ \text{kJ mol}^{-1}$

$\ln (K_P/\text{bar}^{-1}) = \dfrac{281\ 600}{8.3145 \times 2273.15} = 14.90$

$K_P = 2.96 \times 10^6\ \text{bar}^{-1}$

4.29. $\Delta_r G° = 2\Delta_f G°(O_3, g) - 3\Delta_f G°(O_2, g)$

$= 2(163.2) - 3(0) = 326.4 \text{ kJ mol}^{-1}.$

$K = \exp(-\Delta_r G°/RT) = \exp[-326\ 400\ /(8.3145 \times 400)]$

$= 3.371 \times 10^{-43}.$

4.30. a. $\Delta S° = \dfrac{\Delta H° - \Delta G°}{T} = \dfrac{-20\ 100 + 31\ 000}{310.15}$

$= 35.1 \text{ J K}^{-1} \text{ mol}^{-1}$

b. $\ln K_c = \dfrac{31\ 000}{8.3145 \times 310.15} = 12.02$

$K_c = 1.66 \times 10^5 \text{ mol dm}^{-3}$

c. $\Delta G°(25°C) = -20\ 100 - (298.15 \times 35.1)$

$= -30\ 570 \text{ J mol}^{-1}$

$\ln K_c = \dfrac{30\ 570}{8.3145 \times 298.15} = 12.33$

$K_c = 2.27 \times 10^5 \text{ mol dm}^{-3}$

4.31. a. $\Delta H° = -165.98 + 146.44 = -19.54 \text{ kJ mol}^{-1}$

$\Delta S° = 306.4 - 349.0 = -42.6 \text{ J K}^{-1} \text{ mol}^{-1}$

$\Delta G° = -19\ 540 + (42.6 \times 298.15)$

$= -6839 \text{ J mol}^{-1} = -6.84 \text{ kJ mol}^{-1}$

b. $\ln K_P = \dfrac{6839}{8.3145 \times 298.15} = 2.759; \quad K_P = 15.78$

If partial pressure of neopentane = x bar, partial pressure of n-pentane = $(1 - x)$ bar

$\dfrac{x}{1 - x} = 15.78$

$x = 15.78 - 15.78x; \quad x = 0.940$

$1 - x = 0.060$

Thus P(neopentane) = 0.940 bar and P(n-pentane) = 0.060 bar.

4.32. a. Slope of plot of $\ln K_c$ against $1/T$ is

$$\dfrac{\ln 3}{\left(\dfrac{1}{313.15} - \dfrac{1}{298.15}\right)} = \left(\dfrac{1.0986}{1.607 \times 10^{-4}}\right) = -6836$$

From Eq. 4.73,

$\Delta H° = 6836 \times 8.3145 = 56\ 840 \text{ J mol}^{-1}$

$$= 56.8 \text{ kJ mol}^{-1}$$

b. $-56.8 \text{ kJ mol}^{-1}$

4.33. a. Slope of plot of $\ln K_c$ against $1/T$ is

$$-\frac{\ln 1.45}{\dfrac{1}{298.15} - \dfrac{1}{303.15}} = -\frac{0.372}{5.532 \times 10^{-5}} = -6724.6 \ K$$

$$\Delta H° = 6724.6 \times 8.3145 = 55\ 910 \text{ J mol}^{-1}$$

$$= 55.9 \text{ kJ mol}^{-1}$$

At 25 °C, $\Delta G° = -RT \ln 1.00 \times 10^{-14}$

$$= -8.3145 \times 298.15 \times (-32.236)$$

$$= 79\ 912 \text{ J mol}^{-1}$$

$$\Delta S° = \frac{\Delta H° - \Delta G°}{T} = \frac{55\ 910 - 79\ 912}{298.15}$$

$$= -80.5 \text{ J K}^{-1} \text{ mol}^{-1}$$

b. The difference between the reciprocals of the absolute temperatures corresponding to 25 °C and 37 °C is

$$3.3540 \times 10^{-3} - 3.2242 \times 10^{-3} = 1.298 \times 10^{-4}$$

The slope of the plot $\ln K_w$ against $1/T$ was $-6724.6 \ K$, and in going from 25 °C to 37 °C $\ln K_w$ is thus increased by

$$6724.6 \times 1.298 \times 10^{-4} = 0.873$$

At 25°C, $\ln K_w$ is -32.244 and at 37 °C it is therefore

$$-32.244 + 0.873 = -31.371$$

K_w at 37 °C is therefore $2.38 \times 10^{-14} \text{ mol}^2 \text{ dm}^{-6}$.

4.34. a. At 400 °C,

$$\log_{10}(K_P/\text{bar}) = 7.55 - \frac{4844}{673.15} = 0.354$$

$$K_P = 2.259 \text{ bar}$$

$$\Delta G°/\text{kJ mol}^{-1} = -8.3145 \times 673.15 \ln 2.259$$

$$\Delta G° = -4549 \text{ J mol}^{-1} = -4.55 \text{ kJ mol}^{-1}$$

From Figure 4.2 (b),

$$\Delta H° = 4844 \times 8.3145 \times 2.303 \text{ J mol}^{-1}$$

$$= 92\ 750 \text{ J mol}^{-1}$$

$$= 92.75 \text{ kJ mol}^{-1}$$

$$\Delta S° = \frac{92\ 800 + 4549}{673.15} \text{ J K}^{-1} \text{ mol}^{-1}$$

$$= 144.6 \text{ J K}^{-1} \text{ mol}^{-1}$$

b. $K_c = 2.259 \times (0.0831 \times 673.15)^{-1}$

$= 0.0404 \text{ mol dm}^{-3}$

$\Delta G°/\text{kJ mol}^{-1} = -8.3145 \times 673.15 \ln 0.0404$

$\Delta G° = 17\ 960 \text{ J mol}^{-1} = 17.96 \text{ kJ mol}^{-1}$

c. I_2 + cyclopentene $\rightleftarrows$ 2HI + cyclopentadiene

$0.1 - x$ $\qquad 0.1 - x$ $\qquad\qquad 2x$ $\qquad\qquad x$

$$\frac{4x^3}{(0.1 - x)^2} = 0.0404$$

For a very approximate solution, neglect x in comparison with 0.1:

$4x^3 = 0.0404 \times (0.1)^2 = 4.04 \times 10^{-4}$

$x^3 = 1.01 \times 10^{-4}$

$x = 0.0466$

For a better solution, calculate $4x^3/(0.1 - x)^2$ at various x values:

x	0.0466	0.04	0.03	0.0350
$\dfrac{4x^3}{(0.1 - x)^2}$	0.1419	0.0711	0.022	0.0406

$x = 0.0350$

Final concentrations are

$[I_2] = 0.0650\ M$;

$[\text{cyclopentene}] = 0.0650\ M$; $[HI] = 0.0700\ M$;

$[\text{cyclopentadiene}] = 0.0350\ M$

4.35. $\Delta H° = -283.66 + 110.53 = -173.13 \text{ kJ mol}^{-1}$

$\Delta G° = -166.27 + 137.17 = -29.10 \text{ kJ mol}^{-1}$

$\Delta S° = \dfrac{-173\ 130 + 29\ 100}{298.15} = -483.1 \text{ J K}^{-1} \text{ mol}^{-1}$

$\ln K_P = \dfrac{29\ 100}{8.3145 \times 298.15} = 11.74$

$K_P = 1.25 \times 10^5 \text{ bar}^{-3}$

4.36. $\Delta H° = -207.4 + 104.6 = -102.8 \text{ kJ mol}^{-1}$

$\Delta G° = -111.3 + 37.2 = -74.1 \text{ kJ mol}^{-1}$

$\Delta S° = \dfrac{-102\ 800 + 74\ 100}{298.15} = -96.3 \text{ J K}^{-1} \text{ mol}^{-1}$

4.37.　a.　$K_c = 95/5 = 19$

$\Delta G°/\text{J mol}^{-1} = -8.3145 \times 298.15 \ln 19$

$\Delta G° = -7299 \text{ J mol}^{-1} = -7.30 \text{ kJ mol}^{-1}$

　　b.　$\Delta G/\text{J mol}^{-1} = -7299 + 8.3145 \times 298.15 \ln \dfrac{10^{-4}}{10^{-2}}$

$\Delta G = (-7299 - 11\,416) \text{ J mol}^{-1} \quad = -18\,714 \text{ J mol}^{-1}$

$= -18.7 \text{ kJ mol}^{-1}$

The reaction will therefore go from left to right.

4.38.　a.　$\Delta H° = -110.53 - 241.82 + 393.51 = 41.16 \text{ kJ mol}^{-1}$

$\Delta G° = -137.17 - 228.57 + 394.36 = 28.62 \text{ kJ mol}^{-1}$

$\Delta S° = \dfrac{41\,160 - 28\,620}{298.15} = 42.06 \text{ J K}^{-1} \text{ mol}^{-1}$

　　b.　$\ln K_P = -\dfrac{28\,620}{8.3145 \times 298.15} = -11.55$

$K_P = 9.68 \times 10^{-6}$

　　c.　From the data in Table 2.1,

$\Delta d = 28.41 + 30.54 - 44.22 - 27.28$

$= -12.55 \text{ J K}^{-1} \text{ mol}^{-1}$

$\Delta e = 10^{-3}(4.10 + 10.29 - 8.79 - 3.26)$

$= 2.34 \times 10^{-3} \text{ J K}^{-2} \text{ mol}^{-1}$

$\Delta f = 10^4(-4.6 + 0 + 86.2 - 5.0)$

$= 76.6 \times 10^4 \text{ J K}^{-1} \text{ mol}^{-1}$

Then, from Eq. 2.52,

$\Delta H°_T/\text{J mol}^{-1} = 41\,160 - 12.55[(T/\text{K}) - 298.15] + 1.17 \times 10^{-3}[(T/\text{K})^2 - 298.15^2] -$

$76.6 \times 10^4 \left(\dfrac{1}{T/\text{K}} - \dfrac{1}{298.15} \right)$

$= 47\,367 - 12.55(T/\text{K}) + 1.17 \times 10^{-3}(T/\text{K})^2 - 76.6 \times 10^4/(T/\text{K})$

　　d.　$\ln K = \displaystyle\int \dfrac{\Delta H°}{RT^2} dT$

$= \dfrac{1}{8.3145} \displaystyle\int \left[\dfrac{47\,367}{(T/\text{K})^2} - \dfrac{12.55}{T/\text{K}} + 1.17 \times 10^{-3} - \dfrac{76.6 \times 10^4}{(T/\text{K})^3} \right] d(T/\text{K})$

$= \dfrac{1}{8.3145} \left[-\dfrac{47\,367}{(T/\text{K})} - 12.55 \ln (T/\text{K}) + 1.17 \times 10^{-3} (T/\text{K}) + \dfrac{38.3 \times 10^4}{(T/\text{K})^2} \right] + I$

$$= -\frac{5697}{T/K} - 1.51 \ln (T/K) + 1.41 \times 10^{-4} (T/K) + \frac{4.61 \times 10^4}{(T/K)^2} + I$$

I is obtained from the fact that at 298.15, $\ln K = -11.55$

$I = -11.55 + 19.11 + 8.60 - 0.042 - 0.519 = 15.60$

$$\ln K = 15.60 - \frac{5697}{T/K} - 1.51 \ln (T/K) + 1.41 \times 10^{-4} (T/K) + \frac{4.61 \times 10^4}{(T/K)^2}$$

e. $\ln K_P$ at 1000 K

$$= 15.60 - \frac{5697}{1000} - 1.51 \ln 1000 + 0.141 + \frac{4.61 \times 10^4}{1000^2}$$

$$= -0.34$$

$$K_P (1000 \text{ K}) = 0.71$$

4.39. Partial pressures at 1395 K are:

CO: 0.000 140 atm; CO_2: (1 − 0.000 140) atm; O_2: 0.000 070 atm

$$K_P = \frac{(0.000\ 140)^2 \times 0.000\ 070}{(0.9999)^2} = 1.372 \times 10^{-12} \text{ atm} = 1.39 \times 10^{-12} \text{ bar}$$

At 1443 K,

$$K_P = \frac{(0.000\ 250)^2 \times 0.000\ 125}{(0.999\ 75)^2} = 7.814 \times 10^{-12} \text{ atm} = 7.92 \times 10^{-12} \text{ bar}$$

At 1498 K,

$$K_P = \frac{(0.000\ 471)^2 \times 0.000\ 2355}{(0.999\ 529)^2} = 5.227 \times 10^{-11} \text{ atm} = 5.30 \times 10^{-11} \text{ bar}$$

Then

T/K	$10^{12}K_P/$atm	$10^4/(T/K)$	$\ln(10^{12}K_P/$atm$)$
1395	1.372	7.168	0.3163
1443	7.814	6.930	2.0559
1498	52.27	6.676	3.956

From a plot of $\ln (K_P/$atm$)$ against $1/(T/K)$,

slope $= -7.39 \times 10^4$ K   and $\Delta H^\circ = -R \times$ slope $= 609$ kJ mol^{-1}.

$\Delta G^\circ(1395 \text{ K}) = -8.3145 \times 1395 \ln (1.372 \times 10^{-12})$

$= 316.8$ kJ mol^{-1}

$$\Delta S^\circ = \frac{\Delta H^\circ - \Delta G^\circ}{T} = 209 \text{ J K}^{-1} \text{ mol}^{-1}$$

(standard state 1 atm)

4.40. Suppose that there are present x mol of I_2 and y mol of I:

$$x + \frac{y}{2} = 1.958 \times 10^{-3}$$

$$P = \frac{(x + y) \text{ mol } RT}{V}$$

At 800° C

$$\frac{558.0}{760.0} \text{ atm} = \frac{(x + y)0.082\ 05\ \text{dm}^3\ \text{atm K}^{-1}\ \text{mol}^{-1}}{249.8 \times 10^{-3}\ \text{dm}^3} \times 1073.15\ \text{K}$$

$$x + y = 2.0829 \times 10^{-3}$$

$$\frac{y}{2} = 0.1249 \times 10^{-3}; \quad y = 2.498 \times 10^{-4}$$

$$x = 1.833 \times 10^{-3}$$

Degree of dissociation,

$$\alpha = 1 - \frac{x}{1.958 \times 10^{-3}} = 1 - \frac{1.833 \times 10^{-3}}{1.958 \times 10^{-3}} = 0.0638$$

At 1000°C, $x + y = 2.3535 \times 10^{-3}$

$$\frac{y}{2} = 3.955 \times 10^{-4}; \quad y = 7.91 \times 10^{-4}$$

$$x = 1.5625 \times 10^{-3}; \quad \alpha = 0.202$$

At 1200°C, $x + y = 2.7715 \times 10^{-3}$

$$\frac{y}{2} = 8.135 \times 10^{-4}; \quad y = 1.627 \times 10^{-3}$$

$$x = 1.1445 \times 10^{-3}; \quad \alpha = 0.415$$

a. $\alpha = 0.0638, 0.202, 0.415$ at the three temperatures

b. At 800 °C,

$$K_c = \frac{(2.498 \times 10^{-4}\ \text{mol})^2}{1.833 \times 10^{-3}\ \text{mol} \times 0.2498\ \text{dm}^3}$$

$$= 1.363 \times 10^{-4}\ \text{mol dm}^{-3}$$

At 1000 °C,

$$K_c = \frac{(7.91 \times 10^{-4})^2}{1.5625 \times 10^{-3} \times 0.2498}$$

$$= 16.0 \times 10^{-4}\ \text{mol dm}^{-3}$$

At 1200 °C,

$$K_c = \frac{(1.627 \times 10^{-3})^2}{1.1445 \times 10^{-3} \times 0.2498}$$

$$= 92.59 \times 10^{-4} \text{ mol dm}^{-3}$$

c. $K_P = K_c (RT)^{\Sigma v}$ (Eq. 4.26)

$= K_c RT$ since $\Sigma v = 1$

At 800 °C, K_P

$= 1.363 \times 10^{-4} \text{ mol dm}^{-3} \times 100 \text{ dm}^3 \text{ m}^{-3} \times 8.3145 \text{ J K}^{-1} \text{ mol}^{-1} \times 1073.15 \text{ K}$

$= 1.216 \text{ kPa} = 9.12 \text{ Torr} = 0.0122 \text{ bar}$

At 1000 °C, $K_P = 16.97 \text{ kPa}$

At 1200 °C, $K_P = 113.4 \text{ kPa}$

d.

T/K	$10^4 K_c/\text{mol dm}^{-3}$	$10^4/(T/K)$	$\ln \dfrac{10^4 K_c}{\text{mol dm}^{-3}}$
1073.15	1.363	9.318	0.310
1273.15	16.03	7.855	2.774
1473.15	92.59	6.788	4.528

Slope of a plot of $\ln(K_c/\text{mol dm}^{-3})$ against $1/(T/K)$ is $-1.67 \times 10^4 \text{ K}^{-1}$.

$\Delta U° = 139 \text{ kJ mol}^{-1}$

$\Delta H° = \Delta U° + RT = 139\,000 + (8.3145 \times 1273.15)$

$= 149\,586 \text{ J mol}^{-1} = 150 \text{ kJ mol}^{-1}$

e. At 1000°C, $K_P = 16.97 \text{ kPa} = 0.169 \text{ bar}$.

$\Delta G° = -(8.3145 \times 1273.15 \text{ J mol}^{-1}) \ln(0.169/\text{bar})$

$= 18\,819 \text{ J mol}^{-1}$

$= 18.8 \text{ kJ mol}^{-1}$

(standard state: 1 bar)

$$\Delta S° = \frac{\Delta H° - \Delta G°}{T} = 103.0 \text{ J K}^{-1} \text{ mol}^{-1}$$

4.41. $\Delta G° = -RT \ln K° = -8.3145 \times 300 \ln 5.7 \times 10^{-3}$

$= 12\,889 \text{ J mol}^{-1} = 12.9 \text{ kJ mol}^{-1}$

Slope of plot $= \dfrac{\ln (5.7 \times 10^{-3}/7.8 \times 10^{-4})}{(1/300) - (1/340)} = \dfrac{1.989}{3.922 \times 10^{-4}}$

$= 5071 \text{ K}$

$\Delta H° = -8.3145 \times 5071 = -42\,160 \text{ J mol}^{-1} = -42.16 \text{ kJ mol}^{-1}$

$$\Delta S^\circ = (\Delta H^\circ - \Delta G^\circ)/T = -183 \text{ J K}^{-1} \text{ mol}^{-1}$$

4.42.
$$\Delta G^\circ = \Delta H^\circ - T\Delta S^\circ \quad = -85\ 200 + 300 \times 170.2$$
$$= -34\ 140 \text{ J mol}^{-1}$$
$$K_c \quad = \exp(34\ 140/8.3145 \times 300)$$
$$= 8.8 \times 10^5$$

The equilibrium constant is equal to unity when ΔG° is equal to zero.

$$0 = -85\ 200 + T \times 170.2$$
$$T = 500.6 \text{ K}$$

4.43.
$$\Delta G^\circ = -8.3145 \times 310.15 \times \ln 1.66 \times 10^5$$
$$= -30\ 996 \text{ J mol}^{-1}$$
$$\Delta S^\circ \quad = \frac{\Delta H^\circ - \Delta G^\circ}{T} = \frac{-20\ 100 + 31\ 996}{310.15}$$
$$= 35.13 = 35.1 \text{ J K}^{-1} \text{ mol}^{-1}$$

At 25 °C,
$$\Delta G^\circ = \Delta H^\circ - T\Delta S^\circ = -20\ 100 - (298.15 \times 35.1)$$
$$= -30\ 565 \text{ J mol}^{-1}$$
$$= -8.3145 \times 298.15 \ln K_c$$
$$\ln K_c \quad = \frac{30\ 565}{8.3145 \times 298.15} = 12.33$$
$$K_c \quad = 2.26 \times 10^5 \text{ mol dm}^{-3}$$

4.44.
$$\Delta G^\circ = -8.3145 \times 300 \times \ln 7.2 \times 10^{-5}$$
$$= 23\ 793 \text{ J mol}^{-1}$$
$$\Delta S^\circ \quad = \frac{\Delta H^\circ - \Delta G^\circ}{T} = \frac{40\ 000 - 23\ 793}{300}$$
$$= 54.02 \text{ J K}^{-1} \text{ (standard state: 1 mol dm}^{-3}\text{)}$$

The equilibrium constant is unity when ΔG° is zero:

$$0 \quad = \Delta H^\circ - T\Delta S^\circ$$
$$T \quad = \frac{\Delta H^\circ}{\Delta S^\circ} = \frac{40\ 000}{54.02}$$
$$= 740 \text{ K}$$

4.45.
$$\Delta G^\circ = -RT \ln K_c = -8.3145 \times 300 \ln (4.5 \times 10^4)$$
$$= -26\ 725 \text{ J mol}^{-1}$$

$$\Delta S^\circ = \frac{\Delta H^\circ - \Delta G^\circ}{T} = \frac{-40\ 200 + 26\ 725}{300}$$

$$= -44.9 \text{ J K}^{-1} \text{ mol}^{-1}$$

The equilibrium constant is unity when $\Delta G^\circ = 0$.

$$\Delta G^\circ = \quad 0 = -40\ 200 + (44.9 \times T)$$

$$T = 895 \text{ K}$$

4.46. For the equilibrium

$$Br_2(l) = Br_2(g)$$

the vapor pressures are the equilibrium constants. The corresponding ΔG° values (standard state: 1 bar) are

$$\Delta G^\circ(331.35 \text{ K}) = -RT \ln 1 = 0$$

$$\Delta G^\circ(282.45 \text{ K}) = -8.3145 \times 282.45 \text{ (J mol}^{-1}) \ln 0.1334$$

$$= 4730.7 \text{ J mol}^{-1}$$

We can therefore set up two simultaneous equations:

$$0 \quad = \Delta H^\circ - (331.35 \text{ K}) \Delta S^\circ$$

$$4730.7 \text{ J mol}^{-1} \quad = \Delta H^\circ - (282.45 \text{ K}) \Delta S^\circ$$

Subtraction gives

$$\Delta S^\circ = 4730.7 \text{ J mol}^{-1}/48.9 \text{ K} = 96.74 \text{ J K}^{-1} \text{ mol}^{-1}$$

$$\Delta H^\circ = 96.74 \text{ J K}^{-1} \text{ mol}^{-1} \times 331.35 \text{ K} = 32\ 055 \text{ J mol}^{-1}$$

At 25°C,

$$\Delta G^\circ = 32\ 055 - (96.74 \times 298.15) \text{ J mol}^{-1}$$

$$= 3212 \text{ J mol}^{-1}$$

The vapor pressure at 25°C is the corresponding equilibrium constant:

$$P = K_P = \exp(-3212/8.3145 \times 298.15)$$

$$= 0.274 \text{ bar}$$

4.47. For the reaction, $\Delta G^\circ = 2 \times 162.3 = 324.6 \text{ kJ mol}^{-1}$

Then

$$K_P = \exp(-\Delta G^\circ /RT) = \exp(-324\ 600/8.3145 \times 298.15)$$

$$= 1.36 \times 10^{-57} \text{ bar}^{-1} \text{ (The unit arises from the standard state of 1 bar.)}$$

$\Sigma \nu$ for the reaction is $2 - 3 = -1$. From Eq. 4.26,

$$K_c = K_P (RT)^{-\Sigma \nu}$$

$$= \frac{1.36 \times 10^{-57} \text{ bar}^{-1} (8.3145 \times 298.15 \text{ J})}{1.00000 \times 10^5 \text{ Pa bar}^{-1}}$$

$$= 3.37 \times 10^{-59} \text{ m}^3 \text{ mol}^{-1} \text{ (since J Pa}^{-1} = \text{m}^3\text{)}$$

$$= 3.37 \times 10^{-56} \text{ dm}^3 \text{ mol}^{-1}$$

From Eq. 4.32,

$K_x = K_P P^{-\Sigma \nu}$, so that at 2 bar pressure,

$K_x = 1.36 \times 10^{-57} \text{ bar}^{-1} \times 2 \text{ bar} = 2.72 \times 10^{-57}$

This equilibrium constant (Eq. 4.31) is $x(O_3)^2/x(O_2)^3$, and since it is so small, $x(O_2)$ is almost exactly unity. Thus

$$x(O_3)^2 = 2.72 \times 10^{-57}$$

and

$$x(O_3) = 5.22 \times 10^{-29}$$

4.48. At 500 K, by Eq. 2.46,

$$\Delta H^\circ \quad = \Delta H^\circ(300 \text{ K}) + \Delta C_P(500 - 300)$$

$$= -9600 - (7.11 \times 200) = -11\ 022 \text{ J mol}^{-1}$$

For the entropy change at 500 K, the corresponding equation is

$$\Delta S^\circ(T_2) \quad = \Delta S^\circ(T_1) + \int_{T_1}^{T_2} \frac{\Delta C_P\ dT}{T}$$

$$= \Delta S^\circ(T_1) + \Delta C_P \ln (T_2/T_1)$$

At 500 K, ΔS° is therefore

$\Delta S^\circ = 22.18 - 7.11 \ln(500/300) = 18.55 \text{ J K}^{-1} \text{ mol}^{-1}$

Therefore

$\Delta G^\circ = -11\ 022 - 18.55 \times 500 = -20\ 297 \text{ J mol}^{-1}$

The equilibrium constant K_P is therefore

$\exp(20\ 297/8.3145 \times 500) = 131.9$

Since the reaction involves no change in the number of molecules, K_c and K_x also have this value (see Eqs. 4.26 and 4.32), and they are not affected by the pressure.

Let the mole fraction of HI be x; then the mole fractions of H_2 and I_2 are both $(1 - x)/2$, and the expression for K_x is therefore

$K_x = 4x^2/(1 - x)^2 = 131.9$

Taking square roots

$11.48 \quad = 2x/(1 - x)$

whence $x = 0.85$

Pressure has no effect on the above values.

4.49.   Values of $\Delta G°$ and of K[$= \exp(-\Delta G°/RT)^u$] are

<u>Temperature</u>

$\theta/°C$	T/K	$\Delta G°/\text{kJ mol}^{-1}$	K
40.0	313.15	3.98	0.22
42.0	315.15	2.20	0.43
44.0	317.15	0.42	0.85
46.0	319.15	−1.36	1.67
48.0	321.15	−3.14	3.24
50.0	323.15	−4.93	6.27

$\Delta G° = \Delta H° - T\Delta S° = 0$ when $T = 317.6$ and K = 44.4°C

At this temperature there will be equal concentrations of P and D.

■ Binding to Protein Molecules

4.50.   If the concentration of M is [M], that of sites occupied and unoccupied is $n[M]$. The association reactions may be formulated in terms of the sites, S,

$$S + A \underset{\rightleftarrows}{\overset{K_s}{}} SA$$

$$K_s = \frac{[SA]}{[S][A]}$$

where [S] is the concentration of unoccupied sites and [SA] is the concentration of occupied sites. The total concentration of sites, $n[M]$, is

$$n[M] = [S] + [SA] = [SA]\left\{\frac{1}{K_s[A]} + 1\right\}$$

The average number of occupied sites per molecule is the total concentration of occupied sites divided by the total concentration of M:

$$\bar{v} = \frac{[SA]}{[M]} = \frac{n}{\dfrac{1}{K_s[A]} + 1} = \frac{nK_s[A]}{1 + K_s[A]}$$

4.51.   The total concentration of the molecule M is

$$[M]_0 = [M] + [MA] + [MA_2] + \ldots + [MA_n]$$

The total concentration of occupied sites is the total concentration of bound A molecules:

$$[A]_b = [MA] + 2[MA_2] + \ldots + n[MA_n]$$

Expressing every term in terms of [A]:

$$[M]_0 = [M]\{1 + K_1[A] + K_1K_2[A]^2 + \ldots + (K_1K_2 \ldots K_n)[A]^n\}$$

$$[A]_b = [M]\{K_1[A] + 2K_1K_2[A]^2 + \ldots + n(K_1K_2 \ldots K_n)[A]^n\}$$

Thus

$$\bar{\nu} = \frac{[A]_b}{[M]_0} = \frac{K_1[A] + 2K_1K_2[A]^2 + \ldots + n(K_1K_2 \ldots K_n)[A]^n}{1 + K_1[A] + K_1K_2[A]^2 + \ldots + (K_1K_2 \ldots K_n)[A]^n}$$

4.52. With $K_1 = nK_s$, $K_2 = (n-1)K_s/2$,

$K_3 = (n-2)K_s/3 \ldots K_n = K_s/n$, the preceding equation becomes

$$\bar{\nu} = \frac{nK_s[A] + n(n-1)K_s^2[A]^2 + \ldots + nK_s^n[A]^n}{1 + nK_s[A] + n(n-1)K_s^2[A]^2/2 + \ldots + K_s^n[A]^n}$$

$$= \frac{nK_s[A] \ \left\{1 + (n-1)K_s[A] + \ldots + K_s^{n-1}[A]^{n-1}\right\}}{1 + nK_s[A] + n(n-1)K_s^2[A]^2/2 + \ldots + K_s^n[A]^n}$$

The coefficients are the binomial coefficients,

$$\bar{\nu} = \frac{nK_s[A](1 + K_s[A])^{n-1}}{(1 + K_s[A])^n}$$

$$= \frac{nK_s[A]}{1 + K_s[A]}$$

which are the expressions obtained in Problem 50. To test the equation, plot $1/\bar{\nu}$ against $1/[A]$:

$$\frac{1}{\bar{\nu}} = \frac{1}{n} + \frac{1}{nK_s[A]}$$

One of the intercepts is $1/n$. Alternatively, plot $\bar{\nu}$ against $\bar{\nu}/[A]$:

$$\bar{\nu} = n - \frac{\bar{\nu}}{K_s[A]}$$

4.53. If K_n is very much larger than K_1, K_2, and so on, the equation obtained in Problem 51 reduces as follows:

$$\bar{\nu} = \frac{n(K_1K_2 \ldots K_n)[A]^n}{1 + (K_1K_2 \ldots K_n)[A]^n}$$

$$= \frac{nK[A]^n}{1 + K[A]^n}$$

where $K = K_1K_2 \ldots K_n$ is the overall equilibrium constant for the binding of n molecules:

$$nA + M \overset{K}{\rightleftharpoons} MA_n$$

The fraction of sites occupied, θ, is $\bar{v}/n$:

$$\theta = \frac{K[A]^n}{1 + K[A]^n} \quad \text{or}$$

$$\frac{\theta}{1 - \theta} = K[A]^n$$

The slope of a plot of $\ln\{\theta/(1 - \theta)\}$ against [A] is thus n. If the sites are identical and independent (Problem 50), the slope is 1. Intermediate behavior can give nonlinear plots; the maximum slope of a Hill plot cannot be greater than n.

5. PHASES AND SOLUTIONS

■ Thermodynamics of Vapor Pressure

5.1. At equilibrium, $G(\text{graphite}) = G(\text{diamond})$; i.e., $\Delta G_2 = 0$. We are given $\Delta G_1 = -2900 \text{ J mol}^{-1}$.

$$(d\Delta G / \Delta P)_T = \Delta V$$

$$\Delta V = 12.011 \text{ g mol}^{-1} \left(\frac{1}{3.51} - \frac{1}{2.25} \right) \times 10^{-6} \text{ m}^3$$

$$= -1.915 \times 10^{-6} \text{ m}^3 \text{ mol}^{-1}$$

Holding T constant

$$\int_1^2 d\Delta G = \int_{P_1}^{P_2} \Delta V dP = \Delta G_2 - \Delta G_1 = \Delta V(P_2 - P_1)$$

$$P_2 = \frac{\Delta G_2 - \Delta G_1}{\Delta V} + P_1$$

$$= \frac{0 - 2900 \text{ J mol}^{-1}}{-1.92 \times 10^{-6} \text{ m}^3 \text{ mol}^{-1}} + 10^5 \text{ Pa}$$

$$= 1.51 \times 10^9 \text{ Pa} \quad \text{or} \quad 1.51 \times 10^4 \text{ bar}$$

At this pressure the system must also be at approximately 1000 K.

5.2. $$V_l = \frac{\text{Molar mass}}{\text{density}} = \frac{18.01 \times 10^{-3} \text{ kg mol}^{-1}}{0.958 \text{ kg dm}^{-3}} = 18.80 \times 10^{-3} \text{ dm}^3 \text{ mol}^{-1}$$

$$V_v = \frac{18.01 \times 10^{-3} \text{ kg mol}^{-1}}{5.98 \times 10^{-4} \text{ kg dm}^{-3}} = 30.12 \text{ dm}^3 \text{ mol}^{-1}$$

From Eq. 5.8,

$$\frac{dP}{dT} = \frac{\Delta P}{\Delta T} = \frac{\Delta S_{\text{vap}}}{V_v - V_l} = \frac{108.72 \text{ J K}^{-1}}{(30.12 - 0.0188) \text{ dm}^3 \text{ mol}^{-1}} = 3.612 \text{ J dm}^{-3} \text{ K}^{-1}$$

$$\Delta P = 3.612 \text{ J dm}^{-3} \text{ K}^{-1} (1 \text{ K}) = 3.612 \times 10^3 \text{ J m}^{-3}$$

Since $1 \text{ J} = \text{kg m}^2 \text{ s}^{-2}$ or N m   and $1 \text{ Pa} = \text{kg m}^{-1} \text{ s}^{-2}$ or N m^{-2}

$$\Delta P = 3.612 \times 10^3 \text{ Pa}$$

5.3. $$\ln \frac{P_2}{P_1} = \frac{-\Delta_{vap}H_m}{R}\left(\frac{1}{T_2} - \frac{1}{T_1}\right) = \frac{\Delta_{vap}H_m}{R}\left(\frac{T_2 - T_1}{T_2 T_1}\right)$$

$$\ln \frac{101.325}{3.17} = \frac{\Delta_{vap}H_m/\text{J mol}^{-1}}{8.3145}\left(\frac{373.15 - 298.15}{373.15 \cdot 298.15}\right)$$

$$\Delta_{vap}H_m/\text{J mol}^{-1} = 3.4646\,(8.3145)\left(\frac{111\,254}{75}\right)$$

$$= 42.73 \text{ kJ mol}^{-1}$$

$$\Delta_{vap}H_m = 42.7 \text{ kJ mol}^{-1}$$

The CRC Handbook gives 40.57 kJ mol^{-1}.

5.4. Using the Clausius-Clapeyron equation, we get

$$\ln\left(\frac{P_2}{0.00603}\right) = \frac{40\,656}{8.3145}\left(\frac{1}{273.16} - \frac{1}{373.15}\right) = 4.7967.$$

Therefore, $P_2 = 0.00603 \text{ atm} \times e^{4.7967} = 0.73029 \text{ atm}$.

5.5. The molar volume of iodine, I_2, is

$$V_m = \frac{0.253\,81 \text{ kg mol}^{-1}}{4.93 \text{ kg dm}^{-3}} = 51.48 \times 10^{-3} \text{ dm}^3 \text{ mol}^{-1}$$

$$= 51.48 \times 10^{-6} \text{ m}^3 \text{ mol}^{-1}$$

Then from Eq. 5.23,

$$\ln \frac{P_1^g}{P_2^g} = \frac{V_m(P_1 - P_2)}{RT}$$

$$\ln \frac{P_1^g}{P_2^g} = \frac{51.48 \times 10^{-6}(101.3 \times 10^6 - 101\,300)}{8.3145 \times 313.15} = 2.001$$

$$P_1^g / P_2^g = 7.40$$

At 101.3 kPa the pressure is 133 Pa. Therefore, at 101.3×10^3 kPa the pressure is $7.40 \times 133 \text{ Pa} = 984 \text{ Pa}$.

5.6.   The cubic expansion coefficient is a second-order transition since it can be expressed as

$$\alpha = \frac{1}{V}\left[\frac{\partial}{\partial T}\left(\frac{\partial G}{\partial P}\right)_T\right]_P$$

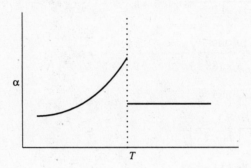

5.7.   From Eq. 5.16,

$$\Delta_{vap}H_m = \left(\frac{1}{T_1} - \frac{1}{T_2}\right)^{-1} R \ln \frac{P_2}{P_1}$$

$$\Delta_{vap}H_m/\text{J mol}^{-1} \quad = (4.975 \times 10^{-4})^{-1}\, 8.3145 \ln \frac{31.86}{1.94}$$

$$= (2010)\,(8.3145)\,(2.7987) = 46\,822$$

$$\Delta_{vap}H_m \quad = 46\,773 \text{ J mol}^{-1} = 46.8 \text{ kJ mol}^{-1}$$

5.8. $\ln \dfrac{P_2}{P_1} = \dfrac{\Delta_{vap}H_m}{R}\left(\dfrac{1}{T_2} - \dfrac{1}{T_1}\right) = \dfrac{\Delta_{vap}H_m}{R}\dfrac{(T_2 - T_1)}{T_2 T_1}$

$$P_2 = \left(14 \text{ Torr}/760\frac{\text{Torr}}{\text{atm}}\right)\left(101\,325\frac{\text{Pa}}{\text{atm}}\right) = 1866.5 \text{ Pa}$$

At $T_1 = 286 + 273.15 = 559.15 \text{ K}$ $P_1 = 101\,325 \text{ Pa}$

At $T_2 = 145 + 273.15 = 418.15$ $P_2 = 1866.5 \text{ Pa}$

$$\ln \frac{1866.5}{101\,325} = \frac{\Delta_{vap}H_m/\text{J mol}^{-1}}{8.3145}\left[\frac{418.15 - 559.15}{(418.15)\,(559.15)}\right]$$

$$\Delta_{vap}H_m/\text{J mol}^{-1} = +3.994\frac{(8.3145)\,(233\,809)}{141} = 55.07 \text{ kJ mol}^{-1}$$

The CRC Handbook value is 71.02 kJ mol^{-1}. The error is large, but considering the relative molecular mass of the compound, its high boiling point, and the wide range of T and P involved in the calculation, it is not surprising that the error is so large.

5.9. Using Trouton's rule,

$$\Delta_{vap}H_m = 88 \text{ J K}^{-1} \text{ mol}^{-1} \, 342.10 \text{ K} = 30.105 \text{ kJ mol}^{-1}$$

$$\% \text{ error} = \frac{31.912 - 30.105}{31.912} \times 100 = 5.7\%$$

5.10. Applying Trouton's rule,

$$\Delta_{vap}H = T\Delta_{vap}S = (88 \text{ J K}^{-1} \text{ mol}^{-1})(383.77 \text{ K})$$

$$= 33.8 \text{ kJ mol}^{-1}$$

$$\ln \frac{P_2}{P_1} = \frac{\Delta_{vap}H}{R}\left(\frac{1}{T_1} - \frac{1}{T_2}\right)$$

$$\ln \frac{P_2}{1 \text{ atm}} = \frac{33.8 \text{ kJ mol}^{-1}}{8.3145 \text{ J K}^{-1} \text{ mol}^{-1}}\left(\frac{1}{383.77} - \frac{1}{353.15}\right)$$

$$= -0.9185$$

$$P_2 = 0.399 \text{ atm}$$

5.11. 2-Propanone is not particularly associated. Therefore, using $\Delta T \approx 9.3 \times 10^{-4} \, T_b\Delta P$, we have

$$\Delta T \approx -9.3 \times 10^{-4}(329.35) \, (2.825) = -0.8653 \text{ K}$$

Thus T_b at 98.5 kPa is $329.35 - 0.87 = 328.48$ K

5.12. At the triple point, the two vapor pressures must be equal since the liquid, solid, and vapor are all in equilibrium with each other. Therefore, we set

$$\frac{-2661}{T_{trp}} + 22.76 = \frac{-3775}{T_{trp}} + 26.88.$$

Solving for T_{trp}, we obtain the triple point temperature: $T_{trp} = 265.5$ K.

Substituting the triple point temperature into one of the two equations given, we calculate the triple point pressure:

$$\ln P_l = \frac{-2661}{265.5} + 22.76 = 12.737.$$

$$P_l = 3.4112 \times 10^5 \text{ Pa}.$$

5.13. Since water is associated, the numerical value 7.5×10^{-4} should be used in Eq. 5.20.
$T_b = 373.15$ K and $\Delta P = 102.7 - 101.3 = 1.4$ kPa, and thus

$$\Delta T = 7.5 \times 10^{-4} \times 373.15 \times 1.4 = 0.39 \text{ K}$$

The calculated boiling point at 101.3 kPa is thus $373.52 - 0.39 = 373.13$ K, an error of only 0.02 K.

5.14. Molar volume of water $= 18.015 \text{ g mol}^{-1}/0.996374 \text{ g cm}^{-3}$

$$= 18.08 \text{ cm}^3 \text{ mol}^{-1}$$

$$\ln \frac{P/\text{Torr}}{27.536} = \frac{18.08 \text{ cm}^3 \text{ mol}^{-1} (760.000 \text{ Torr} - 27.536 \text{ Torr})}{82.056 \text{ cm atm K}^{-1} (760 \text{ Torr atm}^{-1}) (273.15 + 27.5)}$$

$$= \frac{18.08(732.46)}{18\ 749\ 000} = 7.063 \times 10^{-4}$$

$$\ln P/\text{Torr} = 7.063 \times 10^{-4} + \ln 27.536 = 7.063 \times 10^{-4} + 3.3155$$

$$= 3.316$$

$$P = 27.55 \text{ Torr}$$

This is a rather small correction but may be necessary for very accurate work.

5.15. From Eq. 3.115,

$$dG = VdP - SdT.$$

Recognize that a change of state occurs at constant temperature. Therefore, differentiating Eq. 3.115 with respect to P at constant temperature, we obtain (see Eq. 3.119)

$$\left(\frac{\partial G}{\partial P} \right)_T = V.$$

Now, for a change of state where $G_f - G_i = \Delta G$, there will be a corresponding change in volume,

$$\left(\frac{\partial \Delta G}{\partial P} \right)_T = \Delta V.$$

5.16. We start with the Clapeyron equation, Eq. 5.9, and substitute $\dfrac{RT + M}{P}$ for V; this gives, after rearrangement,

$$\frac{dP}{P} = \frac{\Delta_{\text{vap}} H_m dT}{T(RT + M)}$$

Since

$$\frac{1}{T(RT + M)} = \frac{1}{MT} - \frac{R}{M(RT + M)}$$

$$\frac{dP}{P} = \frac{\Delta_{\text{vap}} H_m}{M} \frac{dT}{T} - \frac{R}{M} \frac{\Delta_{\text{vap}} H_m dT}{(RT + M)}$$

Integration with $\Delta_{\text{vap}} H_m$ constant gives

$$\ln \frac{P_2}{P_1} = \frac{\Delta_{\text{vap}} H_m}{M} \ln \frac{T_2}{T_1} - \frac{\Delta_{\text{vap}} H_m}{M} \ln \frac{RT_2 + M}{RT_1 + M}$$

$$\ln \frac{P_2}{P_1} = \frac{\Delta_{\text{vap}} H_m}{M} \ln \frac{T_2}{T_1} \left(\frac{RT_1 + M}{RT_2 + M} \right)$$

5.17. Molar mass of $C_2H_5OH = 46.0695 \text{ g mol}^{-1}$

Molar volume of C_2H_5OH $= \dfrac{46.0695 \text{ g mol}^{-1}}{0.7767 \text{ g dm}^{-3}} = 59.314 \text{ cm mol}^{-1}$

$$= 59.314 \times 10^{-6} \text{ m}^3 \text{ mol}^{-1}$$

Using Eq. (5.23), we get

$$\ln \frac{P/\text{Pa}}{100 \text{ Torr} \times \dfrac{101\,325 \text{ Pa}}{760 \text{ Torr}}}$$

$$= \frac{59.314 \times 10^{-6} \text{ m}^3 \text{ mol}^{-1}}{8.3145 \text{ J mol}^{-1} \text{ K}^{-1} \times 308.15 \text{ K}} \left(1.000 \times 10^7 \text{ Pa} - \frac{100 \text{ Torr} (101\,325 \text{ Pa})}{760 \text{ Torr}} \right)$$

$$\ln P/\text{Pa} = 0.23119 + \ln 1.3332 \times 10^4$$

$$= 0.23119 + 9.49792 = 9.72911$$

$$= 1.67996 \times 10^4 \text{ Pa}$$

$$= \frac{1.67996 \times 10^4 \text{ Pa}}{101\,325 \text{ Pa}} \times 760 \text{ Torr} = 126 \text{ Torr}$$

5.18.   The volume change is

$$\Delta V = \left(\frac{1}{3.5155} - \frac{1}{2.2670} \right) \text{cm}^3 \text{ g}^{-1} = -0.1567 \times 10^{-3} \text{ m}^3 \text{ kg}^{-1}.$$

From the expression derived in Problem 5.15, we have

$$d\Delta G = \Delta V dP.$$

Integrating both sides, we obtain

$$\int_1^2 d\Delta G = \Delta V \int_{P_1}^{P_2} dP.$$

Let state 1 be the standard state and state 2 be at equilibrium. Then, the expression is

$$\Delta G - \Delta G^\circ = \Delta V (P - P^\circ).$$

Now, at equilibrium, the Gibbs energy change ΔG is zero. Also, using 10^5 Pa for the standard pressure P°, we get

$$-240 \text{ kJ kg}^{-1} = -0.1567 \times 10^{-3} \text{ m}^3 \text{ kg}^{-1} (P - 10^5) \text{ Pa}.$$

Solving for P gives $P = 1.53 \times 10^9$ Pa.

5.19.   From Raoult's law (Eq. 5.26) we find, for toluene,

$$P_{\text{tol}} = x_{\text{tol}} P_{\text{tol}}^* = 0.60 \times 0.185 \text{ bar} = 0.111 \text{ bar}.$$

and for benzene,

$$P_{\text{ben}} = x_{\text{ben}} P_{\text{ben}}^* = 0.40 \times 0.513 \text{ bar} = 0.205 \text{ bar}.$$

Therefore, the total pressure is $P_{\text{tot}} = 0.111 \text{bar} + 0.205 \text{ bar} = 0.316 \text{ bar}.$

5.20.   The molar heat of vaporization is

$$\Delta_{\text{vap}} H_m = 801 \text{ J g}^{-1} \times 62.058 \text{ g mol}^{-1} = 49\,708 \text{ J mol}^{-1}$$

The Clausius-Clapeyron equation is $\ln\left(\dfrac{P_2}{P_1}\right) = -\dfrac{\Delta_{vap}H}{R}\left(\dfrac{1}{T_2} - \dfrac{1}{T_1}\right)$. Substituting, we get

$$\ln\left(\frac{760}{30}\right) = -\frac{49\,708}{8.3145}\left(\frac{1}{470} - \frac{1}{T_1}\right),$$

which yields, upon simplification,

$$\ln 25.3 = -5978\left(\frac{1}{470} - \frac{1}{T_1}\right) \text{ or } 3.23 = -5978(2.128 \times 10^{-3} - 1/T_1),$$

from which, we get

$$T_1 = 375 \text{ K or } 102\ ^\circ C.$$

■ Raoult's Law, Equivalence of Units, and Partial Molar Quantities

5.21. $P_T = 0.60(3.572) + 0.40(9.657)$

$= 2.143 + 3.863 = 6.006$ kPa

$x^{vap}_{toluene} = \dfrac{2.143}{6.006} = 0.357$

5.22. The molality m_2 is the amount of solute divided by the mass of solvent. If W_1 is the mass of solvent, the solution contains $m_2 W_1$ mol of solute and W_1/M_1 mol of solvent. The mole fraction is thus

$$x_2 = \frac{m_2 W_1}{W_1/M_1 + m_2 W_1} = \frac{m_2 M_1}{1 + m_2 M_1}$$

If we divide each term by its SI unit,

$$x_2 = \frac{(m_2/\text{mol kg}^{-1})\,(M_1/\text{kg mol}^{-1})}{1 + (m_2/\text{mol kg}^{-1})\,(M_1/\text{kg mol}^{-1})}$$

The customary unit for molar mass M_1 is g mol^{-1}, and we then obtain

$$x_2 = \frac{(m_2/\text{mol kg}^{-1})(M_1/1000 \text{ g mol}^{-1})}{1 + (m_2/\text{mol kg}^{-1})(M_1/1000 \text{ g mol}^{-1})}$$

$$= \frac{(m_2/\text{mol kg}^{-1})(M_1/\text{g mol}^{-1})}{1000 + (m_2/\text{mol kg}^{-1})(M_1/\text{g mol}^{-1})}$$

If the solution is sufficiently dilute, the expression approximates to

$$x_2 = (m_2/\text{mol kg}^{-1})\,(M_1/\text{g mol}^{-1})/1000$$

5.23. Rearranging Eq. 5.122, and substituting gives

$$W_2 = \frac{M_2 \Delta_{fus} T W_1}{K_f} = \frac{0.06202 \text{ kg mol}^{-1} \ (20.0 \text{ K}) \ (1.00 \text{ kg})}{1.86 \text{ K kg mol}^{-1}} = 0.667 \text{ kg}.$$

Converting the mass to a volume using density, and assuming that 1.00 kg of water has a volume of 1.00 dm³, we get

$$V_2 = 0.667 \text{ kg}/(1.1088 \text{ kg dm}^{-3}) = 0.602 \text{ dm}^3.$$

So, the volume ratio is 0.602 dm³ antifreeze to 1.0 dm³ water (or approximately 3:5 volume ratio).

The elevation of boiling point for this solution will be $\Delta_{vap}T = K_b m_2 = 0.51 \times (0.667/0.06202) = 5.48$ K. This means that the solution will boil at 378.6 K or 105 °C.

5.24. If V is the volume of solution, concentration of solute = c_2; amount of solute = Vc_2; mass of solute = Vc_2M_2 where M_2 is the molar mass of solute.

   Mass of solution = $V\rho$;   mass of solvent = $V\rho - Vc_2M_2$;
   amount of solvent = $(V\rho - Vc_2M_2)/M_1$.

The mole fraction is therefore

$$x_2 = \frac{Vc_2}{(V\rho - Vc_2M_2)/M_1 + Vc_2} = \frac{c_2M_1}{\rho + c_2(M_1 - M_2)}$$

Dividing each term by its SI unit,

$$x_2 = \frac{(c_2/\text{mol m}^{-3}) \ (M_1/\text{kg mol}^{-1})}{\rho/\text{kg m}^{-3} + (c_2/\text{mol m}^{-3}) \ [(M_1 - M_2)/\text{kg mol}^{-1}]}$$

The more customary units are, for concentration, mol dm⁻³; for molar mass, g mol⁻¹; for density, kg dm⁻³ ≡ g cm⁻³.

Then

$$x_2 = \frac{(1000 \ c_2/\text{mol dm}^{-3}) \ (M_1/1000 \text{ g mol}^{-1})}{1000 \ \rho/\text{g cm}^{-3} + (1000 \ c_2/\text{mol dm}^{-3}) \dfrac{(M_1 - M_2)}{1000 \text{ g mol}^{-1}}}$$

$$= \frac{(c_2/\text{mol dm}^{-3}) \ (M_1/\text{g mol}^{-1})}{1000 \ \rho/\text{g cm}^{-3} + (c_2/\text{mol dm}^{-3}) \dfrac{(M_1 - M_2)}{\text{g mol}^{-1}}}$$

If the solution is sufficiently dilute, the density of the solution is approximately that of the pure solvent, ρ_1, and the second term in the denominator can be neglected.

$$x_2 = c_2M_1/\rho_1 = \frac{(c_2/\text{mol dm}^{-3}) \ (M_1/\text{g mol}^{-1})}{1000 \ \rho_1/\text{g cm}^{-3}}$$

5.25. It is difficult to work with the two laws in their normal form. Cast Henry's law into the form

$$\mu_2 = \mu_2^0 + RT \ln x_2$$

using the arguments of Section 5.7 and Eq. 5.101. Then

$$d\mu_2/dx_2 = RT/x_2$$

$$d\mu_2 = \frac{RT}{x_2} \, dx_2$$

Using the Gibbs-Duhem equation,

$$x_1 \, d\mu_1 + x_2 \, d\mu_2 = 0$$

$$d\mu_1 = -x_2 \, d\mu_2/x_1 = -RT/x_1 \, dx_2$$

Since $x_1 + x_2 = 1$,

$$dx_2 = -dx_1$$

$$d\mu_1 = RT/x_1 \, dx_1$$

Integration gives

$$\mu_1 = RT \ln x_1 + \text{constant}$$

Since $x_1 = 1$ as μ_1 goes to μ_1^0

$$\mu_1 = RT \ln x_1 + \mu_1^0$$

5.26. Using Eq. 5.121, we get $\Delta_{fus}T = 1.86$ K kg mol^{-1} × 1.0 mol kg^{-1} = 1.86 K, which is not what is observed.

NaCl completely ionizes in solution. Since colligative properties are, to a large extent, determined by the number of particles in solution rather than the actual identity of the species, we should realize that 1.0 mol of solid NaCl results in a solution containing 2.0 mol of particles. Therefore, we calculate that the temperature change is

$$\Delta_{fus}T = 1.86 \text{ K kg mol}^{-1} \times 2.0 \text{ mol kg}^{-1} = 3.72 \text{ K},$$

which accounts for the observation.

5.27 From Problem 5.24,

amount of solute = Vc_2;   mass of solvent = $V\rho - Vc_2M_2$

Thus, the molality is

$$m_2 = \frac{c_2}{\rho - c_2M_2}$$

and the concentration in terms of m_2 is

$$c_2 = \frac{m_2\rho}{1 + m_2M_2}$$

Dividing throughout by SI units,

$$c_2/\text{mol m}^{-3} = \frac{(m_2/\text{mol kg}^{-1})\,(\rho/\text{kg m}^{-3})}{1 + (m_2/\text{mol kg}^{-1})\,(M_2/\text{kg mol}^{-1})}$$

In terms of more usual units,

$$1000\,c_2/\text{mol dm}^{-3} = \frac{(m_2/\text{mol kg}^{-1})\,(1000\,\rho/\text{g cm}^{-3})}{1 + (m_2/\text{mol kg}^{-1})\left(M_2/1000\,\frac{\text{g}}{\text{mol}}\right)}$$

$$c_2/\text{mol dm}^{-3} = \frac{1000(m_2/\text{mol kg}^{-1})\,(\rho/\text{g cm}^{-3})}{1000 + (m_2/\text{mol kg}^{-1})\,(M_2/\text{g mol}^{-1})}$$

In dilute solution,

$$c_2/\text{mol dm}^{-3} \approx (m_2/\text{mol kg}^{-1})\,(\rho_1/\text{g cm}^{-3})$$

where ρ_1 is the density of the solvent. For aqueous solutions $\rho_1 \approx 1$ g cm^{-3} and therefore the numerical values of the concentration and the molality, in the above units, are very similar.

5.28. Assuming that this is an ideal solution, we use Raoult's law with $P_{\text{Hg}} = (1 - x_M)P_{\text{Hg}}^*$, where $P_{\text{Hg}}^* = 768.8$ Torr at the boiling point, and $P_{\text{Hg}} = 754.1$ Torr. Therefore, we calculate

$$x_M = 1 - \frac{P_{\text{Hg}}}{P_{\text{Hg}}^*} = 1.9121 \times 10^{-2}.$$

Now, since

$$1.9121 \times 10^{-2} = \frac{1.152/\text{MW}}{(1.152/\text{MW} + 100.0/200.59)},$$

we can solve for the molecular mass of the element, which gives MW = 118.54 g mol^{-1}. Comparison with the masses of elements shows that the metal is tin.

5.29. From Eq. 5.31,

$$V_{\text{NaCl}}/\text{cm}^3\,\text{mol}^{-1} = \left(\frac{\partial V}{\partial n_{\text{NaCl}}}\right)_{n_{\text{H}_2\text{O}}} = \frac{\partial V}{\partial m}$$

$$= 17.8213 + 1.74782\,m - 0.141675\,m^2$$

From Eq. 5.37,

$$dV_{\text{H}_2\text{O}} = -\left(\frac{n_{\text{NaCl}}}{n_{\text{H}_2\text{O}}}\right) dV_{\text{NaCl}} = -\left(\frac{m}{55.508}\right) dV_{\text{NaCl}}$$

Integration from $m = 0$ to m gives

$$\int dV_{\text{H}_2\text{O}} = -\int \left(\frac{1.74782}{55.508}m - \frac{0.28335}{(55.508)}m^2\right) dm$$

$$V_{\text{H}_2\text{O}} - V_{\text{H}_2\text{O}}^* = \frac{1.74782}{2(55.508)}m^2 + \frac{0.28335}{3(55.508)}m^3$$

With $V_{H_2O}^* = 18.068 \text{ cm}^3$,

$$V_{H_2O}/\text{cm}^3 \text{ mol}^{-1} = 18.068 - 0.015\ 744\ m^2 + 0.001\ 7016\ m^3$$

5.30.   We first develop an expression for $(\partial\rho/\partial n_2)n_1$. Since

$$x_2 = n_2/(n_1 + n_2),$$

$$\left(\frac{\partial x_2}{\partial n_2}\right)_{n_1} = \frac{n_1}{(n_1 + n_2)^2}$$

and $\left(\frac{\partial\rho}{\partial n_2}\right)_{n_1} = \frac{d\rho}{dx_2}\left(\frac{\partial x_2}{\partial n_2}\right)_{n_1} = \frac{n_1}{(n_1 + n_2)^2}\frac{d\rho}{dx_2}$

Substitution and division by $(n_1 + n_2)$ gives

$$V_2 = \frac{M_2}{\rho} - (M_1 x_1 + M_2 x_2)\frac{x_1}{\rho^2}\frac{d\rho}{dx_2}$$

5.31.   Substitution of $x_2 = 0.100$ into the expression in Problem 5.23 gives $\rho = 0.970\ 609 \text{ kg dm}^{-3}$. Differentiating the ρ equation with respect to x_2 gives

$$\frac{d\rho}{dx_2} = -0.289\ 30 + 0.598\ 14\ x_2 - 1.826\ 28\ x_2^2 + 2.377\ 52\ x_2^3 - 1.029\ 05\ x_2^4$$

$$= -0.245\ 47 \text{ kg dm}^{-3}$$

with $M_1(H_2O) = 0.018\ 016 \text{ kg mol}^{-1}$ and $M_2(\text{methanol}) = 0.032\ 043 \text{ kg mol}^{-1}$, substitution with $x_1 = 0.900$ gives $V_2 = 0.037\ 56 \text{ dm}^3 \text{ mol}^{-1}$.

5.32.   Using Dalton's Law of partial pressures, $P_c = y_c P_{tot}$, where y_c is the mole fraction of chloroform in the vapor phase.

$$P_c = (1 - 0.818)220.5 = 40.1 \text{ Torr}.$$

Activity and activity coefficient: $a_c = P_c / P_c^* = 40.131/221.8 = 0.181$; $f_c = a_c/x_c = 0.181/(1 - 0.713) = 0.630$.

5.33.   Using Eq. 5.80,

$$\frac{56.18 \text{ Torr} - 55.24 \text{ Torr}}{56.18 \text{ Torr}} = \frac{12.5 \text{ g } (46.0695 \text{ g mol}^{-1})}{M_2 (520.8 \text{ g})}$$

$$0.01673 = 1.10577 \text{ g mol}^{-1}/M_2$$

$$M_2 = 66.1 \text{ g mol}^{-1}$$

5.34. We prepare a graph of the data from which we see that the solution exhibits positive deviation from Raoult's law.

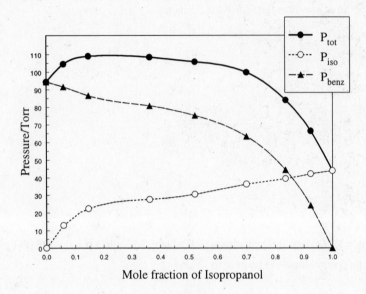

From this graph, we construct the following table. $a_i = P_i/P_i^*, f_i = a_i/x_i$.

x_I	P_I (Torr)	$a_I = \dfrac{P_I}{44.0}$	$f_I = \dfrac{P_I}{x_I}$	P_B (Torr)	$a_B = \dfrac{P_B}{94.4}$	$f_B = \dfrac{P_B}{x_B}$
0.20	25.0	0.568	2.84	84.5	0.895	1.12
0.50	30.0	0.682	1.36	76.0	0.805	1.61
0.80	39.0	0.886	1.11	50.0	0.530	2.65

5.35. From Raoult's law, we have for propylene dibromide (p),

$$P_p = x(l)P_p^* \qquad = 0.600(128) \text{ Torr } (133.33)\text{Pa/Torr}$$

$$= 10.239 \text{ kPa}$$

For ethylene dibromide (e),

$$P_e = x(l)P_e^* \qquad = 0.40(172 \text{ Torr}) (133.32)\text{Pa/Torr}$$

$$= 9.172 \text{ kPa}$$

The total pressure is $P_{total} = 10.239 \text{ kPa} + 9.172 \text{ kPa} = 19.411 \text{ kPa}$. The mole fraction in the vapor phase is given by

$$x = \frac{P}{P_{total}}$$

For propylene dibromide,

$$x_p(v) = \frac{10.239 \text{ kPa}}{19.411 \text{ kPa}} = 0.527$$

For ethylene dibromide,

$$x_e(v) = \frac{9.172 \text{ kPa}}{19.411 \text{ kPa}} = 0.473$$

5.36. First, calculate the moles of each present. For A, $n_A = 75.0 \text{ g}/89.5 \text{ g/mol}^{-1} = 0.83799 \text{ mol}$.

For B, $n_B = 1000 \text{ g}/185 \text{ g mol}^{-1} = 5.4054 \text{ mol}$. Then mole fractions are calculated.

For A, $x_A = 0.83799/6.24339 = 0.1342$. For B, $x_B = 5.4054/(0.83799 + 5.4054) = 0.8658$.

$$P_{\text{total}} = 520 \text{ Torr} = 430 \text{ Torr} + P_A$$

$$P_A = 90 \text{ Torr}$$

Then, from Henry's law,

$$P_A = k' x_A$$

$$k' = \frac{90 \text{ Torr}}{0.1342} = 671 \text{ Torr}$$

5.37.   a.   Henry's law applies to the individual gas components, and we require the partial pressures of N_2 and O_2. Since the partial pressure is directly proportional to the mole fraction, for N_2 we have $P_{N_2} = 0.80 \text{ atm}$, and for O_2 we have $P_{O_2} = 0.20 \text{ atm}$. From Eq. 5.28,

$$x(N_2) = \frac{P_{N_2}}{k'_{N_2}} = \frac{0.80 \text{ atm}}{7.58 \times 10^4 \text{ atm}} = 1.06 \times 10^{-5}$$

$$x(O_2) = \frac{P_{O_2}}{k'_{O_2}} = \frac{0.20 \text{ atm}}{3.88 \times 10^4 \text{ atm}} = 5.15 \times 10^{-6}$$

b.   Since the mole fractions are so small, if we use 1 mol of water as a reference, then there are (to a very good approximation) 1.06×10^{-5} mol of N_2 and 5.15×10^{-6} mol of O_2. The concentration calculation requires the volume of the solution, which is obtained from the density:

$$c(N_2) = \frac{1.06 \times 10^{-5} \text{ mol } (N_2)}{1 \text{ mol}(H_2O)[0.018 \text{ kg }(H_2O)/\text{mol}(H_2O)]} \times \frac{0.9982 \text{ kg } H_2O}{1 \text{ dm}^3} = 5.88 \times 10^{-4} M$$

$$c(O_2) = \frac{5.15 \times 10^{-6} \text{ mol } (O_2)}{1 \text{ mol}(H_2O)[0.018 \text{ kg }(H_2O)/\text{mol}(H_2O)]} \times \frac{0.9982 \text{ kg } H_2O}{1 \text{ dm}^3} = 2.86 \times 10^{-4} M$$

5.38. Assume that the total vapor pressure of pure benzene is present in the total pressure of 750.0 Torr.

$$P = P_T - P_1 = 750.0 - 94.6 = 655.4 \text{ Torr}$$

$$P_2 - k'x_2 \quad \text{and} \quad k' = 4.27 \times 10^5 \text{ Torr}$$

$$655.4 \text{ Torr} = 4.27 \times 10^5 \text{ Torr } x_2$$

$$x_2 = 1.535 \times 10^{-3}$$

$$x_1 = 0.99846$$

Using Raoult's law

$$P_1 = x_1 P_1^* = 0.99846 \times 94.6 = 94.45$$

Thus, the assumption is valid. Next, determine m_2. In 1000 g of benzene,

$$n_1 = \frac{1000}{78.11} = 12.80$$

$$x_2 = \frac{n_2}{n_2 + 12.80} = 1.535 \times 10^{-3}$$

Solving for n_2 gives 0.0200 mol.

Therefore, the molality is 0.0200 m.

■ Thermodynamics of Solutions

5.39. From Raoult's law, $x_1 = P_1/P_1^*$ and $x_2 = 1 - x_1 = 1 - (P_1/P_1^*)$, from which we get

$$x_2 = \frac{P_1^* - P_1}{P_1^*} = \frac{0.0410}{2.332} = \frac{W_2/M_2}{(W_1/M_1) + (W_2/M_2)}$$

$$= \frac{18.04/M_2}{100.0/18.015 + 18.04/M_2}$$

$$0.0176 = \frac{18.04/M_2}{5.551 + 18.04/M_2}$$

$$M_2 = \frac{18.04 - 0.32}{0.0977} = 181.4 \text{ g mol}^{-1}.$$

The correct value is 182.18 g mol^{-1}.

5.40. The vapor pressure has been reduced from 40.00 to 26.66 kPa, so that the mole fraction of the solvent is given by Raoult's law as

$$x_1 = \frac{P_1}{P_1^*} = \frac{26.66 \text{ kPa}}{40.00 \text{ kPa}} = \frac{2}{3}$$

Let the amount of solute be n_2 mol; since $n_1 = 1$ mol, the mole fraction of solvent is

$$x_1 = \frac{1}{1 + n_2} = \frac{2}{3}$$

Consequently, $n_2 = 1/2$. Since 0.80 kg is half a mole, the molar mass is 0.160 kg mol^{-1} = 160 g mol^{-1}.

5.41. From Eq. 5.78,

$$P_1^* - P_1 = x_2 P_1^*$$

$$13.3 - 12.6 = x_2(13.3) \qquad x_2 = 0.053$$

From Eq. 5.79,

$$x_2 = \frac{W_2/M_2}{W_1/M_1 + W_2/M_2} = \frac{1}{(M_2 W_1/M_1 W_2) + 1}$$

$$\frac{M_2 W_1}{M_1 W_2} = \frac{1}{x_2} - 1 = 18.0; \qquad \frac{M_2}{M_1} = 18.0 \times \frac{1.00}{10.00} = 1.80$$

5.42. $\Delta_f T = K_f m_2 = 1.5$ K

$$1.5 = 7.0 \, m_2/\text{mol kg}^{-1}; \quad m_2 = 0.2143 \text{ mol kg}^{-1}$$

$$\text{mol \% impurity} = \frac{0.2143 \times 100}{0.2143 + 1000/128} = \frac{21.43}{8.027}$$

$$= 2.7\%$$

$$100.0\% - 2.7\% = 97.3 \text{ mol \% pure}$$

5.43. $a_1 = \dfrac{6.677}{9.657} = 0.6914 \qquad a_2 = \dfrac{1.214}{3.572} = 0.3399$

$\gamma_1 = \dfrac{0.6914}{0.670} = 1.03 \qquad \gamma_2 = \dfrac{0.3399}{0.330} = 1.03$

5.44. Amount of NaCl = 11.5 g/58.5 g mol^{-1} = 0.197 mol

Amount of H$_2$O = 100.0 g/18.015 g mol^{-1} = 5.551 mol

$$x_1 = \frac{n_1}{n_1 + n_2} = \frac{5.551}{5.748} = 0.966, \quad x_2 = 0.034$$

$$a_1 = \frac{95.325}{101.325} = 0.941 \quad \gamma_1 = \frac{0.941}{0.965} = 0.975$$

5.45. From Figure 5.13, the maximum value of the entropy is 5.76 J K^{-1} mol^{-1}. Then $\Delta_{\text{mix}}G^{\text{id}} = -300(5.76) = -1728$ J mol^{-1}. Therefore, the Gibbs energy of mixing ranges from 0 to -1.73 kJ mol^{-1} for a 50/50 mixture. Since this is a rather small driving force, in a nonideal solution where $\Delta_{\text{mix}}H = 0$, the value of $\Delta_{\text{mix}}H$ must be negative or only slightly positive for mixing to occur.

5.46. $a_{\text{H}_2\text{O}} = \dfrac{P_{\text{H}_2\text{O}}}{P^*_{\text{H}_2\text{O}}} = \dfrac{2.269 \text{ kPa}}{2.339 \text{ kPa}} = 0.9701.$

Since $a_{\text{H}_2\text{O}} = \gamma_{\text{H}_2\text{O}} x_{\text{H}_2\text{O}}$

$\gamma_{\text{H}_2\text{O}} = \dfrac{a_{\text{H}_2\text{O}}}{x_{\text{H}_2\text{O}}} = \dfrac{0.9701}{0.990} = 0.980$

5.47. Find the value of μ_A such that $\mu_A + \mu_B \ (= G)$ is equal to the expression given in the problem.

a. $\mu_A = \left(\dfrac{\partial G}{\partial n_A}\right)_{n_B, T, P} = \mu^*_A + RT \ln x_A + RT n_A \left(\dfrac{\partial \ln x_A}{\partial n_A}\right)_{n_B} +$

$RT n_B \left(\dfrac{\partial \ln x_B}{\partial n_A}\right)_{n_B} + \dfrac{C n_B (n_A + n_B)}{(n_A + n_B)^2} - \dfrac{C n_A n_B}{(n_A + n_B)^2}$

Since $\left(\dfrac{\partial \ln x_A}{\partial n_A}\right)_{n_B} = \left[\dfrac{\partial\left(\dfrac{n_A}{n_A + n_B}\right)}{\partial n_A}\right]_{n_B} = \dfrac{1}{n_A} - \dfrac{1}{(n_A + n_B)}$

and $\left(\dfrac{\partial \ln x_A}{\partial n_A}\right)_{n_B} = \left[\dfrac{\partial\left(\dfrac{n_B}{n_A + n_B}\right)}{\partial n_A}\right]_{n_B} = - \dfrac{1}{(n_A + n_B)}$

$\mu_A = \mu^*_A + RT \ln x_A + RT\left(1 - \dfrac{n_A}{n_A + n_B} - \dfrac{n_B}{n_A + n_B}\right) + C\left[\dfrac{n_A n_B + n_B^2 - n_A n_B}{(n_A + n_B)^2}\right]$

$\mu = \mu^*_A + RT \ln x_A + RT(0) + \dfrac{C n_B^2}{(n_A + n_B)^2}$

$= \mu^*_A + RT \ln x_A + C x_B^2$

b. Write

$\mu_A = \mu^*_A + RT \ln x_A = \mu^*_A + RT \ln \gamma_A$

By comparison,

$RT \ln \gamma_A = C x_B^2 \quad \text{or} \quad \ln \gamma = \dfrac{C x_B^2}{RT}$

Thus $\gamma = e^{C x_B^2 / RT} = 1$ when $x_B \to 0.$

This corresponds to a pure A. In a very dilute solution of A in B we also expect $\gamma_A \to 1$. In that case

$$\overset{*}{\mu_A}{}' = \underset{x_A \to 0}{\lim} (\mu_A - RT \ln x_A)$$

Substitution from above

$$\overset{*}{\mu_A}{}' = \underset{x_B \to 1}{\lim} (\overset{*}{\mu_A} + Cx_B^2) = \overset{*}{\mu_A} + C$$

Therefore,

$$\mu = \overset{*}{\mu_A}{}' + RT \ln x_A + C(x_B^2 - 1) = \overset{*}{\mu_A} + RT \ln x_A + RT \ln \gamma_A$$

$$\ln \gamma_A = C(x_B^2 - 1)/RT = 0 \quad \text{when} \quad x_B \to 1$$

■ Colligative Properties

5.48. From Eq. 5.115,

$$
\begin{aligned}
\ln x_1 \quad &= \frac{19\,000}{8.3145}\left(\frac{1}{353.35} - \frac{1}{298.15}\right) \\
&= 2.285 \times 10^3 (2.830 \times 10^{-3} - 3.354 \times 10^{-3}) \\
&= 6.467 - 7.664 = -1.198 \\
x_1 \quad &= 0.302
\end{aligned}
$$

5.49. Rewriting Eq. 5.28 as $P_2(k'')^{-1} = c_2$, the values for N_2 and O_2 may be substituted directly. We make the assumption that N_2 gives rise to 80% of the pressure and O_2 is responsible for the other 20%. Then,

$$c(N_2) \quad = 0.80(101\,325) \text{ Pa } 2.17 \times 10^{-8} \text{ mol dm}^{-3} \text{ Pa}^{-1}$$

$$= 1.76 \times 10^{-3} \text{ mol dm}^{-3}$$

$$c(O_2) \quad = 0.20(101\,325) \text{ Pa } 1.02 \times 10^{-8} \text{ mol dm}^{-3} \text{ Pa}^{-1}$$

$$= 2.07 \times 10^{-4} \text{ mol dm}^{-3}$$

The total concentration is 1.967×10^{-3} mol dm^{-3}. This value of the concentration approaches the value of the molality. We may then use the molal freezing point depression expression. The result is

$$\Delta_{fus}T = -(1.86)\,(0.001\,967) = -0.003\,66 \text{ K}$$

5.50. Since the molecular weight is 60.06, $c = 0.100$ mol dm^{-3}

$$\pi = cRT = \frac{n}{V} RT = 0.100\,(8.3145)\,(300.15) \text{ kPa} = 249.6 \text{ kPa}$$

5.51. $\Delta_{fus}T = K_f m_2 = 2.17 \times 1.50 = 3.255$ K

From Eq. 5.115, substituting a_1 for x_1 as discussed in the paragraph after Eq. 5.100, we have

$$\ln a = \frac{\Delta_{fus}H_m}{R}\left(\frac{1}{T_f^*} - \frac{1}{T}\right) = -\frac{\Delta_{fus}H\theta}{R\,T_f^*}$$

$$\ln a_1 = -\frac{6009.5(3.255)}{8.3145(273.15)^2} = -0.0315; \quad a = 0.969$$

$$x_1 = \frac{55.6}{55.6 + 1.5} = 0.9737$$

$$f_1 = \frac{a}{x} = \frac{0.9690}{0.9737} = 0.995$$

5.52. The molality of the solute is calculated from

$$\Delta_{vap}T \quad = K_b m_2$$

$$m \quad = \frac{\Delta_{fus}T}{K_b} = \frac{0.9 \text{ K}}{6.26 \text{ K m}^{-1}} = 0.144 \text{ m}$$

$$= 0.144 \text{ mol kg}^{-1}$$

The mass of solute per kilogram of solvent is

$$\frac{0.000\ 85 \text{ kg solute}}{0.150 \text{ kg bromobenzene}} = 0.005\ 67$$

Then for the solute,

$$0.005\ 67 \text{ kg} = 0.144 \text{ mol}$$

and

$$M = \frac{0.005\ 67 \text{ kg}}{0.144 \text{ mol}} = 0.0394 \text{ kg mol}^{-1} = 39.4 \text{ g mol}^{-1}$$

5.53. For the dissociation:

	A_xB_y	$\rightleftharpoons$	xA^{z+}	yB^{z-}
Initial molalities:			m	0	0 mol kg^{-1}
Molalities after dissociation:			$m - \alpha m$	$x\alpha m$	$y\alpha m$ mol kg^{-1}

Total molality $= m(1 - \alpha + x\alpha + y\alpha)$ mol kg^{-1}

Then $\quad i = \dfrac{\text{total molality}}{\text{initial molality}} = 1 - \alpha + x\alpha + y\alpha$

If $v = x + y$, $i = 1 - \alpha + \alpha v = 1 - \alpha(1 - v)$

Then $\quad \alpha = \dfrac{i - 1}{v - 1}$

From our problem, $\Delta_{fus}T = 273.150 - 273.114 = 0.036$ K

$$i = \Delta_{fus}T/K_f m = 0.036 \text{ K}/[1.86 \text{ (K } m^{-1})0.010(m)] = 1.94$$

Since complete dissociation gives $v = 2$ for HCl,

$$\alpha = \frac{i-1}{v-1} = \frac{0.94}{1} = 0.94$$

The electrolyte therefore appears to be 94% dissociated. This extent of dissociation is only apparent because of the nonideality of the solution.

5.54. From Eq. 5.134,

$$\pi = \frac{n_2 RT}{V}$$

Thus

$$\pi = \frac{m_2/M_2}{V} RT$$

where m_2 is the mass of solute dissolved in volume V and M_2 is the molar mass. Since the equations are good only for dilute solutions, we try to find a limiting value of M_2.

From the preceding, $M_2 = (RT/\pi)(m_2/V)$ and so $\lim \dfrac{m_2}{V} \to 0$ and $\lim \dfrac{m_2}{\pi V} \to 0$.

A plot of $(m_2/\pi V)/\text{g dm}^{-3} \text{ atm}^{-1}$ against $(m_2/V)/\text{g dm}^{-3}$ gives a limiting value of $m_2/\pi V$.

The values of $(m_2/\pi V)/\text{g dm}^{-3} \text{ atm}^{-1}$ corresponding to the listed values are

12.93   12.98   12.68   12.16   11.53   10.92

From the plot, the limiting value of $m_2/\pi V$ is about 12.9 g dm^{-3} atm^{-1}. Thus

$$M_2 = RT\left(\frac{m_2}{V_0}\right) = 0.0821 \text{ atm dm}^3 \text{ K}^{-1} \text{ mol}^{-1} \text{ 293.15 (K)} \times$$

$$12.9 \text{ (g dm}^{-3} \text{ atm}^{-1}) = 310 \text{ g mol}^{-1}$$

The molar mass of sucrose is 342 g mol^{-1}, so that there is an error of about 9%. Ignoring the lowest concentration point and extrapolating leads to an error of about 4%. However, since the slope is expected to be fairly close to zero at infinite dilution, a better result cannot be obtained without more low-concentration data.

5.55. From Eq. 5.122,

$$M_2 = \frac{1.856(\text{K kg mol}^{-1}) \times 3.78 \text{ g}}{0.646(\text{K}) \times 300.0 \text{ g}} = 0.0362 \text{ kg mol}^{-1}$$

$$= 36.2 \text{ g mol}^{-1}$$

5.56. $m_2 = 6.09 \text{ g}/187.4 \text{ g mol}^{-1} = 0.0325 \text{ mol}$

$m_1 = 250.0/18.015 = 13.88 \text{ mol}$

$$x_1 = \frac{13.88}{0.0325 + 13.88} = 0.997\ 66$$

From Eq. 5.125,

$$\ln x_1 = \ln 0.997\,66 = \frac{40\,660}{8.3145}\,(1/T - 1/373.15)$$

$$= \frac{-0.002\,339 \times 8.3145}{40\,660} = 1/T - 1/373.15$$

$$1/T = 2.6794 \times 10^{-3}$$

$$T = 373.218 \text{ K}$$

$$\Delta T = 373.218 - 373.15 = 0.068 \text{ K}$$

From Eq. 5.126,

$$\Delta_{vap}T = K_b m_2 = 0.541 \text{ K kg mol}^{-1} \times 0.0325 \text{ mol}/0.2500 \text{ kg} = 0.0703 \text{ K}$$

The agreement is good. The difference between the two values indicates that the solution is sufficiently dilute for Eq. 5.126 to apply.

5.57. The Clapeyron equation (Eq. 5.9) may be used for this problem. Since the boiling point is given at a pressure for 1 atm, it is appropriate to have the pressure of the gas given in atmospheres. Since 1 mmHg = 1 Torr,

$$\log P/\text{Torr} = \log [P/\text{atm} \times 1 \text{ atm } (760 \text{ Torr})^{-1}]$$

$$= \log P/\text{atm} + \log 1 \text{ atm } (760 \text{ Torr})^{-1}$$

The derivative of this expression shows that it does not matter how pressure is expressed as long as we are considering only a change in pressure.

Since $\dfrac{d \log P}{dT} = \log_{10} e \dfrac{1}{P}\dfrac{dP}{dT} = \dfrac{0.434\,29}{P}\dfrac{dP}{dT}$,

$$\frac{dP}{dT} = \frac{P}{0.43429}\frac{d}{dT}\,[5.4672 - 1427.3\,T^{-1} - 3169.3\,T^{-2}]$$

We now can use the Clapeyron equation (Eq. 5.9). At the boiling point, $P = 1$ atm, $T_b = 398.4$ K.

$$\frac{dP}{dT} = \frac{(1 \text{ atm})}{0.43429}\,[1427.3/(398.4)^2 + 6338.6/((398.4)^3] = 2.094 \times 10^{-2} \text{ atm K}^{-1}$$

$$V_m(\text{liquid}) = \frac{63.9 \times 10^{-3} \text{ kg mol}^{-1}}{0.819 \text{ kg dm}^{-3}} = 7.80 \times 10^{-2} \text{ dm}^3 \text{ mol}^{-1}$$

$$V_m(\text{vapor}) = \frac{63.9 \times 10^{-3} \text{ kg mol}^{-1}}{3.15 \times 10^{-4} \text{ kg dm}^{-3}} = 202.86 \text{ dm}^3 \text{ mol}^{-1}$$

$$\Delta V = V_m(\text{vapor}) - V_m(\text{liquid}) = 202.86 - 0.0780 = 202.78 \text{ dm}^3 \text{ mol}^{-1}$$

Rearranging and substituting into Eq. 5.9, we have

$$\Delta_{vap}H_m = T\Delta V_m\frac{dP}{dT} = 398.4 \text{ K}(202.78 \text{ dm}^3 \text{ mol}^{-1})(2.094 \times 10^{-2} \text{ atm K}^{-1})$$

$$= 1692 \text{ dm}^3 \text{ atm mol}^{-1}$$

$$= 1692 \text{ dm}^3 \text{ atm mol}^{-1} \times 10^{-3} \text{ m}^3 \text{ dm}^{-3} \times 101.325 \text{ kJ atm}^{-1} \text{ m}^{-3}$$

$$= 171.4 \text{ kJ mol}^{-1}$$

5.58. Addition of the moles per liter of the salts listed in Table 5.5 gives a 1.0989 molar solution. Assuming that there is sufficient positive charge to cancel the negative charges present, no calculation need be made to determine van't Hoff's i factor. Although the temperature of the ocean is variable, we can assume an average value of 25°C near the surface. Then

$$\pi = cRT$$

$$= 1.0989 \text{ mol dm}^{-3}(8.3145 \text{ J K}^{-1} \text{ mol}^{-1})(298 \text{ K})$$

$$= 2.72 \times 10^3 \text{ J dm}^{-3}$$

Since $J = \text{kg m}^2 \text{ s}^{-2}$ and $Pa = \text{kg m}^{-1} \text{ s}^{-2}$

$$\pi = 2.72 \times 10^3 \text{ kg m}^2 \text{ s}^{-2} \text{ dm}^{-3} \times 10^3 \text{ dm}^3/\text{m}^3 = 2.72 \times 10^6 \text{ kg m}^{-1} \text{ s}^{-2}$$

$$= 2.72 \text{ MPa}$$

6. PHASE EQUILIBRIA

■ Number of Components and Degrees of Freedom

6.1. The region marked *orthorhombic* is a single-phase region. Since this is the phase diagram for pure sulfur, there is only one component. From the phase rule $f = c - p + 2$, with $c = 1$ and $p = 1$, the value of f is: $f = 1 - 1 + 2 = 2$. The two degrees of freedom are pressure and temperature. There is only one phase in the region marked *monoclinic*.

6.2. The compositions of the two phases at a particular temperature are: water saturated with nicotine, and nicotine saturated with water.

The number of degrees of freedom is

$$f = c - p + 2 = 2 - 2 + 2 = 2$$

This means that within the two-phase region, the temperature and weight percent may be varied without changing the two-phase character of the region.

6.3. a. For KCl and H_2O at the equilibrium pressure,

$$f = c - p + 2 = 2 - 1 + 2 = 3$$

Since the equilibrium pressure is specified, this reduces the number of degrees of freedom to 2.

b. Here, NaCl, KCl, and H_2O are present. This is actually a three-component system since the solution contains Na^+, K^+, Cl^-, and H_2O. The first three compositions are reduced to two independent ones by the electroneutrality condition. Therefore, $f = c - p + 2 = 3 - 1 + 2 = 4$, but with the restriction of constant pressure, the variance is reduced by 1, and is therefore 3.

c. Ice, water, and alcohol are only two components. Consequently, $f = c - p + 2 = 2 - 2 + 2 = 2$.

6.4. Aqueous sodium acetate is a two-component system even though the hydrolysis

$$\text{Acetate}^- + H_2O \rightleftarrows OH^- + HAc$$

takes place, since the equilibrium constant defines the concentration of OH^- and HAc if the concentration of sodium acetate is given.

6.5. Starting with pure $CaCO_3$, we have only one component present. When two of the three species are present, the third species is also present; but because of the equilibrium $CaCO_3 \rightleftarrows CaO + CO_2$, there are only two components.

6.6. There are four individual gases, and the equilibrium equation reduces the number of independent components to three.

6.7.

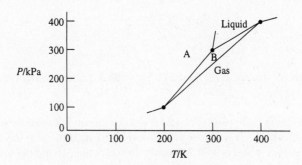

T/K	P/kPa	Phases in Equilibrium
200	100	A, B, gas
300	300	A, B, liquid
400	400	B, liquid, gas

■ Use of the Lever Rule; Distillation

6.8. a. Wt % of B = $\dfrac{99 \times 100}{33 + 99}$ = 75%

From the graph at 75% B, the first vapor appears at 60°C.

b. The composition of the vapor is given by the intersection of the tie line at the vapor curve. In this case, the vapor has a composition of 88% B.

c. The intersection of the liquid line at 65°C corresponds to 53% B in the liquid.

d. Using the average composition of the distillates as the value midway between initial and final composition of the distillates, we have

$$\frac{1}{2}(88\% + 70\%) = 79\% \text{ B in distillate}$$

Let W_R = mass of residue; W_D = mass of distillate

$$W_R + W_D = 132$$

Then, applying the condition that B is distributed through residue and distillate, we have

mass of B in residue + mass of B in distillate = 99 g.

$$0.53W_R + 0.79W_D = 99$$

$$0.53(132 - W_D) + 0.79W_D = 99$$

$$0.26W_D = 29$$

$$W_D = 111.5 \text{ g};   W_R = 132 - 111.5 = 20.5 \text{ g}$$

In the distillate, therefore, 79% of 111.5 g or 88.1 g is component B, and $111.5 - 88.1 = 23.4$ g is component A.

6.9.   From Figure 6.14, the composition at 350 K at equilibrium between the single-phase water-rich region and the two-phase region is approximately 10% nicotine. For the equilibrium value on the nicotine-rich side, the value is approximately 75% nicotine. Using the lever rule, we have

$$\frac{\text{Mass of water-rich layer}}{\text{Mass of nicotine-rich layer}} = \frac{75 - 40}{40 - 10} = \frac{35}{30} = 1.2$$

6.10.   From Eq. 6.21, since $w = nM$, we write

$$\frac{w_A}{w_B} = \frac{n_A M_A}{n_B M_B} = \frac{P_A^* M_A}{P_B^* M_B}   \text{or}   M_A = \frac{P_B^*}{P_A^*} \frac{w_A M_B}{w_B}$$

The vapor pressure of pure chlorobenzene is $56.434 - 43.102 = 13.332$ kPa. Substitution gives

$$M_{\text{chlorobenzene}} = \frac{43.102}{13.334} (1.93) 18.02 = 112.4.$$

The value obtained by addition of relative atomic masses is 112.6.

6.11.   a.   The total pressure is: $P = P_1 + P_2$

From Raoult's law

$$P_1 = x_1 P_1^*$$

$$P = x_1 P_1^* + P_2   \text{but}   P_2 = (1 - x_1)P_2^*$$

$$= x_1 P_1^* + (1 - x_1)P_2^* = x_1 P_1^* + P_2^* - x_1 P_2^*$$

$$= P_2^* + (P_1^* - P_2^*)x_1$$

This is Eq. 6.9.

b. From Eq. 6.8, $y_1 = P_1/P = x_1 P_1^* /P$, upon solution of P just derived,

$$y_1 = \frac{x_1 P_1^*}{P_2^* + (P_1^* - P_2^*)x_1}$$

or solving for x_1 gives

$$x_1 P_1^* = y_1 P_2^* + y_1(P_1^* - P_2^*)x_1$$

$$y_1 P_2^* = [P_1^* - (P_1^* - P_2^*)]x_1$$

$$x_1 = \frac{y_1 P_2^*}{P_1^* + (P_2^* - P_1^*)y_1}$$

Then substitution of x_1 into Eq. 6.9 gives

$$P = P_2^* + (P_1^* - P_2^*)$$

$$= \frac{P_2^*[P_1^* + (P_2^* - P_1^*)y_1] + y_1 P_1^* P_2^* - y_1 P_2^* P_2^*}{P_1^* + (P_2^* - P_1^*)y_1}$$

Upon expanding and cancellation, we have Eq. 6.11.

6.12.   To determine the masses of the material distilled, the numerator of Eq. 6.21 is multiplied by M_A and the denominator by M_B. Since $w = nM$,

$$\frac{w_A}{w_B} = \frac{n_A M_A}{n_B M_B} = \frac{P_A^* M_A}{P_B^* M_B}$$

6.13.   The vapor pressure of water at 372.4 K is 98.805 kPa. The vapor pressure of naphthalene is therefore 101.325 kPa – 98.805 kPa = 2.52 kPa. The molar mass of naphthalene is 128.19 g mol^{-1}. From the modified Eq. 6.21, we have

$$\frac{w_{H_2O}}{w_{napth}} = \frac{P_{H_2O}^* M_{H_2O}}{P_{napth}^* M_{napth}}$$

$$w_{H_2O} = \frac{98.805 \text{ kPa } (18.02 \text{ g}) \text{ 1 kg}}{2.52 \text{ kPa } (128.19 \text{ g})} = 5.51 \text{ kg}$$

6.14.   The vapor pressure of chlorobenzene is 56.434 – 43.102 = 13.332 kPa. From Problem 6.12,

$$\frac{\text{Mass of chlorobenzene}}{\text{Mass of water}} = \frac{13\,332 \times 0.1125}{43\,102 \times 0.01802} = 1.93$$

The sample contains 1.93 g of chlorobenzene for each 1.00 g of water.

6.15.   Using Eq. 6.18 and letting isobutyl alcohol be component 1, we have

$$y_{IAA} = \frac{0.4 \times 2330}{0.4 \times 2330 + 0.6 \times 7460} = 0.172$$

$$y_{IBA} = 1 - y_{IAA} = 1.000 - 0.172 = 0.828$$

6.16. Take the logarithms of both sides of $\rho = m/V$: $\ln \rho = \ln m - \ln V$. Partial differentiation with respect to T gives

$$\left(\frac{\partial \ln \rho}{\partial T}\right)_P = -\left(\frac{\partial \ln V}{\partial T}\right)_P = -\frac{1}{V}\left(\frac{\partial V}{\partial T}\right)_P = -\alpha$$

6.17. From the previous problem,

$$\alpha = \frac{1}{V}\left(\frac{\partial V}{\partial T}\right)_P \approx \frac{1}{V}\left(\frac{\Delta V}{\Delta T}\right)_P$$

for small changes in V and T. For an arbitrary quantity of water, say 1 gram exactly, the equation $V = m/\rho$ gives $V = 1.001\ 769$ cm^3 at 20 °C and $V = 1.001\ 982$ cm^3 at 21 °C and 1 atm.

Using these values we have

$$\alpha \approx \frac{1}{1.001\ 769}\left(\frac{1.001\ 982 - 1.001\ 769}{1\ K}\right) = \frac{0.000\ 213}{1.001\ 769\ K}$$

$$\approx 0.002\ 126\ K^{-1} = 2.126 \times 10^{-4}\ K^{-1}$$

6.18. There are six full horizontal steps and two small fractional steps; therefore, there are approximately six theoretical plates.

6.19. From $\left(\frac{\partial P}{\partial T}\right)_V \left(\frac{\partial T}{\partial V}\right)_P \left(\frac{\partial V}{\partial P}\right)_T = -1$, $\quad \frac{\alpha}{\kappa} = \left(\frac{\partial P}{\partial T}\right)_V \approx \left(\frac{\Delta P}{\Delta T}\right)_V$

Therefore,

$$\Delta P \approx \frac{\alpha}{\kappa}\Delta T = \frac{2.85 \times 10^{-4}\ K^{-1}}{4.49 \times 10^{-5}\ atm^{-1}} \times 6\ K = 38.1\ atm.$$

■ Construction of Phase Diagrams from Physical Data

6.20. Application of the lever rule gives

$$\frac{\text{Mass of solid layer}}{\text{Mass of solid + liquid layer}} = \frac{\overline{se}}{\overline{be}} = \frac{0.18}{0.31} = 0.58$$

or 58% solid and 42% liquid in the two-phase region. The overall composition of the liquid above the liquidus line is $x_{Si} = 0.31$.

6.21. The temperature at which solid solvent is in equilibrium with liquid solvent of mole fraction x_1 is given by Eq. 5.115:

$$\ln x_1 = \frac{\Delta_{fus}H_m}{R}\left(\frac{1}{T_f^*} - \frac{1}{T}\right)$$

Values of x_1 and T determined from this equation for each component give the desired liquidus lines in the regions near large values of x_1. Several values are:

x_1	T/K	x_1	T/K
0.945	1650	0.969	1300
0.863	1600	0.924	1250
0.784	1550	0.879	1200
0.708	1500	0.783	1100
0.564	1400	0.681	1000

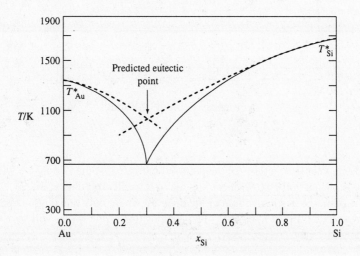

A plot is shown on which the points represent the data points and the solid curves are experimental curves of Figure 6.16. The dotted lines intersect at about $x_{Si} = 0.28$ compared to the actual $x_{Si} = 0.31$. However, the eutectic temperature is approximately 400 K too high.

6.22. We graph the data to generate a phase diagram of the phenol-water system:

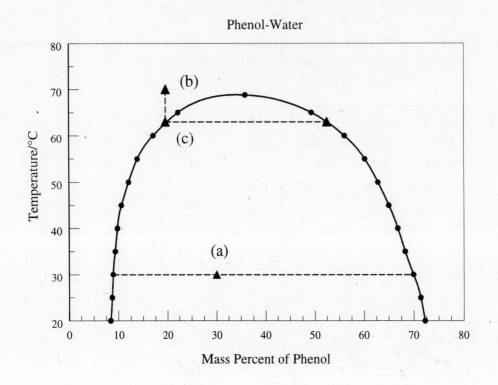

Phenol-Water

a. The tie lines drawn through the point (a) indicate that the two layers will have compositions 8.9% phenol and 70.0 % phenol by mass, respectively (data at 30°C).
b. The point (b) lies in a region of the diagram where only one phase is present.
c. Two phases appear at 63.0°C. The composition of the layers will be 19.6% phenol and 52.5% phenol by mass, respectively.

6.23. Each halt corresponds to a line of three-phase equilibrium and each break to a boundary between a one- and a two-phase region. At 50% Y_2O_3, a compound is formed and may be written as $Fe_2O_3 \cdot Y_2O_3$ or $YFeO_3$.

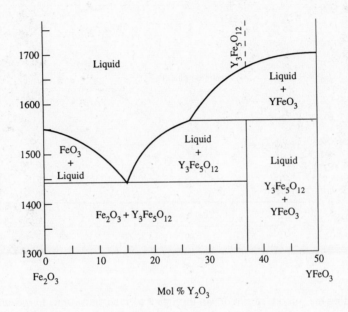

A compound unstable above 1575°C is indicated between 30 and 40% Y_2O_3. This might be taken to be $2Fe_2O_3 \cdot Y_2O_3$ at 33% Y_2O_3, but actually the formula is $Y_3Fe_5O_{12}$, corresponding to $3Y_2O_3 + 5Fe_2O_3$ at 37% Y_2O_3.

6.24.

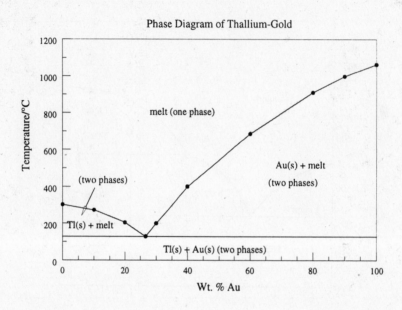

Phase Diagram of Thallium-Gold

By extending the smooth curves drawn through the given data points to the eutectic temperature, the composition of the eutectic (marked with a symbol on the graph) is identified as 26 wt. % Au

6.25.   a.

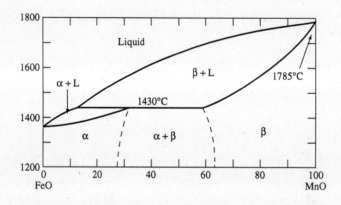

b.   As liquid of 28 mass % MnO cools, β first forms along with liquid. At 1430 °C, β converts to α and $\alpha + L$ remain only briefly as the temperature is lowered about 50 °C, at which point all the liquid is reconverted to the α phase. The compositions are given by the lever rule. As 1200 °C is approached, some β may again make an appearance.

6.26.

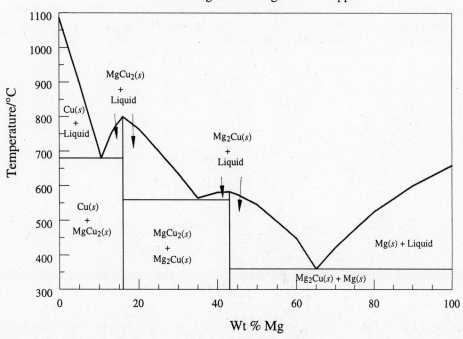

Phase Diagram of Magnesium-Copper

From the graph, the first eutectic (680 °C) has a composition of 10.5 wt. % Mg, the second (560 °C), a composition of 35.0 wt. % Mg while the third (360 °C) has a composition of 65 wt. % Mg.

6.27 Assuming that there are 100 g of the first compound, we have 16.05 g Mg and 83.95 g Cu. The mole ratios of Mg : Cu are

$$\frac{16.05}{24.305} : \frac{83.95}{63.546} \text{ or } 0.660 : 1.321.$$

Dividing both sides by the smallest number (i.e., 0.660), we get

$$1 : \frac{1.321}{0.660} = 1 : 2.001.$$

Therefore, the empirical formula of the first compound is $MgCu_2$. Proceeding in a similar manner, we find that the second compound has an empirical formula Mg_2Cu.

6.28. The diagrams are self-explanatory. The coexistence of the three phases is a clear indication of a peritectic-type diagram. A note of caution is in order here, however. In the range 0–10 mol % Au at 1430°C to 1536°C, still another phase, called δ, exists; and this would not be detected using only the compositions listed. One must be careful to use enough compositions to ensure that all the phases are identified. Also, the equilibrium between liquid and γ is not a simple curve and must be determined by careful experimentation.

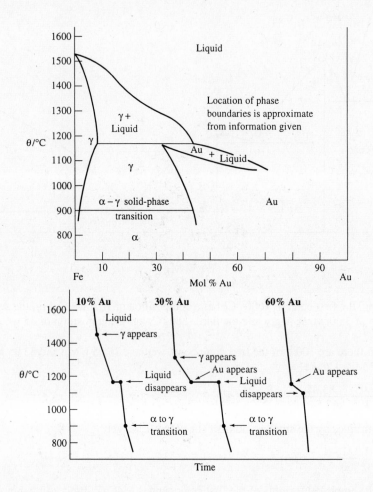

6.29.

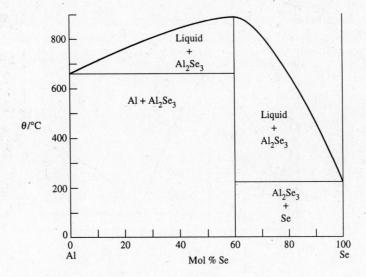

6.30. The diagram is self-explanatory. The lower phase field of the α-phase is less than 1%.

a.

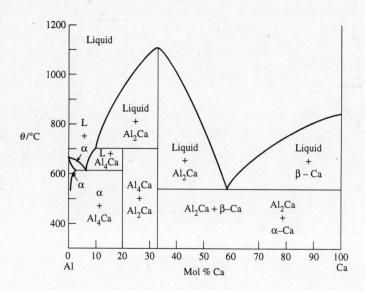

b.

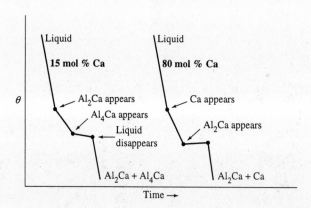

6.31.

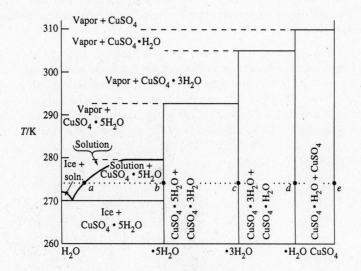

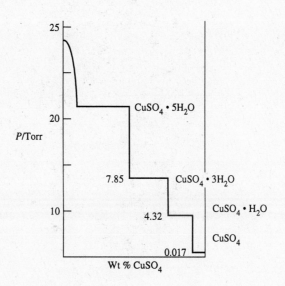

In the upper figure, the dilute $CuSO_4$ begins in a solution single-phase region. Pure
$CuSO_4 \cdot 5H_2O$ precipitates out as the first phase boundary at a is crossed. Water continues to be
removed as more $CuSO_4 \cdot 5H_2O$ precipitates until only pure $CuSO_4 \cdot 5H_2O$ is present at b. In the

next two-phase region, $CuSO_4 \cdot 5H_2O$ dehydrates, forming progressively more $CuSO_4 \cdot 3H_2O$ until all of the pentahydrate is gone at c. The process repeats, the trihydrate forming the monohydrate until only monohydrate is present at d. The monohydrate dehydrates until at e only pure $CuSO_4$ is present.

In the lower figure, the vapor pressure of water drops as the amount of $CuSO_4$ increases (Raoult's law) until the solution is saturated with respect to the pentahydrate. The system is invariant since three phases, vapor, saturated solution, and solid $CuSO_4 \cdot 5H_2O$, are present at the constant temperature of 298.15 K. As the concentration of $CuSO_4$ increases (water is removed), the pressure remains constant until only $CuSO_4 \cdot 5H_2O$ is present. Removal of additional water causes some trihydrate to form, and the pressure drops. Again the system is invariant; three phases are present, vapor, $CuSO_4 \cdot 5H_2O$, and $CuSO_4 \cdot 3H_2O$. The process is continued as before at the other stages.

6.32.   The system $CaO–MgO–SnO_2$:

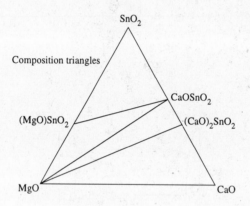

6.33.

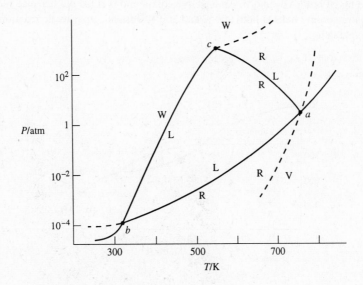

a. Stable triple point R (red phosphorus, solid), liquid (*L*), vapor (*V*).

b. Metastable triple point W (white phosphorus, solid), *L*, *V*. (The vapor pressure of the white form is greater than that of the red form.)

c. Stable triple point W, R, *L*. If we assume that a solid cannot be superheated, the triple point W, R, *L* is totally unstable since it probably lies above the melting point of the liquid.

6.34.

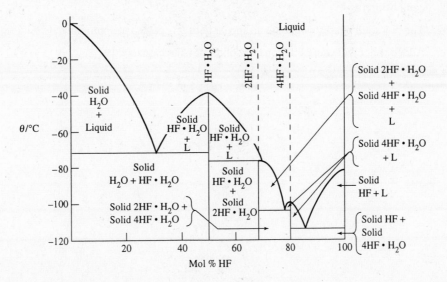

6.35. The 500 K equilibrium line probably contains a eutectic since the temperature is below the melting points of both AB_2 and B. An unstable compound is ruled out because such a reaction would require cooling halts at both 900 K and 500 K. Instead, a peritectic reaction shown is the simplest explanation.

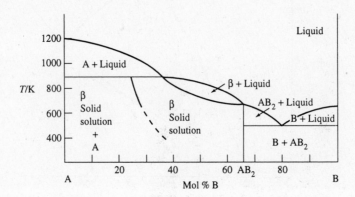

■ Data Derived from Phase Diagrams of Condensed Systems

6.36. a. As water is added, the saturated liquid of composition b would be in equilibrium with two solids A and B. At approximately 20% C, when the composition crosses the line bB , the solid A disappears and only solid B will be present in equilibrium with liquid of composition b.

   b. The two solid phases would not disappear until b is passed at approximately 50% liquid.

   c. Added water will cause dilution and solid salt will cease to exist.

6.37. a. Peritectic point

b. Eutectic point

c. Melting point

d. Incongruent melting

e. Phase transition

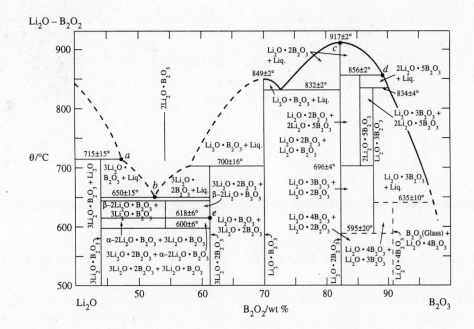

6.38. As liquid is cooled, solid spinel first appears at about 1950 K in equilibrium with liquid. At approximately 1875 K, all of the liquid converts to solid spinel, the composition of which varies according to the lever rule. As the temperature falls to about 1400 K, a two-phase region appears that is Mn_3O_4 + spinel. Below about 1285 K, the spinel converts to corundum and Mn_3O_4 + corundum coexist.

6.39. The composition of the system is 30 mol % acetic acid, 50 mol % water, and 20 mol % toluene. The system point is practically on the $p''q''$ tie line, and there are therefore two liquids present. The ends of this line and thus the concentrations of the two liquids are approximately:

a. 95.5% toluene, 4% acetic acid, 0.5% H_2O and

b. 1% toluene, 37% acetic acid, and 62% H_2O.

The relative amounts of the two liquids are given by the lever rule:

$$\frac{15.4\,B}{3.8\,A} \quad \text{or} \quad 4B \text{ to } 1A$$

6.40.

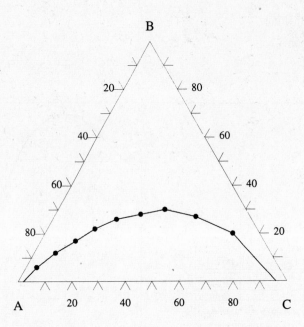

It is clear from the phase diagram that A and C are only slightly soluble in each other in the absence of B. As B is added, two layers are formed, one rich in A and the other, rich in C. As B is added, the mutual solubility of A and C increases until at 30 mole % in B, the three liquids become miscible in all proportions.

6.41.

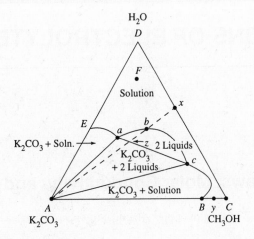

a.

Region	System
AEa	K_2CO_3 in equilibrium with water-rich saturated solution
Aac	K_2CO_3 in equilibrium with conjugate liquids a and c
abc	Two conjugate liquids joined by tie lines
AcB	K_2CO_3 in equilibrium with alcohol-rich saturated solution

b. The state of the system will move along a line joining x and A. Initially solution is formed; as more K_2CO_3 is added two layers a and c form, and once beyond point z, K_2CO_3 ceases to dissolve so that solid K_2CO_3 and the two liquids a and c coexist.

c. As long as two liquids exist, liquid with composition in the region AcB is the alcohol-rich layer and may be separated from the water-rich layer by separatory funnel.

d. When water is added to an unsaturated solution of K_2CO_3 in alcohol, the state of the system moves along the line joining y and D. Some K_2CO_3 will precipitate as the state moves into the ABc region and then redissolves as it moves into the solution region again.

e. On evaporation of F, the system composition follows a line drawn from the water corner through F to the Ac line. At the first composition line, two liquids form and the compositions of the solutions move toward a and c. When the system composition reaches the ac line, K_2CO_3 begins to precipitate and is in equilibrium with the conjugate liquids a and c. Further reduction of water moves the ratio of liquid a to liquid c in favor of c until the line Ac is crossed, at which time solid K_2CO_3 is in equilibrium with a single solution.

7. SOLUTIONS OF ELECTROLYTES

■ Faraday's Laws, Molar Conductivity, and Weak Electrolytes

7.1.   96 500 C deposits (63.5/2) g of copper; the quantity passed is therefore

$$\frac{96\,500 \times 0.04 \times 2}{63.5} \text{ C}$$

The current was passed for 3600 s; the current is therefore

$$\frac{96\,500 \times 0.04 \times 2}{63.5 \times 3600} \text{ A} = 0.03377 \text{ A} = 33.8 \text{ mA}$$

7.2.   Quantity of electricity passed $= \dfrac{96\,500 \times 0.007\,19}{107.9} \text{ C}$

$$\text{Current} = \frac{96\,500 \times 0.007\,19}{107.9 \times 45 \times 60} \text{ A} = 2.38 \times 10^{-3} \text{ A} = 2.38 \text{ mA} \cong 2.4 \text{ mA}$$

7.3.   The chemical reactions involved are

$$C_6H_5OH + Br_2(g) \rightarrow C_6H_4(Br)OH + HBr$$
$$2Br^- \rightarrow Br_2(g) + 2e^-$$
$$2K^+ + 2e^- \rightarrow 2K(s)$$

Two moles of electrons are involved in the generation of each mole of bromine gas, which reacts with one mole of phenol. Each batch consists of 500.0 kg (5.00×10^5 g) or 5313 moles of phenol (molecular weight = 94.114 g mol^{-1}), which requires 5313 mol. of bromine, or 10 626 mol. of electrons for the reaction. Therefore, since $It = nF$, where n is the number of moles of electrons exchanged in the reaction, we get

$$t = \frac{2 \times 5313 \text{ mol} \times 96\,485 \text{ C mol}^{-1}}{20\,000 \text{ C s}^{-1}} \times \frac{1 \text{ h}}{3600 \text{ s}} = 14.3 \text{ h.}$$

7.4.

$c/10^{-4}M$	$\dfrac{\Lambda}{\Omega^{-1}\,cm^2\,mol^{-1}}$	α	$1-\alpha$	$\dfrac{K = c\alpha^2/(1-\alpha)}{10^{-3}\,M}$
625	53.1	.147	.853	1.583
312.5	72.4	.200	.800	1.563
156.3	96.8	.267	.733	1.520
78.1	127.7	.353	.647	1.504
39.1	1.64	.453	.547	1.467
19.6	205.8	.569	.431	1.472
9.8	249.2	.688	.312	1.487

The values are reasonably constant; average $K = 1.51 \times 10^{-3}$ mol dm^{-3}.

7.5. $\Lambda_{AgCl} = 61.9 + 76.4 = 138.3\ \Omega^{-1}\ cm^2\ mol^{-1}$

Solubility $= \dfrac{1.26 \times 10^{-6}}{138.3}$ mol cm^{-3}

$= 9.11 \times 10^{-9}$ mol cm^{-3}

$= 9.11 \times 10^{-6}$ mol dm$^{-3} = 9.11\ \mu M$

7.6.   The increase in conductivity, $4.4 \times 10^{-4}\ \Omega^{-1}\ cm^{-1}$, is due to the $CaSO_4$ present; thus

$$\Lambda\left(\tfrac{1}{2}Ca^{2+} + \tfrac{1}{2}SO_4^{2-}\right) = \frac{4.4 \times 10^{-4}\ \Omega^{-1}\ cm^{-1}}{2c}$$

where c is the concentration of $CaSO_4$; $2c$ is the concentration of $\tfrac{1}{2}CaSO_4$. The value of $\lambda\left(\tfrac{1}{2}SO_4^{2-}\right)$ is obtained from the conductivity of the Na_2SO_4 solution:

$$\Lambda\left(Na^+ + \tfrac{1}{2}SO_4^{2-}\right) = \frac{2.6 \times 10^{-4}\ \Omega^{-1}\ cm^{-1}}{2.0 \times 10^{-6}\ mol\ cm^{-3}}$$

(Note that since the concentration of Na_2SO_4 is 0.001 M, that of $\tfrac{1}{2}Na_2SO_4$ is 0.002 M.) Thus

$$\Lambda\left(Na^+ + \tfrac{1}{2}SO_4^{2-}\right) = 130.0\ \Omega^{-1}\ cm^2\ mol^{-1}$$

Thus, since

$\lambda(Na^+) = 50.1\ \Omega^{-1}\ cm^2\ mol^{-1}$

$\lambda\left(\tfrac{1}{2}SO_4^{2-}\right) = 79.9\ \Omega^{-1}\ cm^2\ mol^{-1}$

Then

$$\Lambda\left(\tfrac{1}{2}Ca^{2+} + \tfrac{1}{2}SO_4^{2-}\right) = 139.4\ \Omega^{-1}\ cm^2\ mol^{-1}$$

and

$$c = \frac{4.4 \times 10^{-4} \ \Omega^{-1} \ cm^{-1}}{2 \times 139.4 \ \Omega^{-1} \ cm^2 \ mol^{-1}} = 1.578 \times 10^{-3} \ mol \ dm^{-3}$$

Thus

$$c_{Ca^{2+}} = 1.578 \times 10^{-3} \ mol \ dm^{-3}$$

$$c_{SO_4^{2-}} = (1.0 \times 10^{-3} + 1.578 \times 10^{-3}) \ mol \ dm^{-3}$$

$$= 2.578 \times 10^{-3} \ mol \ dm^{-3}$$

$$K_{sp} = 4.07 \times 10^{-6} \ mol^2 \ dm^{-6}$$

7.7. From Eq. (7.9) we have $\Lambda = \kappa/c$; therefore the electrolytic conductance κ is

$$\kappa = \Lambda c \ = 128.96 \ S \ cm^2 \ mol^{-1} \times 0.10 \times 10^{-3} \ mol \ cm^{-3}$$

$$= 1.2896 \times 10^{-2} \ S \ cm^{-1}.$$

Now, using Eq. (7.8), the cell constant is

$$l/A = \kappa/G \ = \frac{1.2896 \times 10^{-2} \ S \ cm^{-1}}{0.01178 \ S} = 1.0947 \ cm^{-1}.$$

To find the equivalent conductance of the electrolyte, we use

$$\Lambda = \frac{G(l/A)}{c} = \frac{1.0947 \ cm^{-1} \times 0.00824 \ S}{0.01178 \ S}$$

$$= 180 \ S \ cm^2 \ mol^{-1}.$$

7.8. $\Lambda(KCl) = (73.5 + 76.4) \ \Omega^{-1} \ cm^2 \ mol^{-1}$

The electrolytic conductivity at 0.01 M is

$$\kappa(KCl) \ = 149.9 \ \Omega^{-1} \ cm^2 \ mol^{-1} \times 10^{-5} \ mol \ cm^{-3}$$

$$= 1.50 \times 10^{-3} \ \Omega^{-1} \ cm^{-1}$$

Recall that conductance is inversely proportional to the resistance. The electrolytic conductivity of the ammonia solution is thus

$$\kappa(NH_4OH) = 1.50 \times 10^{-3} \times \frac{189}{2460} = 1.15 \times 10^{-4} \ \Omega^{-1} \ cm^{-1}$$

The molar conductivity of $NH_4^+ + OH^-$ is

$$\Lambda(NH_4^+ + OH^-) = (73.4 + 198.6) \ \Omega^{-1} \ cm^2 \ mol^{-1}$$

If $c = [NH_4^+] = [OH^-]$,

$$272.0 \ \Omega^{-1} \ cm^2 \ mol^{-1} = \frac{1.15 \times 10^{-4} \ \Omega^{-1} \ cm^{-1}}{c}$$

$$c = 4.23 \times 10^{-7} \ mol \ cm^{-3} = 4.23 \times 10^{-4} \ mol \ dm^{-3}$$

The concentrations of NH_4OH, NH_4^+, and OH^- are thus

$$NH_4OH \quad \rightleftarrows \quad NH_4^+ \quad + \quad OH^-$$

$$0.01 - 4.23 \times 10^{-4} \qquad 4.23 \times 10^{-4} \qquad 4.23 \times 10^{-4} \text{ mol dm}^{-3}$$

$$K_b = 1.87 \times 10^{-5} \text{ mol dm}^{-3}$$

7.9. From the conductivity and concentration, we get

$$\Lambda = \frac{1.53 \times 10^{-4} \text{ S cm}^{-1}}{0.0312 \times 10^{-3} \text{ mol cm}^{-3}} = 4.90 \text{ S cm}^2 \text{ mol}^{-1}.$$

For the weak base, we write

$$B + H_2O \rightarrow BH^+ + OH^-$$
$$c(1-\alpha) \qquad \alpha c \qquad \alpha c$$

so that

$$K_b = \frac{[BH^+][OH^-]}{[B]} = \frac{\alpha^2 c}{(1-\alpha)}.$$

Since $\alpha = \Lambda/\Lambda° = 2.07 \times 10^{-2}$, from which with $c = 0.0312$ mol dm^{-3}, we calculate

$$K_b = \frac{(2.07 \times 10^{-2})^2 \times 0.0312 \text{ mol dm}^{-3}}{(1 - 2.07 \times 10^{-2})} = 1.37 \times 10^{-5} \text{ mol dm}^{-3}.$$

7.10. Note that each number in the first row (concentrations) must be multiplied by 10^{-3} M to yield the molar concentration. Using the model suggested by the Debye-Huckel Onsager equation (Equation 7.53), we assign equivalent conductance as the dependent variable and $\sqrt{c}$ as the independent variable. The result of the linear regression is

$$\Lambda = 151.41 - 83.303\sqrt{c}.$$

In the limit as $\sqrt{c} \rightarrow 0$, we have $\Lambda° = 151.41$ S cm^2 mol^{-1}.

7.11. Equation 7.20 can be rearranged to

$$c\Lambda^2 = K\Lambda_0^2 - K\Lambda_0\Lambda$$

$c\Lambda^2$ could therefore be plotted against Λ. Alternatively, since

$$c\Lambda = K\Lambda_0^2 \cdot \frac{1}{\Lambda} - K\Lambda_0$$

$c\Lambda$ can be plotted against $1/\Lambda$. The slope and intercepts are as shown below:

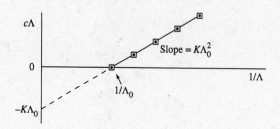

Λ values are obtained by the use of Eq. 7.9; for the lowest concentration, 1.566×10^{-4} mol dm^{-3},

$$\Lambda = \frac{1.788 \times 10^{-6} \ \Omega^{-1} \ cm^{-1} \times 1000 \ dm^{-3} \ cm^{-3}}{1.566 \times 10^{-4} \ mol \ dm^{-3}}$$

$$= 11.4 \ \Omega^{-1} \ cm^2 \ mol^{-1}$$

Similarly, for the other concentrations:

$c/10^{-4}$ mol dm^{-3}	1.566	2.600	6.219	10.441
Λ/Ω^{-1} cm^2 mol^{-1}	11.4	9.30	6.45	5.11
$c\Lambda/10^{-6}$ Ω^{-1} cm^{-1} mol^{-1}	1.785	2.418	4.011	5.335
$1/(\Lambda/\Omega^{-1}$ cm^2 mol^{-1})	0.0877	0.1075	0.155	0.196

In a plot of $c\Lambda$ against $1/\Lambda$, the intercepts are

$$-K\Lambda_0 = -1.15 \times 10^{-6} \ \Omega^{-1} \ cm^{-1} \ mol^{-1}$$

$$1/\Lambda_0 = 0.035 \ \Omega^{-1} \ cm^2 \ mol^{-1}; \quad \Lambda_0 = 30 \ \Omega^{-1} \ cm^2 \ mol^{-1}$$

$$K = 4.0 \times 10^{-8} \ mol \ cm^{-3} = 4.0 \times 10^{-5} \ mol \ dm^{-3}$$

7.12.   The concentration of the acid in water = 1500 ppm = $\dfrac{1500 \ g \ acid}{10^6 \ g \ solution}$

$$= \frac{1.500 \ g \ acid}{10^3 \ g \ solution} = \frac{1.500 \ g \ acid/60.05 \ g \ mol^{-1}}{1.00 \ kg \ solution}$$

$$= 0.0250 \ m.$$

Since the solution has the same density as water, 1.00 kg of solution has a volume of 1.0 dm^3. In other words, we may assume the solution to have a concentration of 0.0250 M.

Now, for a weak acid whose degree of dissociation is α and the concentration is c M, Eq. (7.18) gives

$$K_a = \frac{\alpha^2 c}{(1-\alpha)}, \ \text{or} \ \alpha^2 c + K_a \ \alpha - K_a = 0.$$

Solving this equation for the degree of dissociation α with $c = 0.0250$ M gives $\alpha = 2.6548 \times 10^{-2}$ (the other solution is negative). Since $\alpha = \Lambda/\Lambda°$, we have

$$\Lambda = \alpha\Lambda° = 2.6548 \times 10^{-2} \times 390.7 \ S \ cm^2 \ mol^{-1} = 10.372 \ S \ cm^2 \ mol^{-1}.$$

Therefore, the conductance measured by the cell cannot exceed

$$\Lambda c = 10.372 \text{ S cm}^2 \text{ mol}^{-1} \times 0.0250 \times 10^{-3} \text{ mol cm}^{-3}$$

$$= 2.59 \times 10^{-4} \text{ S cm}^{-1}.$$

7.13. The $\Lambda°$ value for H_2O is calculated as

$$\Lambda°(HCl) - \Lambda°(KCl) + \Lambda°(KOH) = 550.6 \text{ S cm}^2 \text{ mol}^{-1}.$$

In pure water, the only species conducting electricity are H^+ and OH^- ions, each of which have concentrations of $\sqrt{1.008 \times 10^{-14}} = 1.004 \times 10^{-7}$ mol dm^{-3}. Since this is a very low concentration, we may assume that $\Lambda \approx \Lambda°$. Therefore,

$$\kappa = \Lambda c = 1.004 \times 10^{-10} \text{ mol cm}^{-3} \times 550.6 \text{ S cm}^2 \text{ mol}^{-1}$$

$$= 5.528 \times 10^{-8} \text{ S cm}^{-1}.$$

■ Debye-Hückel Theory and Transport of Electrolytes

7.14. From Eq. 7.50,

Thickness $\propto c^{-1/2}$

Thickness $\propto \varepsilon^{1/2}$

Therefore,

a. At 0.0001 M, thickness $= 0.964 \times \sqrt{1000} = 30.5$ nm

b. At $\varepsilon = 38$, thickness $= 0.964 \times \sqrt{\dfrac{38}{78}} = 0.673$ nm

7.15. $\Lambda_{1/2 Na_2SO_4} = \Lambda_{NaCl} + \Lambda_{1/2 K_2SO_4} - \Lambda_{KCl}$

$$= 126.5 + 153.3 - 149.9 = 129.9 \ \Omega^{-1} \text{ cm}^2 \text{ mol}^{-1}$$

7.16. $\Lambda°_{NH_4OH} = \Lambda°_{NH_4Cl} - \lambda°_{Cl^-} + \lambda°_{OH^-}$

$$= 129.8 - 65.6 + 174.0 = 238.2 \text{ cm}^2 \ \Omega^{-1} \text{ mol}^{-1}$$

$$\alpha = \frac{9.6}{238.2} = 0.0403$$

7.17.   a.   Quantity of electricity $= 2\,h \times 3600\,s\,h^{-1} \times 0.79\,A$

$= 5688\,C$

Amount deposited $= 5688/96\,465 = 0.05895$ mol

Loss of LiCl in anode compartment $= \dfrac{0.793\,g}{42.39\,g\,mol^{-1}}$

$= 0.01871$ mol

Anode reaction: $Cl^- \rightarrow \frac{1}{2}Cl_2 + e^-$

0.05895 mol Cl^- is removed by electrolysis.

Net loss $= 0.01871$ mol $Cl^- = 0.05895 - 0.01871 = 0.04024$ mol Cl^- have migrated into the anode compartment.

$$t_{Cl^-} = \frac{0.04024}{0.05895} = 0.683$$

$$t_{Li^+} = 1 - 0.683 = 0.317$$

b. $\lambda_{Li^+}^{\circ} = 0.317 \times 115.0 = 36.5\ \Omega^{-1}\,cm^2\,mol^{-1}$

$\lambda_{Cl^-}^{\circ} = 78.5\ \Omega^{-1}\,cm^2\,mol^{-1}$

Then, from Eq. 7.64,

$$u_+ = 36.5/96\,485 = 3.78 \times 10^{-4}\,cm^2\,V^{-1}\,s^{-1}$$

$$u_- = 78.5/96\,485 = 8.14 \times 10^{-4}\,cm^2\,V^{-1}\,s^{-1}$$

7.18.   Molecular weight of $CdI_2 = 366.21$

$96\,500\,C$ deposits $\frac{1}{2}$ mol $Cd^{2+} = 56.205$ g of Cd^{2+}

∴ current passed is $\dfrac{0.034\,62 \times 96\,485}{56.205} = 59.43\,C$

Anode compartment (152.64 g) originally contained

$$\frac{7.545 \times 10^{-3} \times 152.64}{1000} = 1.1517 \times 10^{-3}\ mol$$

It finally contains

$$\frac{0.3718}{366.21} = 1.0153 \times 10^{-3}\ mol$$

Loss in anode compartment $= 1.364 \times 10^{-4}$ mol

96 500 C would have brought about a loss of

$$\frac{1.364 \times 10^{-4} \times 96\,485}{59.43} = 0.2214 \text{ mol of } CuI_2$$

$$= 0.4428 \text{ mol of } \frac{1}{2} CuI_2$$

$$\therefore t_+ = 0.4428; \quad t_- = 0.5572$$

7.19.   The individual ionic conductivities are

$$\lambda_+ = 0.821 \times 426.16 = 349.9 \ \Omega^{-1} \text{ cm}^2 \text{ mol}^{-1}$$

$$\lambda_- = 0.179 \times 426.16 = 76.3 \ \Omega^{-1} \text{ cm}^2 \text{ mol}^{-1}$$

Then, by Eq. 7.64, the ionic mobilities are

$$u_+ = \frac{349.9 \ \Omega^{-1} \text{ cm}^2 \text{ mol}^{-1}}{96\,485 \text{ C}} = 3.63 \times 10^{-3} \text{ cm}^2 \text{ V}^{-1} \text{ s}^{-1}$$

$$u_- = \frac{76.3 \ \Omega^{-1} \text{ cm}^2 \text{ mol}^{-1}}{96\,485 \text{ C}} = 7.91 \times 10^{-4} \text{ cm}^2 \text{ V}^{-1} \text{ s}^{-1}$$

7.20.   The ionic mobilities are (Eq. 7.64)

$$u_+ = \frac{50.1 \ \Omega^{-1} \text{ cm}^2 \text{ mol}^{-1}}{96\,485 \text{ C}} = 5.19 \times 10^{-4} \text{ cm}^2 \text{ V}^{-1} \text{ s}^{-1}$$

$$u_- = \frac{76.4 \ \Omega^{-1} \text{ cm}^2 \text{ mol}^{-1}}{96\,485 \text{ C mol}^{-1}} = 7.92 \times 10^{-4} \text{ cm}^2 \text{ V}^{-1} \text{ s}^{-1}$$

The velocities in a gradient of 100 V cm^{-1} are thus

Na^+: 5.19×10^{-2} cm s^{-1}

Cl^-: 7.92×10^{-2} cm s^{-1}

7.21.   The molar conductivity of LiCl is

$$\Lambda = (38.6 + 76.4) \ \Omega^{-1} \text{ cm}^2 \text{ mol}^{-1}$$

The specific conductivity of a 0.01 M solution is this quantity multiplied by 10^{-4} mol cm^{-3}:

$$\kappa = 115.0 \times 10^{-5} \ \Omega^{-1} \text{ cm}^{-1}$$

The resistance of a 1-cm length of tube is thus

$$R = \frac{1 \text{ cm/5 cm}^2}{115.0 \times 10^{-5} \ \Omega^{-1} \text{ cm}^{-1}} = 173.9 \ \Omega$$

The potential required to produce a current of 1 A is

$173.9 \ \Omega \times 1 \text{ A} = 173.9$ V

The potential gradient is thus 173.9 V cm^{-1}.

The mobilities of the ions are (Eq. 7.64)

$$\text{Li}^+: \quad \frac{38.6 \ \Omega^{-1} \ \text{cm}^2 \ \text{mol}^{-1}}{96 \ 485 \ \text{C} \ \text{mol}^{-1}} = 4.00 \times 10^{-4} \ \text{cm}^2 \ \text{V}^{-1} \ \text{s}^{-1}$$

$$\text{Cl}^-: \quad \frac{76.4 \ \Omega^{-1} \ \text{cm}^2 \ \text{mol}^{-1}}{96 \ 485 \ \text{C} \ \text{mol}^{-1}} = 7.92 \times 10^{-4} \ \text{cm}^2 \ \text{V}^{-1} \ \text{s}^{-1}$$

The velocities are

$\text{Li}^+: \ 0.070 \ \text{cm s}^{-1}; \quad \text{Cl}^-: \ 0.138 \ \text{cm s}^{-1}$

7.22. The work is given by $dw = Fdr$, where the force of attraction is

$$F = -Q_1 Q_2 / r^2$$

Therefore

$$w = \int_{r_1}^{\infty} -\frac{Q_1 Q_2}{4\pi\varepsilon_0 r^2} \ dr = \frac{Q_1 Q_2}{4\pi\varepsilon_0} \left(\frac{1}{\infty} - \frac{1}{r_1} \right)$$

a. $\varepsilon_0 = 8.854 \times 10^{-12} \ \text{C}^2 \ \text{J}^{-1} \ \text{m}^{-1}; \quad r_1 = 10^{-9} \ \text{m}$

$$w = -\frac{(1.6 \times 10^{-19} \ \text{C})^2}{4\pi \ 8.85 \times 10^{-12} \ \text{C}^2 \ \text{J}^{-1} \ \text{m}^{-1}} \left(\frac{-1}{10^{-9} \ \text{m}} \right)$$

$$= 2.30 \times 10^{-19} \ \text{J}$$

b. $w = -2.30 \times 10^{-28} \ (1/\infty - 1/10^{-3} \ \text{m})$

$= 2.30 \times 10^{-25} \ \text{J}$

c. $w = -2.30 \times 10^{-28} \ (1/0.10 \ \text{m} - 1/10^{-9} \ \text{m})$

$= -2.30 \times 10^{-28} \ (10 - 10^9) = 2.30 \times 10^{-19} \ \text{J}$

7.23. The exponential is shown as curve a, $4\pi r^2$ as curve b, and their product as curve c in the accompanying diagram. With $z_c = 1$ and $z_i = -1$, the function to be differentiated is

$$f = e^{e^2/4\pi\varepsilon_0 \varepsilon r k_B T} \ 4p\pi r^2$$

Differentiation gives

$$\frac{df}{dr} = 8\pi r \ e^{e^2/4\pi\varepsilon_0 \varepsilon r k_B T} - 4\pi r^2 \cdot \frac{e^2}{4\pi\varepsilon_0 \varepsilon r^2 k_B T} \ e^{e^2/4\pi\varepsilon_0 \varepsilon r k_B T}$$

Setting this equal to zero leads to

$$r^* = \frac{e^2}{8\pi\varepsilon_0 \varepsilon k_B T}$$

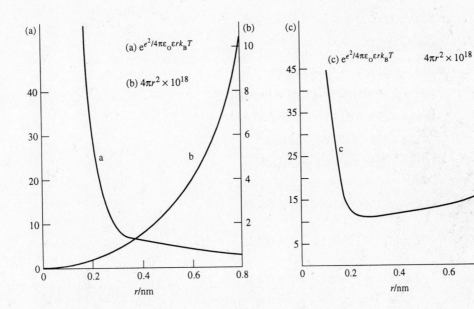

The value of this at 25.0°C, with $\varepsilon = 78.3$, is

$$3.58 \times 10^{-10} \text{ m} = 0.358 \text{ nm}$$

With $z_c = 1$, the potential energy for two univalent ions, from Eq. 7.47, is

$$E_p = \frac{e^2}{4\pi\varepsilon_0\varepsilon r}$$

Introduction of the expression for r^* gives

$$E_p = 2k_B T$$

At 25.0°C, $E_p = 8.23 \times 10^{-21} \text{ J} = 4.96 \text{ kJ mol}^{-1}$

■ Thermodynamics of Ions

7.24. NaCl: $-239.7 - 167.4 = -407.1 \text{ kJ mol}^{-1}$

CaCl$_2$: $-543.1 - 334.8 = -877.9 \text{ kJ mol}^{-1}$

ZnBr$_2$: $-152.3 - (2 \times 120.9) = -394.1 \text{ kJ mol}^{-1}$

7.25. H^+: -1051.4 kJ mol^{-1}

   Na^+: $679.1 - 1051.4 = -372.3$ kJ mol^{-1}

   Mg^{2+}: $274.1 - (2 \times 1051.4) = -1828.7$ kJ mol^{-1}

   Al^{3+}: $-1346.4 - (3 \times 1051.4) = -4500.6$ kJ mol^{-1}

   Cl^-: $-1407.1 + 1051.4 = -355.7$ kJ mol^{-1}

   Br^-: $-1393.3 + 1051.4 = -341.9$ kJ mol^{-1}

7.26. KNO_3: $I = \frac{1}{2}(0.1 \times 1^2 + 0.1 \times 1^2) = 0.1\ M$

   K_2SO_4: $I = \frac{1}{2}(0.2 \times 1^2 + 0.1 \times 2^2) = 0.3\ M$

   $ZnSO_4$: $I = \frac{1}{2}(0.1 \times 2^2 + 0.1 \times 2^2) = 0.4\ M$

   $ZnCl_2$: $I = \frac{1}{2}(0.1 \times 2^2 + 0.2 \times 1^2) = 0.3\ M$

   $K_4Fe(CN)_6$: $I = \frac{1}{2}(0.4 \times 1^2 + 0.1 \times 4^4) = 1.0\ M$

7.27.   Ionic strength of solution.

$$I = \frac{1}{2}(0.4 \times 1^2 + 0.2 \times 2^2) = 0.6\ M$$

$$\log_{10}\gamma_\pm = -z_+ |z_-|\, 0.51\sqrt{I}$$

$$= -2 \times 2 \times 0.51\sqrt{0.6}$$

$$= -2.04 \times 0.775 = -1.58$$

$$\gamma_\pm = 0.026$$

7.28.   a. $s = 1.274 \times 10^{-5}\ M$

$$\log_{10}\gamma_\pm = -0.51 \times (1.274 \times 10^{-5})^{1/2}$$

$$= -1.82 \times 10^{-3}$$

$$\gamma_\pm = 0.996$$

$$K_s = \gamma_\pm^2 s^2 = (0.996 \times 1.274 \times 10^{-5})^2$$

$$= 1.609 \times 10^{-10}$$

$$\Delta G^\circ = -RT \ln K_s$$

$$= -8.3145 \times 298.15 \ln 1.609 \times 10^{-10}$$

$$= 55.90\ \text{kJ mol}^{-1}$$

b. $I = \frac{1}{2}(0.01 \times 0.005 \times 2^2) = 0.015 \, M$

$\log_{10}\gamma_{\pm} = -0.51\sqrt{0.015} = -0.0625$

$\gamma_{\pm} = 0.866$

$s = \dfrac{K_s^{1/2}}{\gamma_{\pm}} = \dfrac{(1.609 \times 10^{-10})^{1/2}}{0.866} = 1.46 \times 10^{-5} \, M$

7.29. $\log_{10}\gamma_{\pm} = \dfrac{-z_+ \, |z_-| \, 0.51\sqrt{I}}{1 + a \, (0.33 \times 10^{10}) \sqrt{I}}$

For $a = 0$ and z_+ and $|z_-| = 1$

$\log_{10}\gamma_{\pm} = -0.51\sqrt{I}$

I	0.01	0.10	0.50	1.0	1.5	2.0
$\log_{10}\gamma_{\pm}$	−0.051	−0.16	−0.36	−0.51	−0.62	−0.72

For $a = 0.1$

$\log_{10}\gamma_{\pm} = \dfrac{-0.51\sqrt{I}}{1 + 0.33\sqrt{I}}$

I	0.01	0.10	0.50	1.0	1.5	2.0
$\log_{10}\gamma_{\pm}$	−0.49	−0.15	−0.29	−0.38	−0.44	−0.49

For $a = 0.2$

$\log_{10}\gamma_{\pm} = \dfrac{-0.51\sqrt{I}}{1 + 0.66\sqrt{I}}$

I	0.01	0.10	0.50	1.0	1.5	2.0
$\log_{10}\gamma_{\pm}$	−0.48	−0.13	−0.24	−0.31	−0.35	−0.35

For $a = 0.4$

$\log_{10}\gamma_{\pm} = \dfrac{-0.51\sqrt{I}}{1 + 1.32\sqrt{I}}$

I	0.01	0.10	0.50	1.0	1.5	2.0
$\log_{10}\gamma_{\pm}$	−0.045	−0.11	−0.19	−0.22	−0.24	−0.25

For $a = 0.8$

$\log_{10}\gamma_{\pm} = \dfrac{-0.51\sqrt{I}}{1 + 2.64\sqrt{I}}$

I	0.01	0.10	0.50	1.0	1.5	2.0
$\log_{10}\gamma_{\pm}$	−0.040	−0.088	−0.13	−0.14	−0.15	−0.15

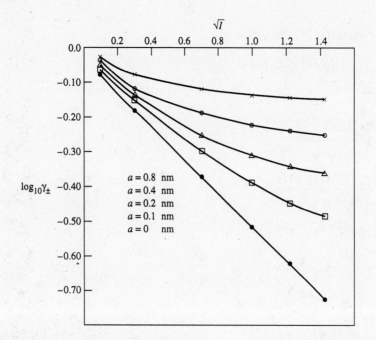

7.30. The electrostatic contribution to the Gibbs energy (Eq. 7.87) is, per mole of ions,

$$G_{es}^\circ = \frac{z^2 e^2 L}{8\pi\varepsilon_0 \varepsilon r}$$

$$= \frac{(1.602 \times 10^{-19})^2 \times 6.022 \times 10^{23}}{8\pi \times 8.854 \times 10^{-12} \times 0.133 \times 10^{-9}\, \varepsilon} \; \text{J mol}^{-1}$$

$$= \frac{5.22 \times 10^5}{\varepsilon} \; \text{J mol}^{-1}$$

In the membrane,

$$G_{es}^\circ = 130.5 \; \text{kJ mol}^{-1}$$

In water,

$$\Delta G_{es}^\circ = 6.7 \; \text{kJ mol}^{-1}$$

$$\Delta G_{es}^\circ \,(\text{water} \to \text{membrane}) = 124 \; \text{kJ mol}^{-1}$$

7.31. $\Lambda^\circ\left(\frac{1}{2}\text{CaF}_2\right) = 51.1 + 47.0 = 98.1 \; \Omega^{-1} \, \text{cm}^2 \, \text{mol}^{-1}$

Observed κ due to salt $= 3.86 \times 10^{-5} - 1.5 \times 10^{-6}$

$$= 3.71 \times 10^{-5} \; \Omega^{-1} \, \text{cm}^{-1}$$

Solubility $= \dfrac{3.71 \times 10^{-5}}{98.1}$ mol cm^{-3}

$= 3.782 \times 10^{-4}$ mol dm^{-3} $\left(\text{of } \frac{1}{2}CaF_2\right)$

1 mol of $\frac{1}{2}CaF_2$ has a mass of $20.04 + 19.00 = 39.04$ g.

Solubility $= 0.0148$ g dm^{-3}

Solubility product $= [Ca^{2+}][F^-]^2$

$= (0.5 \times 3.782 \times 10^{-4}) \times (3.782 \times 10^{-4})^2$

$= 2.70 \times 10^{-11}$ mol^3 dm^{-9}

7.32. $x\,M\,CuSO_4$: $\quad I = \frac{1}{2}(2^2 + 2^2)\,x = 4x\,M$

$I = 0.1\,M \quad \text{if} \quad x = 0.025$

$x\,M\,Ni(NO_3)_2$: $\quad I = \frac{1}{2}(2^2 + 2)\,x = 3x$

$I = 0.1\,M \quad \text{if} \quad x = 0.033$

$x\,M\,Al_2(SO_4)_3$: $\quad I = \frac{1}{2}(2 \times 3^2 + 3 \times 2^2)\,x = 15x\,M$

$I = 0.1\,M \quad \text{if} \quad x = 0.006\ 67$

$x\,M\,Na_3PO_4$: $\quad I = \frac{1}{2}(3 + 3^2)\,x = 6x\,M$

$I = 0.1\,M \quad \text{if} \quad x = 0.0167$

7.33. a. First, neglect the effect of activity coefficients: if s is the solubility

$s(2s)^2 = 4.0 \times 10^{-9}$ mol^3 dm^{-9}

$s = 1.0 \times 10^{-3}$ mol dm^{-3}

The ionic strength is

$\frac{1}{2}(1 \times 2^2 + 2 \times 1)\,1.0 \times 10^{-3} = 3.0 \times 10^{-3}$ mol dm^{-3}

By the Debye-Hückel limiting law

$\log_{10}\gamma_{\pm} = -0.51 \times 2 \times \sqrt{3.0 \times 10^{-3}} = -0.0559$

$\gamma_{\pm} = 0.88$

If now the true solubility is s, the activities of the ions are

Pb^{2+}: $\quad \gamma_+ s$; $\quad F^-$: $\quad 2\gamma_- s$

Then

$$(\gamma_+ s)(2\gamma_- s)^2 = 4.0 \times 10^{-9} \text{ mol}^3 \text{ dm}^{-3}$$

$$\gamma_+ \gamma_-^2 \, 4s^3 = 4.0 \times 10^{-9} \text{ mol}^3 \text{ dm}^{-9}$$

$$\gamma_\pm^3 \, 4s^3 = 4.0 \times 10^{-9} \text{ mol}^3 \text{ dm}^{-9} \quad \text{(from Eq. 7.105)}$$

Thus

$$s^3 = \frac{4.0 \times 10^{-9} \text{ mol}^3 \text{ dm}^{-9}}{(0.88)^3 \times 4}$$

$$s = 1.14 \times 10^{-3} \text{ mol dm}^{-3}$$

We could proceed to further approximations as necessary.

b. In 0.01 M NaF, the ionic strength is essentially 0.01 mol dm^{-3} and

$$\log_{10}\gamma_\pm = -2 \times 0.51 \times \sqrt{0.01} = -0.102$$

$$\gamma_\pm = 0.791$$

If s is the solubility,

$$s = [Pb^{2+}]; \quad [F^-] = 0.01 \text{ mol dm}^{-3}$$

Then

$$s\gamma_+ \times (0.01 \, \gamma_-)^2 = 4.0 \times 10^{-9} \text{ mol dm}^{-3}$$

$$\gamma_+ \gamma_-^2 \, s \times 0.0001 = 4.0 \times 10^{-9} \text{ mol dm}^{-3}$$

$$\gamma_\pm^3 \, s \times 0.0001 = 4.0 \times 10^{-9} \text{ mol dm}^{-3}$$

$$s = \frac{4.0 \times 10^{-9} \text{ mol dm}^{-3}}{0.0001(0.791)^3} = 8.08 \times 10^{-5} \text{ mol dm}^{-3}$$

7.34. We proceed by successive approximations, first taking the activity coefficients to be unity. Then, if s is the solubility,

$$s^2 = 4.0 \times 10^{-3} \text{ mol}^2 \text{ dm}^{-6}$$

$$s = 0.0632 \text{ mol dm}^{-3}$$

This is the ionic strength, thus

$$\log_{10}\gamma_\pm = -0.51\sqrt{0.0632} = -0.128$$

$$\gamma_\pm = 0.744$$

To a second approximation,

$$\gamma_\pm^2 s^2 = (0.744)^2 s^2 = 4.0 \times 10^{-3} \text{ mol}^2 \text{ dm}^{-6}$$

$$s = 0.085 \text{ mol dm}^{-3}$$

To a third approximation,

$$\log_{10}\gamma_\pm = -0.51\sqrt{0.085} \; ; \quad \gamma_\pm = 0.71$$

$$(0.71)^2 s^2 \quad = 4.0 \times 10^{-3} \text{ mol}^2 \text{ dm}^{-6}$$

$$s \quad = 0.089 \text{ mol dm}^{-3}$$

To a fourth approximation,

$$\log_{10}\gamma_\pm = -0.51\sqrt{0.089} \; ; \quad \gamma_\pm = 0.704$$

$$(0.704)^2 s^2 \quad = 4.0 \times 10^{-3}$$

$$s \quad = 0.090 \text{ mol dm}^{-3}$$

7.35.   For Problem 7.24 it was found that

$$G_{es}^\circ = \frac{5.22 \times 10^5}{\varepsilon} \text{ J mol}^{-1}$$

For the transfer from water (ε_1) to lipid (ε_2)

$$\Delta G_{es}^\circ / \text{J mol}^{-1} = 5.22 \times 10^5 \left(\frac{1}{\varepsilon_2} - \frac{1}{\varepsilon_1}\right)$$

$$\Delta S_{es}^\circ = -\left(\frac{\partial \Delta G_{es}^\circ}{\partial T}\right)_P \qquad\qquad \text{(from Eq. 3.119)}$$

Since ε_2 is temperature independent, this leads to

$$\Delta S_{es}^\circ / \text{J K}^{-1} \text{ mol}^{-1} = 5.22 \times 10^5 \frac{\partial}{\partial T}\left(\frac{1}{\varepsilon_1}\right)$$

$$= -5.22 \times 10^5 \frac{1}{\varepsilon_1^2} \frac{\partial \varepsilon}{\partial T}$$

$$= -5.22 \times 10^5 \frac{1}{\varepsilon_1} \cdot \frac{\partial \ln \varepsilon}{\partial T}$$

$$= \frac{5.22 \times 10^5 \times 0.0046}{78} = 31 \text{ J K}^{-1} \text{ mol}^{-1}$$

The entropy increases because of the release of bound water molecules when the K$^+$ ions pass into the lipid.

7.36. a.   At infinite dilution the work of charging an ion is given directly by (Eq. 7.86):

$$w_{rev} = \frac{z^2 e^2}{8\pi\varepsilon_0 \varepsilon r}$$

For 1 mol of Na$^+$

$$w_{rev} = \frac{(1.602 \times 10^{-19}\text{ C})^2\, 6.022 \times 10^{23}\text{ mol}^{-1}}{8\pi \times 8.854 \times 10^{-12}\text{ C}^2\text{ N}^{-1}\text{ m}^{-2} \times 78} \times \frac{1}{95 \times 10^{-12}\text{ m}}$$

$$= 9373\text{ J mol}^{-1}$$

For 1 mol of Cl$^-$,

$$w_{rev} = 4920\text{ J mol}^{-1}$$

For 1 mol of Na$^+$Cl$^-$ at infinite dilution,

$$w_{rev} = 14\,293\text{ J mol}^{-1} = 14.3\text{ kJ mol}^{-1}$$

b.   These values are reduced when the electrolyte is at a higher concentration, the work of charging the ionic atmosphere being negative and equal to $kT \ln \gamma_i$. Thus, for 1 mol of Na$^+$ ions, of activity coefficient γ_+, the work of charging the atmosphere is

$$RT \ln \gamma_+$$

Similarly for the chloride ion, the work per mole is

$$RT \ln \gamma_-$$

For 1 mol of Na$^+$Cl$^-$

$$w_{rev}(\text{atm}) = RT(\ln \gamma_+ + \ln \gamma_-)$$

$$= RT \ln \gamma_+\gamma_- = 2RT \ln \gamma_\pm$$

If $\gamma_\pm = 0.70$

$$w_{rev}(\text{atm}) = 2(8.3145 \times 298.15\text{ J mol}^{-1}) \ln 0.70$$

$$= -1768\text{ J mol}^{-1}$$

The net work of charging is thus

$$w_{rev} = 14\,293 - 1768 = 12\,525\text{ J mol}^{-1}$$

$$= 12.5\text{ kJ mol}^{-1}$$

7.37. The ionic strength of the solution is

$$I = \frac{1}{2} [0.1 + 0.1 + (0.2 \times 4) + 0.4] = 0.70 \; M$$

The mean activity coefficient γ_i of the barium and sulfate ions is given by

$$\log_{10} \gamma_i = -2^2 \times 0.51 \times \sqrt{0.70}$$

$$= -1.707$$

$$\gamma_i = 0.0196$$

If the solubility in the solution is s,

$$9.2 \times 10^{-11} = s^2 (0.0196)^2$$

whence $s = 4.88 \times 10^{-4} \; M$

7.38. The ionic strength of the solution is

$$I = \frac{1}{2}[0.02 + (0.01 \times 2^2)] = 0.03 \; M$$

By the DHLL,

$$\log_{10} \gamma_i = -0.51 \times \sqrt{0.03} = -0.0883$$

$$\gamma_i = 0.816$$

The solubility product is therefore

$$K_s = (1.561 \times 10^{-5})^2 \times (0.816)^2$$

$$= 1.623 \times 10^{-10} \; M^2$$

The solubility in pure water is thus

$$(1.623 \times 10^{-10})^{1/2} = 1.27 \times 10^{-5} \; M$$

7.39. The enthalpy change ΔH_{neut} for the neutralization of HCN by NaOH is less than the value 55.90 kJ mol^{-1} because of the energy required for the dissociation of HCN, ΔH_{diss},

$$\Delta H_{neut} = 55.90 \text{ kJ mol}^{-1} - \Delta H_{diss}$$

Thus

$$\Delta H_{diss} = 55.90 \text{ kJ mol}^{-1} - \Delta H_{neut}$$

$$= 55.90 - 12.13 = 43.77 \text{ kJ mol}^{-1}$$

7.40. $I = \frac{1}{2}(0.004 \times 1^2 + 0.004 \times 1^2 + 0.004 \times 2^2) = 0.012$

From Eq. 7.104,

$$\log \gamma_i = -z_i^2 \, B \sqrt{I}$$

for Na^+

$$\log \gamma_{Na^+} = -1^2 \times 0.51 \times \sqrt{0.012} = -0.05587$$

$$\gamma_{Na^+} = 0.879$$

for SO_4^{2-}

$$\log \gamma_{SO_4^{2-}} = -2^2 \times 0.51 \times \sqrt{0.012} = -0.2235$$

$$\gamma_{SO_4^{2-}} = 0.598$$

From Eq. 7.111,

$$\gamma_\pm = -0.51 \, |z_+| \, |z_-| \sqrt{I}$$

$$= -0.51 \, |1| \, |-2| \sqrt{0.012} = -0.1117$$

$$= 0.773$$

■ Ionic Equilibria

7.41.

Palmitate side	Other side
Initial concentrations:	
$[Na^+] = 0.1 \, M$	$[Na^+] = 0.2 \, M$
$[P^-] = 0.1 \, M$	$[Cl^-] = 0.2 \, M$
Final concentrations:	
$[Na^+] = (0.1 + x)M$	$[Na^+] = (0.2 - x)M$
$[P^-] = 0.1 \, M$	$[Cl^-] = (0.2 - x)M$
$[Cl^-] = x \, M$	

Then

$$(0.2 - x)^2 = (0.1 + x)$$

$$0.04 - 0.4x + x^2 = x^2 + 0.1x$$

$$x = \frac{0.04}{0.5} = 0.08$$

Final concentrations are thus, on the palmitate side,

$$[Na^+] = 0.18\ M; \quad [Cl^-] = 0.08\ M$$

On the other side,

$$[Na^+] = [Cl^-] = 0.12\ M$$

7.42. $$\frac{[H_2NCH_2COOH]\,[H^+]}{[H_3N^+CH_2COOH]} = 1.5 \times 10^{-10}\ M$$

$$\frac{[H_3N^+CH_2COO^-]\,[H^+]}{[H_3N^+CH_2COOH]} = 4.0 \times 10^{-3}$$

Dividing the first by the second gives

$$\frac{[H_2NCH_2COOH]}{[H_3N^+CH_2COO^-]} = \frac{1.5 \times 10^{-10}}{4.0 \times 10^{-3}} = 3.8 \times 10^{-8}$$

This is convincing evidence for the predominance of the zwitterion $H_3N^+CH_2COO^-$.

7.43.   a.   pH = 1; H_3PO_4 predominant

b.   pH = 2.7; $H_2PO_4^-$ predominant

c.   pH = 4.3; $H_2PO_4^-$ predominant

d.   pH = 11.4; HPO_4^{2-} predominant

e.   pH = 14; PO_4^{3-} predominant

7.44.   Let the final concentrations be

Left-hand Compartment	Right-hand Compartment
$[K^+]/M = 0.05 - x$	$[K^+]/M = 0.15 + x$
$[Cl^-]/M = 0.05 - x$	$[Cl^-]/M = x$
	$[P^-]/M = 0.1$

At equilibrium,

$$(0.05 - x)^2 = x\,(0.15 + x)$$

whence $x = 0.01$

The final concentrations are therefore

Left-hand Compartment	Right-hand Compartment
$[K^+] = 0.04\ M$	$[K^+] = 0.16\ M$
$[Cl^-] = 0.04\ M$	$[Cl^-] = 0.01\ M$
	$[P^-] = 0.1\ M$

It is easy to check that the product $[K^+][Cl^-]$ is the same on each side of the membrane.

8. ELECTROCHEMICAL CELLS

■ Electrode Reactions and Electrode Potentials

8.1. a. $H_2 \rightarrow 2H^+ + 2e^-$

$Cl_2 + 2e^- \rightarrow 2Cl^-$

$H_2 + Cl_2 \rightarrow 2H^+ + 2Cl^-; \quad z = 2$

$$E = E^\circ - \frac{RT}{2F} \ln \left(a_{H^+}^2 a_{Cl^-}^2 \right)^u$$

b. $2Hg(l) + 2Cl^- \rightarrow Hg_2Cl_2 + 2e^-$

$2H^+ + 2e^- \rightarrow H_2$

$2Hg + 2H^+ + 2Cl^- \rightarrow Hg_2Cl_2 + H_2; \quad z = 2$

$$E = E^\circ + \frac{RT}{2F} \ln \left(a_{H^+}^2 a_{Cl^-}^2 \right)^u$$

c. $Ag + Cl^- \rightarrow AgCl(s) + e^-$

$2e^- + Hg_2Cl_2(s) \rightarrow 2Hg + 2Cl^-$

$2Ag(s) + Hg_2Cl_2(s) \rightarrow 2AgCl(s) + 2Hg(s)$

$E = E^\circ$ (no concentration dependence)

d. $\frac{1}{2} H_2(g) \rightarrow H^+ + e^-$

$AuI(s) + e^- \rightarrow Au(s) + I^-$

$AuI(s) + \frac{1}{2} H_2(g) \rightarrow Au(s) + H^+ + I^-; \quad z = 1$

$$E = E^\circ - \frac{RT}{F} \ln (a_{H^+} a_{I^-})^u$$

e. $Ag(s) + Cl^-(a_1) \rightarrow AgCl(s) + e^-$

$AgCl(s) + e^- \rightarrow Ag(s) + Cl^-(a_2)$

$Cl^-(a_1) \rightarrow Cl^-(a_2); \quad z = 1$

$$E = \frac{RT}{F} \ln \frac{a_1}{a_2}$$

8.2. $E°$

a. $A + 2H^+ + 2e^- \rightarrow AH_2$ -0.60 V

   $B + 2H^+ + 2e^- \rightarrow BH_2$ -0.16 V

       $AH_2 + B \rightarrow A + BH_2$

AH_2 is oxidized by B; the half-cell $A + 2H^+ + 2e^- \rightarrow AH_2$ is written as a reduction. In the overall reaction this equation is reversed and therefore represents an oxidation. Because the net reaction has a positive emf (exergonic), the reaction is spontaneous in the forward direction.

b. -0.16 V $- (-0.60$ V$) = 0.44$ V

c. $[H_3O^+]$ does not appear in the equilibrium expression, and the hydrogen-containing entities cancel in the numerator and denominator. Therefore, there is no effect of pH on the equilibrium ratio.

8.3.   The $\Delta G°$ values for the two reactions are

$$Cr^{3+} + 3e^- \rightarrow Cr \quad \Delta G_1^\circ = -zE_1^\circ F = -3 \times -0.74 \times 96\ 485 \text{ J mol}^{-1} \tag{1}$$

$$Cr^{3+} + e^- \rightarrow Cr^{2+} \quad \Delta G_2^\circ = -zE_2^\circ F = -1 \times -0.41 \times 96\ 485 \text{ J mol}^{-1} \tag{2}$$

The reaction $Cr^{2+} + 2e^- \rightarrow Cr$ is obtained by subtracting reaction (2) from reaction (1), and the $\Delta G°$ value for $Cr^{2+} + 2e^- \rightarrow Cr$ is obtained by subtracting ΔG_2° from ΔG_1°:

$$\Delta G° = -3 \times -0.74 \times 96\ 485 - (-1 \times -0.41 \times 96\ 485) \text{ J mol}^{-1}$$

$$= 1.81 \times 96\ 485 \text{ J mol}^{-1}$$

Since $Cr^{2+} + 2e^- \rightarrow Cr$ involves two electrons, and since $\Delta G = -zE°F$, it follows that

$$1.81\ F = -2(E°/\text{V})F$$

or $E° = -0.90$ V

8.4.   Left-hand electrode $H_2 \rightarrow 2H^+(1\ m) + 2e^-$

     Right-hand electrode $2e^- + 2H^+(aq) + F^{2-} \rightarrow S^{2-}$

     Overall reaction $2H^+(aq) + F^{2-} + H_2 \rightarrow 2H^+(1\ m) + S^{2-}$

The expression for the emf of the cell is

$$E = E° - \frac{RT}{2F} \ln \frac{[S^{2-}][1\ m]^2}{[F^{2-}]\ c^2}$$

8.5. a. In writing the representation of the cell, the oxidation reaction occurs at the anode, which is placed at the left-hand position of the cell. In this case Fe^{2+} is losing electrons, and the oxidation process is

$$Fe^{2+} \rightarrow Fe^{3+} + e^-$$

The cathode reaction is written on the right-hand side of the cell and is

$$Ce^{4+} + e^- \rightarrow Ce^{3+}$$

where reduction occurs. The overall reaction is the sum of these two reactions. The cell representation is

$$Fe^{3+}(aq)\ |\ Fe^{2+}(aq)\ \vdots\ Ce^{4+}(aq)\ |\ Ce^{3+}(aq)$$

The symbols for both sides are always written as though they were a reduction process, i.e., oxidized form | reduced form. The voltage of this cell is then the reduction potential of the right-hand electrode minus the reduction potential of the left-hand electrode. Thus

$$E° = E°_{Ce^{4+}\ |\ Ce^{3+}} - E°_{Fe^{3+}\ |\ Fe^{2+}} = 1.44\ V - 0.771\ V = 0.67\ V$$

b. Upon examining the standard reduction potentials in Table 8.1, we see that the following half-cell reactions can be combined to give the cited reaction.

$$Ag^+(aq) + e^- \rightarrow Ag \qquad\qquad 0.7991\ V \qquad\qquad\qquad (1)$$

$$AgCl(s) + e^- \rightarrow Ag + Cl^- \qquad 0.2224\ V \qquad\qquad\qquad (2)$$

Reversal of the second equation and addition gives the desired equation:

$$Ag^+(aq) + Cl^-(aq) \rightarrow AgCl(s)$$

Equation (1) is the reduction reaction and is placed on the right-hand side. The anode reaction is placed on the left-hand side of the cell representation.

$$Ag\ |\ AgCl(s)\ |\ Cl^-(aq)\ \vdots\ Ag^+(aq)\ |\ Ag(s)$$

The voltage of this cell is the right-hand reduction potential minus the left-hand reduction potential.

$$E° = E°_{Ag^+\ |\ Ag} - E°_{AgCl\ |\ Ag} = 0.7991\ V - 0.2224\ V = 0.5777\ V$$

c. HgO undergoes reduction to Hg and is the cathode. H_2 is oxidized and is the anode. The electrode potentials are found in the CRC Handbook.

$$2H_2O + 2e^- \rightarrow H_2 + 2OH^- \qquad\qquad -0.8277\ V \qquad\qquad (3)$$

$$HgO + H_2O + 2e^- \rightarrow Hg + 2OH^- \qquad 0.0977\ V \qquad\qquad (4)$$

Reversing the sense of equation (3) and adding to (4) gives

$$HgO(s) + H_2(g) \rightarrow Hg(l) + H_2O(l)$$

The cell is represented by

$$\text{Pt, } H_2O(l) \mid H_2(g), OH^-(aq) \vdots OH^-(aq), \quad H_2O \mid HgO(s) \mid Hg(l)$$

The cell potential is

$$E° = E°_{HgO \mid Hg} - E°_{H_2O \mid H_2} = 0.0977 \text{ V} - (-0.8277 \text{ V}) = 0.9254$$

■ Thermodynamics of Electrochemical Cells

8.6. $Fe^{3+}/Fe^{2+} = 0.771 \text{ V}; \quad I_2/I^- = 0.5355 \text{ V}$

$$E° = 0.771 - 0.5355$$

$$= 0.2355 \text{ V}$$

$$\ln K_c^\mu = \frac{2 \times 0.2355}{0.0257}$$

$$= 18.3268$$

$$K_c = 9.10 \times 10^7 \text{ dm}^6 \text{ mol}^{-2}$$

8.7. The $E°$ for the process

$$Sn + Fe^{2+} \rightleftarrows Sn^{2+} + Fe$$

is $-0.4402 - (-0.136) = -0.304 \text{ V}$

Since $z = 2$ we have

$$-0.304 = \frac{0.0257}{2} \ln K_c^\mu$$

$$\ln K = -23.658$$

$$K_c = 5.312 \times 10^{-11}$$

8.8. $\Delta G° = -z \times 96\ 485 \times 0.25 \text{ J}; \quad z = 1$

$$= -24\ 100 \text{ J mol}^{-1} = -24.1 \text{ kJ mol}^{-1}$$

8.9. From Table 8.1 the relevant values are

$$O_2 + 4H^+ + 4e^- \rightarrow 2H_2O \qquad E° = 1.23 \text{ V} \tag{1}$$

$$2H^+ + 2e^- \rightarrow H_2 \qquad E° = 0 \tag{2}$$

Subtraction of (2) from $\frac{1}{2}$(1) gives the required equation

$$H_2 + \frac{1}{2}O_2 \rightleftarrows H_2O$$

with $E° = 1.23$ V and $z = 2$. Then,

$$\Delta G° = -2 \times 96\ 485 \times 1.23 \text{ J mol}^{-1}$$

$$= -237\ 400 \text{ J mol}^{-1} = -237.4 \text{ kJ mol}^{-1}$$

8.10.   From Table 8.1,

$$Cu^{2+} + 2e^- \to Cu \qquad E_1^° = 0.337 \text{ V} \qquad\qquad (1)$$

$$Cu^{2+} + e^- \to Cu^+ \qquad E_2^° = 0.153 \text{ V} \qquad\qquad (2)$$

If we subtract $2 \times (2)$ from (1) we obtain

$$2Cu^+ \to Cu^{2+} + Cu \qquad E° = E_1^° - E_2^° = 0.184 \text{ V}; \quad z = 2$$

$$\therefore \ln K^u = \frac{2 \times 0.184\ F}{(8.3145)(298.15)} = \frac{2 \times 0.184}{0.0257} = 14.32$$

$$K = 1.66 \times 10^6 \text{ dm}^3 \text{ mol}^{-1}$$

If Cu_2O is dissolved in dilute H_2SO_4, half will form Cu^{2+} and half Cu.

8.11.   Note that the $\Delta G°$ given is for the reaction of 3 moles of H_2 to form 2 moles of Sb. The electrode reactions may be written as

$$3H_2 \to 6H^+ + 6e^-,$$

$$Sb_2O_3 + 6H^+ + 6e^- \to 2Sb + 3H_2O.$$

Using Eq. (8.2), we get $E° = 83\ 700 \text{ J}/(6 \times 96\ 485) = 0.1446$ V.

Since this reaction is spontaneous, the electron flow is from the hydrogen electrode (negative) to the antimony electrode (positive).

■ Nernst Equation and Nernst Potentials

8.12. $E = \dfrac{RT}{zF} \ln \dfrac{m_1}{m_2}$

$$= 0.0257 \ln 2 = 0.0178 \text{ V}$$

8.13.   The process is

$$Pyruvate^- + 2H^+ + 2e^- \rightleftarrows lactate^-$$

and the Nernst equation is

$$E' = E° - \frac{RT}{2F} \ln \frac{[\text{lactate}^-]}{[\text{pyruvate}^-]}$$

Then

$$E'/V \quad = -0.185 - \frac{0.0257}{2} \ln\left(\frac{10}{90}\right)$$

$$E' \quad = -0.157 \text{ V}$$

8.14. a. From Table 8.1,

$$Fe^{3+} + e^- \rightarrow Fe^{2+} \qquad E_1^\circ = 0.771 \text{ V} \tag{1}$$

$$Fe^{2+} + 2e^- \rightarrow Fe \qquad E_2^\circ = -0.4402 \text{ V} \tag{2}$$

The corresponding ΔG° values are:

$$\Delta G_1^\circ = -0.771 \times 96\ 485 = -74\ 389.9 \text{ J mol}^{-1} \tag{1}$$

$$\Delta G_2^\circ = -2(-0.4402) \times 96\ 485 = 84\ 945.4 \text{ J mol}^{-1} \tag{2}$$

The required reaction is obtained by adding (1) and (2):

$$Fe^{3+} + 3e^- \rightarrow Fe \quad \Delta G^\circ = 10.56 \text{ kJ mol}^{-1}$$

Then, since $z = 3$,

$$E^\circ = -\frac{10\ 560}{3 \times 96\ 485} = -0.0365 \text{ V}$$

b. The electrode reactions are

$$Sn^{2+} \rightarrow Sn^{4+} + 2e^- \qquad E_1^\circ = -0.15 \text{ V} \tag{1}$$

$$3e^- + Fe^{3+} \rightarrow Fe \qquad E_1^\circ = -0.0365 \text{ V} \tag{2}$$

The cell reaction is

$$3Sn^{2+} + 2Fe^{3+} \rightarrow 3Sn^{4+} + 2Fe \quad E^\circ = -0.186 \text{ V}; \quad z = 6$$

From the Nernst equation, Eq. 8.13,

$$E \quad = -0.186 - \frac{0.0257}{6} \ln \frac{(0.01)^3}{(0.1)^3 (0.5)^2}$$

$$= -0.186 + 0.024 = -0.16 \text{ V}$$

8.15. From $K_a = \dfrac{[H^+][CH_3COO^-]}{[CH_3COOH]}$, we obtain

$$[H^+] = K_a \frac{[CH_3COOH]}{[CH_3COO^-]} .$$

The concentration of acetate ions formed from the dissociation of acetic acid will be negligible compared to the acetate concentration from the fully dissociated sodium salt. Therefore,

$$[H^+] = 1.81 \times 10^{-5} \times \frac{0.0100}{0.0358} = 5.056 \times 10^{-6} \text{ } M.$$

The cell reactions are (see Table 8.1):

Anode $H_2(g) \rightarrow 2H^+ + 2e^-$; $E° = 0.0000$ V (by definition)

Cathode $Hg_2Cl_2 + 2e^- \rightarrow 2Hg + 2Cl^-$; $E° = 0.2415$ V

However, since the cathode reaction (reduction) is accounted for by the standard reduction potential of the electrode, the Nernst equation contains only the hydrogen ion concentration. For this concentration, the Nernst equation gives

$$E = (0.0000 + 0.2415) - \frac{0.0257}{2} \ln [5.056 \times 10^{-6}]^2 = 0.5549 \text{ V}.$$

8.16.   The reactions taking place during the electrolysis are

$$Cu^{2+} + 2e^- \rightarrow Cu; \qquad E° = \quad 0.3419 \text{ V}$$

$$2Br^- \rightarrow Br_2 + 2e^-; \qquad E° = \quad -1.0873 \text{ V}$$

If the *reverse* reaction were taking place in a galvanic cell, the initial cell voltage is given by

$$E = E° - \frac{0.0257}{2} \ln [Cu^{2+}][Br^-]^2.$$

$$E = (1.0873 - 0.3419) - \frac{0.0257}{2} \ln (0.0500 \times 0.1000^2)$$

$$= 0.843 \text{ V}.$$

Therefore, a minimum voltage of 0.843 V will have to be applied at the beginning in order for the electrolysis reaction to occur.

At the end of the electrolysis, the concentrations are:

$$[Cu^{2+}] = 0.0500 \text{ } M - \frac{(2.872/63.456)}{1.0000} = 0.00474 \text{ } M.$$

$$[Br^-] = 0.1000 \text{ } M - 2 \times \frac{(2.872/63.456)}{1.0000} = 0.0094 \text{ } M.$$

Therefore, the voltage required will be

$$(1.0873 - 0.3419) - \frac{0.0257}{2} \ln (0.0047 \times 0.0094^2) = 0.934 \text{ V}.$$

8.17.   For the reaction

$$I_2 + I^- \rightarrow I_3^- \qquad E° = -0.0010 \text{ V} \quad \text{and} \quad z = 2$$

$$K_c = \frac{[I_3^-]}{[I^-]} = \frac{[I_3^-]}{0.5 \text{ mol dm}^{-3}}$$

$$-0.0010 = \frac{0.0257}{2} \ln \frac{[I_3^-]}{0.5 \text{ mol dm}^{-3}}$$

$$\ln \quad \frac{[I_3^-]^u}{0.5} = -0.0778$$

$$\frac{[I_3^-]^u}{0.5} = 0.925$$

$$[I_3^-] = 0.463 \; M$$

8.18. $\Delta\Phi/V = \dfrac{8.3145 \times 298.15}{96\,500} \ln\dfrac{0.18}{0.12}$

$\Delta\Phi = 0.0104 \text{ V} = 10.4 \text{ mV}$

8.19. From Table 8.1, we see that Au^+ has a much higher reduction potential than Pb^{2+}. Therefore, the gold will be deposited first.

As the Au^+ concentration falls, the lead begins to be deposited, i.e., we have

$$2Au(s) + Pb^{2+} \rightleftarrows Pb(s) + 2Au^+$$

for which we write (see Equation 8.7)

$$E^\circ = (-1.692 - 0.126) \text{ V} - \frac{0.0257}{2} \ln\frac{[Au^+]^2}{0.0100}$$

or $[Au^+]^2 = 0.0100 e^{(-1.818 \times 2/0.0257)} = 3.603 \times 10^{-64}$.

Therefore, $[Au^+] = 1.898 \times 10^{-32} \; M$.

The conclusion is that only an infinitesimal amount of gold will be left in the solution by the time the lead starts to deposit at the electrode. Therefore, this is an acceptable way of separating the two metals.

8.20. Reaction at the right-hand electrode:

$$e^- + H^+(0.2 \; m) \rightarrow \frac{1}{2} H_2(10 \text{ bar})$$

At the left-hand electrode:

$$\frac{1}{2} H_2(1 \text{ bar}) \rightarrow e^- + H^+(0.1 \; m)$$

Overall reaction:

$$H^+(0.2 \; m) + \frac{1}{2} H_2(1 \text{ bar}) \rightarrow \frac{1}{2} H_2(10 \text{ bar}) + H^+(0.1 \; m); \quad z = 1$$

Cell emf:

$$E = \frac{RT}{F} \ln \frac{0.2 \times (1 \text{ bar})^{1/2}}{0.1 \times (10 \text{ bar})^{1/2}}$$

$E/V = 0.0257 \ln \dfrac{2}{\sqrt{10}} = -0.0118 \text{ V}$

$E = -11.8 \text{ mV}$

8.21. The reactions at the two electrodes are

$$\frac{1}{2} H_2(1 \text{ bar}) \rightarrow H^+(0.1 \ m) + e^-$$

$$H^+(0.2 \ m) + e^- \rightarrow \frac{1}{2} H_2(10 \text{ bar})$$

Every H^+ ion produced in the left-hand solution will have to pass through the membrane to preserve electrical neutrality.

$$H^+(0.1 \ m) \rightarrow H^+(0.2 \ m)$$

The net reaction is $\frac{1}{2} H_2(1 \text{ bar}) \rightarrow \frac{1}{2} H_2(10 \text{ bar})$

$$E = \frac{RT}{F} \ln \frac{1}{\sqrt{10}} = 0.0257 \ln \frac{1}{\sqrt{10}}$$

$$= -0.0296 \text{ V} = -29.6 \text{ mV}$$

8.22. The capacitance of the cell membrane (see Eq. 8.20) is

$$C = \frac{8.854 \times 10^{-12} \times 3 \times 10^{-10}}{10^{-8}} \text{ F} = 2.66 \times 10^{-13} \text{ F}$$

a. With a potential difference of 0.085 V, the net charge on either side of the wall is

$$Q = CV = 2.66 \times 10^{-13} \times 0.085 = 2.26 \times 10^{-14} \text{ C}$$

b. The number of K^+ ions required to produce this charge is

$$\frac{Q}{e} = \frac{2.26 \times 10^{-14}}{1.602 \times 10^{-19}} = 1.41 \times 10^5$$

The number of ions inside the cell is

$$0.155 \times 10^{-12} \times 6.022 \times 10^{23} = 9.33 \times 10^{10}$$

The fraction of ions at the surface is therefore

$$\frac{1.41 \times 10^5}{9.33 \times 10^{10}} = 1.5 \times 10^{-6}$$

8.23. At the left-hand electrode:

$$\frac{1}{2} H_2 \rightarrow H^+ + e^-$$

At the right-hand electrode:

$$CrSO_4(s) + 2e^- \rightarrow SO_4^{2-} + Cr(s)$$

Overall reaction:

$$CrSO_4(s) + H_2 \rightarrow 2H^+ + SO_4^{2-} + Cr(s)$$

$$E° = -0.40 \text{ V} \quad \text{and} \quad z = 2$$

a. $E = E° - \dfrac{RT}{2F} \ln ([H^+]^2[SO_4^{2-}])^u$

$E/V \quad = -0.40 - \dfrac{0.0257}{2} \ln (0.002)^2(0.001)$

$E \quad = -0.152 \text{ V}$

b. The ionic strength is

$I = \dfrac{1}{2}\{0.002 + (0.001 \times 4)\} = 0.003 \ M$

Historically, common logs have been used in activity coefficient calculations where the value of B is easily remembered as 0.51.

$\log_{10}\gamma_{\pm} = -2 \times 0.51 \sqrt{0.003}$

$= -0.0559$

$\gamma_{\pm} = 0.879$

$E = E° - \dfrac{RT}{2F} \left\{ \ln ([H^+]^2[SO_4^{2-}])^u + \ln \gamma_+^2\gamma_- \right\}$

$= E° - \dfrac{RT}{2F} \left\{ \ln ([H^+]^2[SO_4^{2-}])^u + \ln \gamma_{\pm}^3 \right\}$

$E/V = -0.152 - \dfrac{0.0257}{2} \ln (0.879)^3$

$E = -0.152 + 0.050 = -0.147$

8.24. At the left-hand electrode:

$Cu(s) \rightarrow Cu^{2+} + 2e^-$

At the right-hand electrode:

$AgCl(s) + e^- \rightarrow Ag(s) + Cl^-$

Overall reaction:

$2AgCl(s) + Cu(s) \rightarrow 2Ag(s) + Cu^{2+} + 2Cl^-, \quad z = 2$

To a good approximation, it can be assumed that the activity coefficients are unity at $10^{-4} \ M$ (DHLL gives $\gamma_{\pm} = 0.988$). Then

$E = E° - \dfrac{RT}{zF} \ln ([Cu^{2+}][Cl^-]^2)^u$

$0.191 = E°/V - \dfrac{0.0257}{2} \ln \left[10^{-4} (2 \times 10^{-4})^2\right]^u$

$= E°/V + 0.337$

$E° = -0.146 \text{ V}$

Suppose that at 0.20 M the activity coefficients are γ_+ and γ_-. Then

$$-0.074 = -0.146 - \frac{0.0257}{2} \ln [0.20(0.40)^2] - \frac{0.0257}{2} \ln \gamma_+ \gamma_-^2$$

$$-0.074 = -0.146 + 0.044 - 0.01285 \ln \gamma_\pm^3$$

$$\ln \gamma_\pm^3 = -2.179$$

$$\gamma_\pm^3 = 0.113$$

$$\gamma_\pm = 0.48$$

8.25. a. The anode reaction:

$$2Tl(s) + 2Cl^-(0.02\ m) \rightarrow 2TlCl(s) + 2e^-$$

The cathode reaction:

$$Cd^{2+}(0.01\ m) + 2e^- \rightarrow Cd(s)$$

The overall cell reaction:

$$2Tl(s) + Cd^{2+}(0.01\ m) + 2Cl^-(0.02\ m) \rightarrow 2TlCl(s) + Cd(s); \quad z = 2$$

b. This reaction can alternatively be written as

$$(2)\quad 2Tl(s) + Cd^{2+}(0.01\ m) \rightarrow 2Tl^+(\text{in } 0.01\ m\ CdCl_2) + Cd(s); \quad z = 2$$

for which the Nernst equation is

$$E = E°_{Cd^{2+}\ |\ Cd} - E°_{Tl^+\ |\ Tl} - \frac{0.0257\ V}{2} \ln ([Tl^+]^2/[Cd^{2+}])^u$$

$$E° = -0.40 - (-0.34) = -0.06\ V$$

When $a_{CdCl_2} = 1$, E_1 is the value of E, and substituting $[Tl^+] = K_{sp}/[Cl^-]$ we have

$$E_1° / V = -0.06 - 0.01285 \ln (K_{sp}^2 / [Cd^{2+}][Cl^-]^2)^u$$

$$= -0.06 - 0.01285 \ln (1.6 \times 10^{-3})^2$$

$$E_1° = -0.06 - (-0.165) = 0.105\ V$$

When $m = 0.01\ m$,

$$E/V = -0.06 - 0.01285 \ln (1.6 \times 10^{-3})^2/(0.01)(.02)^2$$

$$E = -0.06 - (-0.0057) = -0.054\ V$$

8.26. The diffusible K^+ ions are at a higher potential on the right-hand side of the membrane; there is thus a tendency for a few of them to cross to the left-hand side and create a positive potential there. (The same conclusion is reached by considering the diffusible Cl^- ions; they are at a higher concentration on the left-hand side, and a few tend to cross to the right-hand side and create a negative potential there.)

The Nernst potential is given by Eq. 8.19:

$$\Phi = \frac{RT}{zF} \ln \frac{c_1}{c_2}$$

$$= \frac{8.3145 \times 310.15}{96\ 485} \ln \frac{0.16}{0.04}$$

$$= 0.0370 \text{ V}$$

$$= 37 \text{ mV}$$

8.27. The overall reaction is

Lactate$^-$ + 2 cytochrome c (Fe^{3+}) $\rightarrow$ pyruvate$^-$ + 2 cytochrome c (Fe^{2+}) + 2H$^+$

with $z = 2$ and E$^{\circ\prime}$ = 0.254 + 0.185 = 0.439 V. If K' is the ratio at pH 7,

$$E^{\circ\prime} = 0.439 = \frac{RT}{2F} \ln K' = \frac{0.0257}{2} \ln K'$$

$$\ln K' = 34.16$$

$$K' = 6.87 \times 10^{14}$$

If K'' is the ratio at pH 6,

$$K_{\text{true}} = K' \times (10^{-7})^2 = K'' \times (10^{-6})^2$$

so that

$$K'' = K' \times 10^{-2} = 6.87 \times 10^{12}$$

8.28. The processes are

$$\text{AgCl(s)} + e^- \rightarrow \text{Ag(s)} + \text{Cl}^-(0.1 \ m)$$

$$\text{Ag(s)} + \text{Cl}^-(0.01 \ m) \rightarrow \text{AgCl(s)} + e^-$$

The electrical neutrality is maintained by the passage of H$^+$ ions from right to left:

$$\text{H}^+(0.01 \ m) \rightarrow \text{H}^+(0.10 \ m)$$

The overall process is

$$\text{H}^+(0.01 \ m) + \text{Cl}^-(0.01 \ m) \rightarrow \text{H}^+(0.10 \ m) + \text{Cl}^-(0.10 \ m)$$

The emf is

$$E = -\frac{RT}{F} \ln \frac{0.1 \times 0.1}{0.01 \times 0.01}$$

$$E/\text{V} = -0.0257 \ln 100$$

$$E = -0.118 \text{ V}$$

8.29. a. The processes at the electrodes are

$$\frac{1}{2}\text{H}_2 \rightarrow \text{H}^+(m_1) + e^-$$

$$H^+(m_2) + e^- \rightarrow \frac{1}{2} H_2$$

To maintain electrical neutrality of the solutions, for every mole of H^+ produced in the left-hand solution, t_+ mol of H^+ ions will cross the membrane from left to right, and t_- mol of Cl^- ions will pass from right to left. In the left-hand solution there is therefore a net gain of

$$(1 - t_+) \text{ mol} = t_- \text{ mol of } H^+$$

and of t_- mol of Cl^-.

In the right-hand solution, the net loss is

$$(1 - t_+) \text{ mol} = t_- \text{ mol of } H^+$$

and t_- mol of Cl^-.

The overall process is thus

$$t_- H^+(m_2) + t_- Cl^-(m_2) \rightarrow t_- H^+(m_1) + t_- Cl^-(m_1)$$

The emf is

$$E = -\frac{RT}{F} \ln \frac{m_1^{t_-} m_1^{t_-}}{m_2^{t_-} m_2^{t_-}}$$

$$= \frac{2t_- RT}{F} \ln \frac{m_2}{m_1}$$

b. For $m_1 = 0.01 \, m$ and $m_2 = 0.10 \, m$,

$$0.0190 = 2t_- \times 0.0257 \ln 10$$

$$t_- = 0.161$$

$$t_+ = 0.839$$

8.30. $M \rightarrow M^+ (0.1 \, m) + e^-$

Cell reaction: $$\frac{H^+(0.2 \, m) + e^- \rightarrow \frac{1}{2} H_2(1 \text{ bar})}{H^+(0.2 \, m) + M \rightarrow M^+ (0.1 \, m) + \frac{1}{2} H_2(1 \text{ bar})}$$

$$E^\circ = E^\circ_{H^+ \mid H_2} - E^\circ_{M^+ \mid M}$$

The Nernst equation is

$$-0.4 = E^\circ_{M^+ \mid M}/V - 0.0257 \ln \frac{(0.1)}{(0.2)} = -E^\circ_{M^+ \mid M}/V + 0.018$$

$$E^\circ_{M^+ \mid M} = 0.418 \text{ V}$$

Upon addition of KCl, almost all of the M^+ precipitates, and $0.10\ m$ Cl^- is in excess. The value of M^+ in solution is found from the K_{sp}, namely

$$K_{sp} = [M^+][Cl^-] = (M^+)\,(0.10\ M)$$

and from the Nernst equation

$$-0.1 = -E^\circ{}_{M^+\,|\,M} - 0.0257\ \ln \frac{K_{sp}/(0.10)}{0.2}$$

$$-0.1 = -0.418 - 0.0257\ \ln K_{sp}^{u} + 0.0257\ \ln 0.02$$

$$-0.1 + 0.418 + 0.100 = -0.0257\ \ln K_{sp}^{u}$$

$$\ln K_{sp}^{u} = -\frac{0.418}{0.0257} = -7.08$$

$$K_{sp} = 8.2 \times 10^{-8}\ \text{mol}^2\ \text{kg}^{-2}$$

8.31. The E° values for the half reactions (Table 8.2) are

$$\text{pyruvate}^- + 2H^+ + 2e^- \rightarrow \text{lactate}^- \qquad E^{\circ\prime} = -0.19\ \text{V} \qquad\qquad (1)$$

$$\text{NAD}^+ + H^+ + 2e^- \rightarrow \text{NADH} \qquad E^{\circ\prime} = -0.34\ \text{V} \qquad\qquad (2)$$

The required reaction is obtained by subtracting (2) from (1). Hence

$$E^{\circ\prime} = -0.19 - (-0.34) = 0.15\ \text{V}$$

Since $n = 2$, the corresponding $\Delta G^{\circ\prime}$ is

$$\Delta G^{\circ\prime} = -2 \times 96\ 485 \times 0.15\ \text{J}$$

$$= -28\ 946\ \text{J}$$

a. The equilibrium ratio at pH 7 can be calculated from the relationship

$$\Delta G^{\circ\prime} = -RT \ln K'$$

$$= -8.3145 \times 298.15\ \ln K'$$

$$\ln K' = 28\ 946/(8.3145 \times 298.15)$$

$$= 11.68$$

$$K' = 1.2 \times 10^5\ \text{dm}^3\ \text{mol}^{-1}$$

Alternatively, K' could have been calculated directly from $E^{\circ\prime}$, using Eq. (8.7):

$$0.15 = \frac{0.02569}{2}\ \ln K'$$

$$\ln K' = 11.68$$

$$K' = 1.2 \times 10^5\ \text{dm}^3\ \text{mol}^{-1}$$

This K' at pH 7.0 is related to the true (pH-independent) K by the equation

$$K = \frac{[\text{lactate}^-][\text{NAD}^+]}{[\text{pyruvate}^-][\text{NADH}][H^+]} = \frac{K'}{[H^+]} = \frac{K'}{10^{-7}}$$

b. Similarly the K'' at pH 8.0 is related to K by

$$K = \frac{K''}{10^{-8}}$$

Thus

$$\frac{K''}{10^{-8}} = \frac{K'}{10^{-7}}$$

and

$$K'' = 1.2 \times 10^5 \times \frac{10^{-8}}{10^{-7}}$$

$$= 1.2 \times 10^4$$

■ Temperature Dependence of Cell emfs

8.32. a. We are given that

$$\text{fumarate}^{2-} + 2H^+ + 2e^- \rightarrow \text{succinate}^{2-} \quad E^{\circ\prime} = 0.031 \text{ V} \tag{1}$$

$$\text{pyruvate}^- + 2H^+ + 2e^- \rightarrow \text{lactate}^- \quad E^{\circ\prime} = -0.185 \text{ V} \tag{2}$$

Subtraction of (2) from (1) gives

$$\text{fumarate}^{2-} + \text{lactate}^- \rightarrow \text{succinate}^{2-} + \text{pyruvate}^- \quad E^{\circ\prime} = 0.216 \text{ V}; \quad z = 2$$

(Note that this is also E°, the hydrogen ions having canceled out.)

The Gibbs energy change is

$$\Delta G^\circ = -zFE^\circ = -2 \times 96\ 485 \times 0.216 = -41.7 \text{ kJ mol}^{-1}$$

The entropy change is obtained by use of Eq. 8.23:

$$\Delta S^\circ = 2 \times 96\ 485 \times 2.18 \times 10^{-5} \text{ J K}^{-1} \text{ mol}^{-1}$$

$$= 4.21 \text{ J K}^{-1} \text{ mol}^{-1}$$

b. The enthalpy change can be calculated by use of Eq. 8.25, or more easily from the ΔG° and ΔS° values:

$$\Delta H^\circ = \Delta G^\circ + T\Delta S^\circ$$

$$= -41\ 680 + (298.15 \times 4.207) \text{ J mol}^{-1}$$

$$= -40\ 430 \text{ J mol}^{-1} = -40.4 \text{ kJ mol}^{-1}$$

8.33. a. The individual electrode processes are

$$\text{Cd(Hg)} \rightarrow \text{Cd}^{2+} + 2e^-$$

$$\text{Hg}_2^{2+} + 2e^- \rightarrow 2\text{Hg}$$

and the overall reaction is

$$Cd(Hg) + Hg_2^{2+} \rightarrow Cd^{2+} + 2Hg; \quad z = 2$$

Since the solution is saturated with $Hg_2SO_4 \cdot \frac{8}{3}H_2O$ the overall reaction can be written as

$$Cd(Hg) + Hg_2SO_4(s) + \frac{8}{3}H_2O(l) \rightarrow CdSO_4 \cdot \frac{8}{3}H_2O(s) + 2Hg(l)$$

b. $\Delta G° = -2 \times 96\ 487 \times 1.01832 \quad = -196\ 510\ J\ mol^{-1}$

$$= -196.5\ kJ\ mol^{-1}$$

$\Delta S° = 2 \times 96\ 487 \times (-5.00 \times 10^{-5}) = -9.65\ J\ K^{-1}\ mol^{-1}$

$\Delta H° = -196\ 510 - (9.65 \times 298.15) \quad = -199\ 390\ J\ mol^{-1}$

$$= -199.39\ kJ\ mol^{-1}$$

8.34. Since we are only interested in the slope (i.e., not the intercept) of the straight line that best fits the E vs. t data, we need not bother to convert the temperature to Kelvin.

We perform a linear regression analysis using t as the independent variable and E as the dependent variable. The result is:

$$E = 0.93046 - 2.8834 \times 10^{-4}\ t.$$

Differentiation with respect to t gives

$$\left(\frac{\partial E}{\partial t} \right)_P = -2.8834 \times 10^{-4}\ V\ (°C)^{-1} = \left(\frac{\partial E}{\partial T} \right)_P = -2.8834 \times 10^{-4}\ V\ K^{-1}$$

The entropy change for the conversion of one mole of (½)Br_2 to Br^- ($z = 1$) is

$$96\ 485\ C\ mol^{-1} \times 2.8834 \times 10^{-4}\ V\ K^{-1} = -27.820\ J\ K^{-1}\ mol^{-1}.$$

8.35. The cell reaction is $Mg(s) + Cl_2(g) \rightarrow Mg^{2+}(aq) + 2Cl^-(aq)$; i.e., $z = 2$. From Eq. (8.23), we get

$$\left(\frac{\partial E}{\partial T} \right)_P = \frac{\Delta S}{zF} = \frac{-337.3\ J\ K^{-1}\ mol^{-1}}{2 \times 96\ 485\ C\ mol^{-1}} = -1.7477 \times 10^{-3}\ V\ K^{-1}.$$

8.36. $E°' = E° = 0.031 + 0.197 = 0.228\ V$

$\Delta G° = -2 \times 96\ 485 \times 0.228 = -44\ 000\ J\ mol^{-1}$

a. If $x\ kJ\ mol^{-1} = \Delta G_f^°$ (fumarate),

$$-690.44 - 139.08 - x + 181.75 = -44.00$$

$$x = -603.8$$

$$\Delta G_f^° = -603.8\ kJ\ mol^{-1}$$

b. $\Delta S° = 2 \times 96\ 485 \times (-1.45 \times 10^{-4})$

$$= -28.0\ J\ K^{-1}\ mol^{-1}$$

$\Delta H° = \Delta G° + T\Delta S° = -44\ 000 - (298.15 \times 28.0)$

$$= -52\ 350 \text{ J mol}^{-1}$$

If y kJ mol$^{-1} = \Delta H_f^{\circ}$ (fumarate) ,

$$-908.68 - 210.66 - y + 287.02 = 52.35$$

$$y = -780.0$$

$$\Delta H_f^{\circ} = -780.0 \text{ kJ mol}^{-1}$$

8.37. a. The cell reaction can be written

$$Tl + H^+(a = 1) \rightarrow Tl^+(\text{in HBr}; a = 1) + \tfrac{1}{2} H_2(1 \text{ bar})$$

$$E = E_{H^+|H_2}^{\circ} - E_{Tl^+|Tl}^{\circ} - 0.0257 \ln ([Tl^+]/[H^+])^u$$

$$ 0 \qquad\quad -0.34$$

Since $Tl^+ = K_{sp}/[Br^-]$

$$E/V \ = 0.34 - 0.0257 \ln (K_{sp}/[H^+][Br^-])^u$$

$$= 0.34 - 0.0257 \ln K_{sp}^{u} = 0.34 + 0.236$$

$$E \ = 0.58 \text{ V}$$

b. ΔH for the reaction is ΔH° for the half cell

$$Tl \rightarrow Tl^+ + e^-$$

$$\Delta H^{\circ} \ = -zF(E - TdE/dT)$$

$$= -96\ 485\ [(0.34 - 298.15(-0.003)]$$

$$= -96\ 485\ (1.234) \text{ J mol}^{-1}$$

$$= -119\ 062 \text{ rounded to } -119 \text{ kJ mol}^{-1}$$

$$E^{\circ}$$
$$Tl \rightarrow Tl^+ + e^- \qquad -0.34V$$
$$H^+ + e^- \rightarrow \tfrac{1}{2}H_2 \qquad 0$$

■ Applications of emf Measurements

8.38. Subtraction of the second reaction from the first gives

$$AgBr(s) \rightarrow Ag^+ + Br^- \quad E^{\circ} = -0.7278 \text{ V}; \quad z = 1$$

Then

$$-0.7278 = 0.0257 \ln ([Ag^+][Br^-])^u$$

$$\ln ([Ag^+][Br^-])^u = -28.319$$

$$K_{sp} = [Ag^+][Br^-] = 5.03 \times 10^{-13} \text{ mol}^2 \text{ kg}^{-2}$$

Solubility $= \sqrt{K_{sp}} = 7.09 \times 10^{-7} \text{ mol kg}^{-1}$

8.39. The standard emf of the AgCl|Ag electrode is 0.2224 V and the cell reaction is

$$AgCl(s) + \frac{1}{2} H_2 \rightarrow H^+ + Cl^- + Ag(s); \quad z = 1$$

$$E = E^\circ - \frac{RT}{F} \ln (a_{H^+})^u$$

$$\approx E^\circ - \frac{2RT}{F} \ln (a_{H^+})^u$$

$$0.517 = 0.2224 - 2 \times 0.059\ 16 \log_{10} (a_{H^+})^u$$

$$\log_{10}(a_{H^+})^u = \frac{0.2224 - 0.517}{2 \times 0.059\ 16} = -2.49$$

$$pH = 2.49$$

8.40. The cell is represented as $Ag \mid AgI(s) \mid I^-(aq) \vdots Ag^+(aq) \mid Ag(s)$.

At the two electrodes:

$$e^- + AgI(s) \rightarrow Ag(s) + I^-$$

$$Ag(s) \rightarrow Ag^+ + e^-$$

Overall reaction:

$$AgI(s) \rightarrow Ag^+ + I^-; \quad z = 1$$

$$E = \frac{RT}{F}\ln[m_{Ag^+} \times m_{Cl^-}]^u = \frac{RT}{F}\ln K_{sp}$$

$$\therefore \ -0.9509 = 0.0257 \ln K_{sp}^u$$

$$K_{sp} = 8.53 \times 10^{-17} \ mol^2 \ kg^{-2}$$

$$Solubility = \sqrt{K_{sp}} = 9.24 \times 10^{-9} \ mol \ kg^{-1}$$

8.41. The standard cell potential for the cell shown is $E^\circ = (0.2224 - 0.0254) = 0.1970$ V. Since the molality is exactly 1.0, Eq. (8.43) simplifies to $0.2053 = 0.1970 - 2 \times 0.0257 \ln \gamma_\pm$. Solving for the mean activity coefficient, we calculate

$$\gamma_\pm = \exp[-(0.2053 - 0.1970)/(2 \times 0.0257)] = 0.8509.$$

8.42.　a.　The electrical work is $-\Delta G$.

$$\Delta G^\circ = \Delta H^\circ - T\Delta S^\circ$$

$$\Delta G^\circ/J \ mol^{-1} \quad = -2\ 877\ 000 - 298.15 \times (-432.7)$$

$$= -2\ 784\ 000 \ J \ mol^{-1} = -2750 \ kJ \ mol^{-1}$$

Electrical work available = 2750 kJ mol^{-1}

　　b.　The total work done is $-\Delta A$.

$$\Delta G^\circ = \Delta A^\circ - \Sigma \nu RT$$

$$\Sigma v = 4 - 1 - \frac{13}{2} = -3.5$$

$$\Sigma vRT = -3.5 \times 298.15 \times 8.3145 = -8676 \text{ J mol}^{-1}$$

$$\Delta A° = -2748 + 8.68 = -2739 \text{ kJ mol}^{-1}$$

Total available work $= 2740 \text{ kJ mol}^{-1}$

8.43. a. The overall process is

$$Cd + 2AgCl \rightarrow 2Ag + CdCl_2$$

and is a two-electron process.

$$\Delta G = \Delta G° + \frac{RT}{2F} \ln a_{Cd^{2+}} a_{Cl^-}^2$$

where a is the activity of the respective ions.

$$E = E° - \frac{RT}{2F} \ln a_{Cd^{2+}} a_{Cl^-}^2$$

$$0.7585 = 0.5732 - \frac{RT}{2F} \ln (m_{Cd^{2+}} \gamma_{Cd^{2+}} m_{Cl^-}^2 \gamma_{Cl^-}^2)$$

$$0.1853 = -\frac{RT}{2F} \ln[(0.01\gamma_{Cd^{2+}})(0.02)^2 \gamma_{Cl^-}^2)]$$

$$= -\frac{8.3145(298.15)}{2(96\,485)} \ln (4 \times 10^{-6})\gamma_\pm^3$$

$$\ln \gamma_\pm^3 = -1.161$$

$$\gamma_\pm = 0.679$$

b. The ionic strength of a 0.010 m solution of $CdCl_2$ is

$$I = \frac{1}{2} \Sigma m_i z_i^2 = \frac{1}{2} [(0.010 \times 2^2) + (2 \times 0.010 \times 1^2)]$$

$$= 0.030 \text{ mol kg}^{-1}$$

From the Debye-Hückel limiting law,

$$\log_{10}\gamma_\pm = -Az_+ z_- \sqrt{I}$$

$$= -0.51 \times 2 \times 1 \times \sqrt{0.030}$$

$$= -0.177$$

$$\gamma_\pm = 0.6653 = 0.67 \text{ (to the degree of accuracy allowed by the problem)}$$

The two values are in good agreement.

8.44. The LiOH is required for the hydrogen electrode and the LiCl salt is used to complete the AgCl electrode. Both the Cl^- ion and the H^+ ion will behave according to their activities in solution. Begin by determining the emf of the cell.

$$E_{cell} = E_{AgCl} - E_{H_2} = E^{\circ}_{AgCl} - \frac{RT}{F} \ln a_{Cl^-} - \frac{RT}{F} \ln a_{H^+}$$

This can be rewritten using $K_w = a_{H^+} a_{OH^-}$, and so the desired relationship is established.

$$E_{cell} = E^{\circ}_{AgCl} - \frac{RT}{F} \ln a_{Cl^-} - \frac{RT}{F} \ln K_w + \frac{RT}{F} \ln a_{OH^-}$$

Combining the two activity terms gives

$$E_{cell} = E^{\circ}_{AgCl} - \frac{RT}{F} \ln \frac{a_{Cl^-}}{a_{OH^-}} - \frac{RT}{F} \ln K_w$$

Rewriting this in terms of activity coefficients and molalities gives, after rearrangement,

$$\frac{(E_{cell} - E^{\circ}_{AgCl})F}{RT} + \ln (m_{Cl^-}/m_{OH^-}) = -\ln K_w - \ln \frac{\gamma_{Cl^-}}{\gamma_{OH^-}}$$

The value of m_{OH^-} is 0.01 mol kg^{-1}. With that substitution, plot the left-hand side of the equation with $E^{\circ}_{AgCl} = 0.2224$ V at 298.15 K against the ionic strength, which varies with the concentration, and extrapolate to zero ionic strength. At zero ionic strength the γ_{Cl^-} and γ_{OH^-} approach unity. Then the value of the curve is $-\ln K_w$.

In the following data, I is based on $m + 0.01$ m OH$^-$. The latter is constant. For example, $I = 1/2[0.01 \times 1^2 + 0.01 \times (-1)^2 + 0.01 \times 1^2 + 0.01 \times (-1)^2]$.

m/mol kg^{-1}	0.01	0.02	0.05	0.10	0.20
I/mol kg^{-1}	0.02	0.03	0.06	0.11	0.21

$\dfrac{(E_{cell} - E^{\circ}_{AgCl})F}{RT}$	32.2086	31.5079	30.566	29.834	29.087
$\ln \dfrac{m}{0.01}$	0.000	0.693	1.609	2.303	2.996
Sum of last 2 terms	32.209	32.201	32.175	32.137	32.083

From the indicated plot shown, the value of $-\ln K_w = 32.226$. The value of K_w is 1.010×10^{-14}.

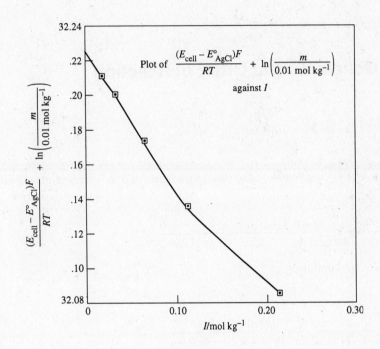

9. CHEMICAL KINETICS I. THE BASIC IDEAS

■ Rate Constants and Order of Reaction

9.1.　a.　3

b.　Both rates are 3.6×10^{-3} mol dm^{-3} s^{-1}

c.　None

d.　Rate of Br$^-$ disappearance decreased by a factor of 8; no effect on rate constant.

9.2.　$\alpha = 1, \beta = 1$

$$k = \frac{5.0 \times 10^{-7} \text{ mol dm}^{-3} \text{ s}^{-1}}{(3.5 \times 10^{-2} \text{ mol dm}^{-3})(2.3 \times 10^{-2} \text{ mol dm}^{-3})}$$

$$= 6.2 \times 10^{-4} \text{ dm}^3 \text{ mol}^{-1} \text{ s}^{-1}$$

9.3.　$\alpha = 2, \beta = 1$

$$k = \frac{7.4 \times 10^{-9} \text{ mol dm}^{-3} \text{ s}^{-1}}{(1.4 \times 10^{-2} \text{ mol dm}^{-3})^2 (2.3 \times 10^{-2} \text{ mol dm}^{-3})}$$

$$= 1.6 \times 10^{-3} \text{ dm}^6 \text{ mol}^{-2} \text{ s}^{-1}$$

9.4.　a.　Half-life:

$$t_{1/2} = \ln 2/k$$

$$= \frac{0.6931}{3.72 \times 10^{-5} \text{ s}^{-1}} = 18\ 632 \text{ s}$$

$$= 5.18 \text{ hours}$$

b.　Fraction undecomposed:

$$\frac{a_0 - x}{a_0} = e^{-kt}$$

$$= e^{-3.72 \times 10^{-5} \times 3 \times 3600} = 0.669$$

9.5.　$\dfrac{a_0 - x}{a_0} = e^{-kt}$

For half life, $t_{1/2}$: $0.5 = \exp(-kt_{1/2})$

For time $t_{99\%}$ to go to 99% completion:

$$0.01 = \exp(-kt_{99\%})$$

$$t_{99\%} = \frac{1}{k} \ln 100; \quad t_{1/2} = \frac{1}{k} \ln 2$$

$$\frac{t_{99\%}}{t_{1/2}} = \frac{\ln 100}{\ln 2} = \frac{4.605}{0.693} = 6.64$$

9.6. For a second-order reaction, $t_{1/2} = 1/a_0 k$ (Table 9.1)

a. $t_{1/2} = \dfrac{1}{1.3 \times 10^{11} \times 10^{-1}} = 7.7 \times 10^{-11}$ s $= 77$ ps

b. $t_{1/2} = \dfrac{1}{1.3 \times 10^{11} \times 10^{-4}} = 7.7 \times 10^{-8}$ s $= 77$ ns

9.7. $\ln \dfrac{a_0}{a_0 - x} = -\ln \text{(fraction remaining)} = kt$

$$t_{1/2} = \frac{\ln 2}{k}; \quad k = \frac{\ln 2}{t_{1/2}} = 0.02467 \text{ years}^{-1}$$

a. After 25 years,

ln (fraction remaining) $= -0.02467 \times 25 = -0.61675$

Fraction remaining $= 0.540$

Quantity remaining $= 0.540$ μg

b. After 50 years,

ln (fraction remaining) $= -0.02467 \times 50 = -1.2335$

Fraction remaining $= 0.291$

Quantity remaining $= 0.291$ μg

c. After 70 years,

ln (fraction remaining) $= -0.02467 \times 70 = -1.7269$

Fraction remaining $= 0.178$

Quantity remaining $= 0.178$ μg

9.8. $n_0 = 0.0500$ g/(62.023 g mol^{-1}) = 8.0604 $\times$ 10^{-4} mol.

Amount reacted in 70.0 min:

$$(n_0 - n) = \frac{0.00659 \text{ dm}^3 \times 1 \text{ bar}}{0.083145 \text{ dm}^3 \text{ bar K}^{-1} \text{ mol}^{-1} \times 288 \text{ K}}$$

$$= 2.7521 \times 10^{-4} \text{ mol}.$$

Therefore, using the integrated rate law, we get

$$k = \frac{1}{70.0} \ln\left(\frac{8.0604 \times 10^{-4}}{8.0604 \times 10^{-4} - 2.7521 \times 10^{-4}}\right)$$

$$= 5.967 \times 10^{-3} \text{ min}^{-1}.$$

Half-life:

$$t_{1/2} = \frac{\ln 2}{5.967 \times 10^{-3} \text{ min}^{-1}} = 116 \text{ min}.$$

9.9.

	2NO	+	Cl$_2$	$\rightarrow$	2NOCl
Initial amounts:	5		2		0 mol
Amounts when half of the Cl$_2$ has reacted:	3		1		2 mol

The rate is then equal to

$$(3/5)^2(1/2) \times 2.4 \times 10^{-3} = 4.32 \times 10^{-4} \text{ mol dm}^{-3} \text{ s}^{-1}$$

9.10. Let n_0 denote the initial amount of NOCl and let $2x$ be the moles reacted in time t. Then, at time t, we have

$$2NOCl(g) \rightarrow 2NO(g) + Cl_2(g)$$
$$n_0 - 2x \qquad 2x \qquad x$$

Examining the quantities present, we see that the total amount of gases at any time t is $(n_0 + x)$. The concentrations can then be calculated.

Since the initial pressure $P_0 = \dfrac{n_0 RT}{V}$, the initial concentration $\left(\dfrac{n_0}{V}\right) = \dfrac{P_0}{RT}$.

At time t, $P_t = (n_0 + x)\dfrac{RT}{V}$. Therefore, $\left(\dfrac{x}{V}\right) = \dfrac{P_t}{RT} - \dfrac{n_0}{V}$.

Substitution gives

$$\left(\frac{x}{V}\right) = \frac{1}{RT}(P_t - P_0), \text{ and } \left(\frac{n_0 - 2x}{V}\right) = \frac{1}{RT}(3P_0 - 2P_t).$$

So, the rate is given by

$$\frac{-d[NOCl]}{dt} = k[NOCl]^2 = \frac{k}{RT}(3P_0 - 2P_t)^2.$$

9.11.

% decomposed	$v / \dfrac{\text{Torr}}{\text{min}}$	$\log_{10}$ (% remaining)	$\log_{10}\left(v\dfrac{\text{Torr}}{\text{min}}\right)$
0	8.53	2.00	0.931
5	7.49	1.98	0.874
10	6.74	1.95	0.827
15	5.90	1.93	0.771
20	5.14	1.90	0.711
25	4.69	1.88	0.671
30	4.31	1.85	0.634
35	3.75	1.81	0.574
40	3.11	1.78	0.493
45	2.67	1.74	0.427
50	2.29	1.70	0.360

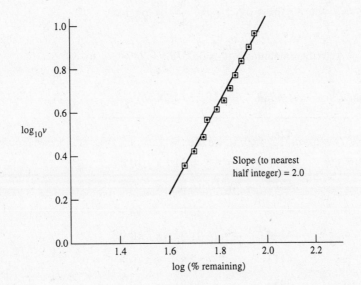

Slope of plot $\log_{10}$ (v/Torr min^{-1}) against $\log_{10}$ (% remaining) (to nearest half integer) = 2.0. This is the order with respect to time.

9.12. Half-life, $t_{1/2} = 14.3 \times 24 \times 60 \times 60$ s $= 1.236 \times 10^6$ s

$$k = \frac{0.693}{t_{1/2}} = 5.61 \times 10^{-7} \text{ s}^{-1}$$

$$= \frac{0.693}{14.3} \text{ days}^{-1} = 0.0485 \text{ days}^{-1}$$

a. After 10 days,

$$\ln \frac{n_0}{n} = 0.0485 \times 10 = 0.485; \quad \frac{n_0}{n} = 1.624$$

Activity remaining $= \dfrac{n}{n_0} = 0.616 = 61.6\%$

b. After 20 days,

$$\ln \frac{n_0}{n} = 0.0485 \times 20 = 0.97; \quad \frac{n_0}{n} = 2.64$$

Activity remaining $= \dfrac{n}{n_0} = 0.379 = 37.9\%$

c. After 100 days,

$$\ln \frac{n_0}{n} = 0.0485 \times 100 = 4.85; \quad \frac{n_0}{n} = 127.7$$

Activity remaining $= \dfrac{n}{n_0} = 0.0078 = 0.78\%$

9.13.

t/days	n/min	$\ln(n_0/n)$
0	4280 (n_0)	–
1	4245	8.21 x 10^{-3}
2	4212	16.0 x 10^{-3}
3	4179	23.9 x 10^{-3}
4	4146	31.8 x 10^{-3}
5	4113	39.8 x 10^{-3}
10	3952	79.7 x 10^{-3}
15	3798	119.5 x 10^{-3}

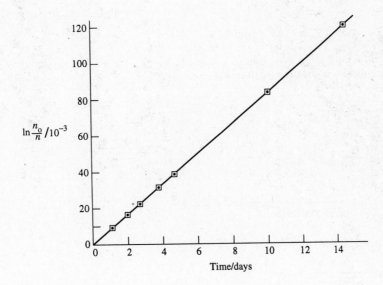

From a plot of $\ln (n_0/n)$ against t, slope $= k$

$$= \frac{119.5 \times 10^{-3}}{15} = 7.97 \times 10^{-3} \text{ days}^{-1}$$

$$= \frac{7.97 \times 10^{-3}}{3600 \times 24} = 9.22 \times 10^{-8} \text{ s}^{-1}$$

$$t_{1/2} = \frac{0.693}{7.97 \times 10^{-3}} = 87.0 \text{ days}$$

a. After 60 days,

$$\ln (n_0/n) = 0.478$$

$$(n_0/n) = 1.613; \quad n = 2653$$

b. After 365 days,

$$\ln (n_0/n) = 2.909$$

$$(n_0/n) = 18.34; \quad n = 233$$

9.14. The rate equation is

$$k_1 t = \frac{x_e}{a_0} \ln \frac{x_e}{x_e - x} \qquad \text{(Eq. 9.48)}$$

The time for half the equilibrium amount of product to be formed is given by putting $x = x_e/2$:

$$k_1 t_{1/2} = \frac{x_e}{a_0} \ln \frac{x_e}{x_e/2} = \frac{x_e}{a_0} \ln 2$$

The equilibrium constant is $x_e/(a_0 - x_e)$:

$$\frac{a_0 - x_e}{x_e} = \frac{1}{0.16}$$

$$\frac{a_0}{x_e} = \frac{1}{0.16} + 1 = 7.25$$

Thus $\quad t_{1/2} = \dfrac{1}{7.25 \times 3.3 \times 10^{-4} s^{-1}} \ln 2 = 290 \text{ s}$

9.15. For a second order reaction, Eqs. (9.32) and (9.33) give

$$kt = \frac{1}{a_0 - 2x} - \frac{1}{a_0},$$

where we have assumed that $2x$ moles of A is consumed in each step. Since the times involved in this problem are multiples of the half-life, $t_{1/2}$, we write

$$k(nt_{1/2}) = \frac{1}{a_0 - 2x} - \frac{1}{a_0} ; n = 1, 2, 3, \text{ or } \infty.$$

Substituting from Eq. (9.40), $t_{1/2} = 1/(a_0 k)$, the above equation becomes

$$\frac{n}{a_0} = \frac{1}{a_0 - 2x} - \frac{1}{a_0}.$$

Therefore,

$$(a_0 - 2x) = \left(\frac{a_0}{n+1}\right), \text{ and } x = \frac{a_0}{2}\left(\frac{n}{n+1}\right). \tag{1}$$

It follows from the stoichiometry that $P_A \propto (a_0 - 2x)$, $P_B \propto 2x$, $P_C \propto x$. From this, and Eq. (1), we derive

$$P_A = \left(\frac{P_{0;NOCl}}{n+1}\right), P_B = P_{0;NOCl}\left(\frac{n}{n+1}\right), \text{ and } P_C = \frac{P_{0;NOCl}}{2}\left(\frac{n}{n+1}\right). \tag{2}$$

Therefore, from Eq. (2), we get

t	P_A/bar	P_B/bar	P_C/bar	P/bar
0	2.00	0.00	0.00	2.00
$t_{1/2}$	1.00	1.00	0.50	2.50
$2\,t_{1/2}$	0.67	1.33	0.67	2.67
$3\,t_{1/2}$	0.50	1.50	0.75	2.75
∞	0.00	2.00	1.00	3.00

9.16. $$\frac{dx}{dt} = k(a_0 - x)^n$$

which integrates to

$$k = \frac{1}{t(n-1)} \frac{1}{(a_0 - x)^{(n-1)}} - \frac{1}{a_0^{(n-1)}}$$

with $x = a_0/2$,

$$t_{1/2} = \frac{1}{k(n-1)} \left[\frac{1}{(a_0/2)^{(n-1)}} - \frac{1}{a_0^{(n-1)}} \right]$$

$$= \frac{2^{(n-1)} - 1}{k \, a_0^{(n-1)} (n-1)}$$

9.17. The reaction is

$$2C_4H_6 \rightarrow C_8H_{12}.$$
$$\quad n_0 - 2x \qquad x$$

Total number of moles at any time $t = n_0 - x$. Therefore,

$$P_0 = n_0 RT/V; \quad P = (n_0 - x)RT/V, \quad \text{and} \quad xRT/V = P_0 - P.$$

The partial pressure of butadiene is, therefore, given by

$$P_{C_4H_6} = (n_0 - 2x)RT/V = 2P - P_0.$$

Therefore, we retabulate the data in terms of the partial pressure of butadiene:

t/min	3.25	12.18	24.55	42.50	68.05
$P_{C_4H_6}$/Torr	605.0	536.4	461.6	386.6	317.2

Now, we need to check whether the reaction follows first-order or second-order kinetics. Plotting $P_{C_4H_6}$ against t, we obtain a curved line, whereas plotting $1/P_{C_4H_6}$ against t yields a straight line. This indicates that the reaction follows second-order kinetics. From the linear regression results, the integrated rate law is found to be

$$1/P_{C_4H_6} = 1.5865 \times 10^{-3} + 2.3182 \times 10^{-5} t,$$

from which we get (see Eq. 9.33)

$$k = 2.3182 \times 10^{-5} \, (M \, s)^{-1}.$$

9.18. When the concentration has reached nc, where c is the concentration produced by one dose, the concentration will fall to $(n-1)c$ during the interval between successive doses. The next dose restores the concentration to $(n-1)c + c = nc$. The "steady state" has therefore been reached.

9.19.

	2A	+	B	→	Z
Initially:	a_0		b_0		0
After time t:	$a_0 - 2x$		$b_0 - x$		x

$$\frac{dx}{dt} = k(a_0 - 2x)(b_0 - x)$$

$$kdt = \frac{dx}{(a_0 - 2x)(b_0 - x)}$$

$$= \left[\frac{1}{a_0 - 2x} - \frac{1}{2(b_0 - x)}\right]\frac{dx}{2b_0 - a_0}$$

$$kt = \frac{1}{2b_0 - a_0}\left[-\frac{1}{2}\ln(a_0 - 2x) + \frac{1}{2}\ln 2(b_0 - x)\right] + I$$

When $t = 0$, $x = 0$, and therefore

$$I = \frac{1}{2b_0 - a_0}\left(\frac{1}{2}\ln a_0 - \frac{1}{2}\ln 2b_0\right)$$

$$kt = \frac{1}{2b_0 - a_0}\left[\frac{1}{2}\ln\frac{a_0}{a_0 - 2x} - \frac{1}{2}\ln\frac{2b_0}{2(b_0 - x)}\right]$$

$$= \frac{1}{2(2b_0 - a_0)}\left[\ln\frac{a_0(b_0 - x)}{b_0(a_0 - 2x)}\right]$$

9.20.

	2A	+	B	→	Z
Initially:	$2a_0$		a_0		0
After time t:	$2a_0 - 2x$		$a_0 - x$		x

$$\frac{dx}{dt} = k(2a_0 - 2x)^2(a_0 - x)$$

$$= 4k(a_0 - x)^3$$

$$\frac{dx}{(a_0 - x)^3} = 4k\, dt$$

$$\frac{1}{2(a_0 - x)^2} = 4kt + I$$

$x = 0$ when $t = 0$; $I = 1/(2a_0^2)$

$$\frac{1}{2(a_0 - x)^2} - \frac{1}{2a_0^2} = 4\,kt$$

$$\frac{2a_0 x - x^2}{a_0^2 (a_0 - x)^2} = 8kt$$

The half-life $t_{1/2}$ is when $x = a_0/2$ (Note that at this time half of A and half of B have been consumed; if all reactants were not present in stoichiometric proportions, the half-lives would be different for the two reactants, and we could not speak of the half-life of the reaction.)

$$\frac{2a_0^2/2 - 2a_0^2/4}{a_0^2 (a_0/2)^2} = 8kt_{1/2}$$

which reduces to

$$t_{1/2} = \frac{3}{8a_0^2 k}$$

9.21. $\dfrac{d[Y]}{dt} = k_1 [A]; \quad \dfrac{d[Z]}{dt} = k_2 [A]$

$$\frac{d[Y]}{d[Z]} = \frac{k_1}{k_2}$$

Integrating

$$[Y] = \frac{k_1}{k_2} [Z] + I$$

At $t = 0$, $[Y] = [Z]$ and therefore $I = 0$. Thus

$$\frac{[Y]}{[Z]} = \frac{k_1}{k_2}$$

9.22. The rate of consumption of A is given by

$$-\frac{d[A]}{dt} = k_1 [A] \tag{1}$$

The rates of formation of B and C are

$$\frac{d[B]}{dt} = k_1 [A] - k_2 [B] \tag{2}$$

$$\frac{d[C]}{dt} = k_2 [B] \tag{3}$$

The first equation may be integrated at once to give

$$[A] = [A]_0 e^{-kt} \tag{4}$$

Insertion of this in Eq. (2) gives

$$\frac{d[B]}{dt} = k_1 [A]_0 e^{-k_1 t} - k_2 [B] \tag{5}$$

With the boundary condition $t = 0$, $[B] = 0$, this integrates to

$$[B] = [A]_0 \frac{k_1}{k_2 - k_1} (e^{-k_1 t} - e^{-k_2 t}) \qquad (6)$$

$$[A]_0 = [A] + [B] + [C] \qquad (7)$$

and thus

$$[C] = [A]_0 - [A] - [B] \qquad (8)$$

and insertion of Eqs. (4) and (6) leads to

$$[C] = [A]_0 \left(1 + \frac{k_2 e^{-k_1 t} - k_1 e^{-k_2 t}}{k_1 - k_2} \right)$$

9.23. a. The rate equation, with the catalyst concentration incorporated in the rate constants, is

$$\frac{dx}{dt} = k_1(a_0 - x) - k_{-1}x_e^2$$

At equilibrium

$$k_1(a_0 - x_e) - k_{-1}x_e^2 = 0$$

where x_e is the concentration of Y and Z at equilibrium. Hence

$$k_{-1} = \frac{k_1(a_0 - x_e)}{x_e^2}$$

Insertion of this into the first equation leads, after rearrangement, to

$$\frac{x_e^2 \, dx}{(a_0 - x)x_e^2 - (a_0 - x_e)x^2} = k_1 \, dt$$

Integration of the left-hand side of this equation can be carried out after resolution into partial fractions:

$$\text{L.H.S.} = \frac{p}{x_0 - x} + \frac{q}{a_0 x_e + a_0 x - x_e x}$$

and it is found that

$$p = \frac{x_e}{2a_0 - x_e} \qquad q = \frac{x_e(a_0 - x_e)}{2a_0 - x_e}$$

The integration is then straightforward (but fairly lengthy); with the boundary condition $t = 0$, $x = 0$, the result is

$$k_1 = \frac{x_e}{(2a_0 - x_e)t} \ln \frac{a_0 x_e + x(a_0 - x_e)}{a_0(x_e - x)}$$

Readers wishing further mathematical details are referred to C. Capellos and B. H. T. Bielski, *Kinetic Systems* (New York: Wiley, Interscience, 1972) pp. 41–43.

b. The equilibrium constant $K = k_1/k_{-1}$ is given by

$$K = \frac{a_0 - x_e}{x_e^2}$$

and this quadratic can be solved to give x_e in terms of K. This expression can then be substituted into the equation for k_1.

c. To deal with the numerical data, which are in terms of percent hydrolysis, it is convenient to define

$$r \equiv \frac{x}{a_0} \quad \text{and} \quad r_e = \frac{x_e}{a_0}$$

The integrated equation then becomes

$$k_1 = \frac{x_e}{(2 - r_e)t} \ln \frac{r_e + r(1 - r_e)}{r_e - r}$$

From the data we then have $r_e = 0.90$ and

t/s	r	$r_e + r(1 - r_e)$	$r_e - r$	$\ln \dfrac{r_e + r(1 - r_e)}{r_e - r}$
1350	0.212	0.9212	0.688	0.292
2070	0.307	0.9307	0.593	0.451
3060	0.434	0.9434	0.466	0.705
5340	0.595	0.9595	0.315	1.114
7740	0.7345	0.97345	0.1655	1.77

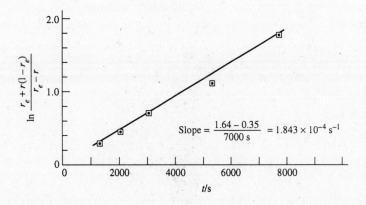

Slope of a plot of $\ln \dfrac{r_e + r(1 - r_e)}{r_e - r}$ against t is 1.843×10^{-4} s^{-1}.

This slope is

$$\frac{k_1(2 - r_e)}{r_e} = 1.222\, k_1$$

Thus $k_1 = 1.51 \times 10^{-4}$ s^{-1}

This is the pseudo-first-order rate constant for the reaction in the presence of 0.05 M HCl. The second-order rate constant is

$$\frac{1.51 \times 10^{-4}\ \text{s}^{-1}}{0.05\ \text{mol dm}^{-3}} = 3.02 \times 10^{-3}\ \text{dm}^3\ \text{mol}^{-1}\ \text{s}^{-1}$$

For the equilibrium

A	$\rightleftarrows$	Y	+	Z
0.05×0.1		0.05×0.9		0.05×0.9 mol dm^{-3}

and the equilibrium constant k_1/k_{-1} is

$$\frac{(0.05 \times 0.9)^2}{0.05 \times 0.1} = 0.405\ \text{mol dm}^{-3}$$

Thus

$$k_{-1} = \frac{1.51 \times 10^{-4}\ \text{s}^{-1}}{0.405\ \text{mol dm}^{-3}} = 3.73 \times 10^{-4}\ \text{dm}^3\ \text{mol}^{-1}\ \text{s}^{-1}$$

The corresponding third-order rate constant is

$$\frac{3.73 \times 10^{-4}\ \text{dm}^3\ \text{mol}^{-1}\ \text{s}^{-1}}{0.05\ \text{mol dm}^{-3}} = 7.46 \times 10^{-3}\ \text{dm}^6\ \text{mol}^{-2}\ \text{s}^{-1}$$

9.24. If a_0 is the initial concentration of A, and x is the concentration of ions at equilibrium,

$$\frac{dx}{dt} = k_1(a_0 - x) - k_{-1}x^2 \tag{1}$$

At equilibrium

$$k_1(a_0 - x_e) - k_{-1}x_e^2 = 0 \tag{2}$$

The deviation Δx from equilibrium, $x - x_e$, is given by

$$\frac{d\Delta x}{dt} = \frac{dx}{dt} = k_1(a_0 - x) - k_{-1}x^2 \tag{3}$$

Subtraction of Eq. 2 leads to

$$\frac{d\Delta x}{dt} = -k_1 \Delta x - k_{-1}(\Delta x)^2 - 2k_{-1}x_e\Delta x \tag{4}$$

Since Δx is very small, the term in $(\Delta x)^2$ can be neglected:

$$\frac{d\Delta x}{dt} = -(k_1 + 2k_{-1}x_e)\,\Delta x \tag{5}$$

Integration gives

$$\ln \Delta x = -(k_1 + 2k_{-1}x_e)t + I \tag{6}$$

The boundary condition is $t = 0$, $\Delta x = (\Delta x)_0$, where $(\Delta x)_0$ is the initial value of Δx. Thus

$$\ln \frac{(\Delta x)_0}{\Delta x} = (k_1 + 2k_{-1}x_e)t \tag{7}$$

By definition, the relaxation time t^* is the time corresponding to

$$\frac{(\Delta x)_0}{\Delta x} = e \tag{8}$$

Thus

$$t^* = \frac{1}{k_1 + 2k_{-1}x_e} \tag{9}$$

■ Temperature Dependence

9.25. $T_1 = 293.15$ K $1/T_1 = 3.4112 \times 10^{-3}$ K^{-1}

$T_2 = 303.15$ K $1/T_2 = 3.2987 \times 10^{-3}$ K^{-1}

Slope of a plot of $\ln k$ against $1/T$ is

$$\frac{\ln 2}{(3.2987 - 3.4112) \times 10^{-3}\ \text{K}^{-1}} = \frac{0.6931\ \text{K}}{-0.1125 \times 10^{-3}}$$

$$= 6161\ \text{K}$$

$$E = -8.3145\ \text{J K}^{-1}\ \text{mol}^{-1} \times \text{slope} = 51\,200\ \text{J mol}^{-1}$$

$$= 51.2\ \text{kJ mol}^{-1}$$

9.26. $T_1 = 493.15$ K $1/T_1 = 2.0278 \times 10^{-3}$ K^{-1}

$T_2 = 503.15$ K $1/T_2 = 1.9875 \times 10^{-3}$ K^{-1}

Slope of a plot of $\ln k$ against $1/T$ is

$$-\frac{0.6931\ \text{K}}{0.0403 \times 10^{-3}} = -17\,200\ \text{K}$$

$$E = 143\,000\ \text{J mol}^{-1} = 143.0\ \text{kJ mol}^{-1}$$

9.27. We perform a linear regression analysis using the dependent variables $\ln(k)$ and independent variables $1/T(\text{K})$. The result is

$\ln(k) = 32.217 - 11\ 704/T.$

Therefore,

$$E_a = 11\ 704 \times 8.3145 \text{ J mol}^{-1} = 9.73 \text{ kJ mol}^{-1}.$$

$$A = e^{(32.217)} = 9.81 \times 10^{13} \text{ min}^{-1} = 1.63 \times 10^{12} \text{ s}^{-1}.$$

9.28. Rate constant ratio $= e^{20\ 000/(8.3145 \times T/\text{K})}$

   a. If $T = 273.15$ K, ratio $= e^{20\ 000/(8.3145 \times 273.15)}$

$$= 6.68 \times 10^3$$

   b. If $T = 1273.15$ K, ratio $= e^{20\ 000/(8.3145 \times 1273.15)}$

$$\doteq 6.62$$

9.29. Taking the natural log of both sides, we get

$$\ln(k) = \ln(a) + n\ln(T) - E/(\text{RT}).$$

Therefore,

$$\left(\frac{d\ln k}{dT} \right) = \frac{n}{T} + \frac{E}{RT^2} \, .$$

From the definition of E_a,

$$E_a = RT^2 \left(\frac{d\ln k}{dT} \right) = nRT + E.$$

9.30. a. From the equation obtained by Mahmud, *et al.*,

$$E = 1766 \times 8.3145 \text{ J K}^{-1} \text{ mol}^{-1} = 14\ 683 \text{ J mol}^{-1}.$$

Therefore, the activation energy is

$$E_a = 2.87 \times 8.3145 \text{ J K}^{-1} \text{ mol}^{-1} \times 900 \text{ K} + 14\ 683 \text{ J mol}^{-1} = \ 36\ 159 \text{ J mol}^{-1}.$$

   b. Experimental rate constant $= 5.6 \times 10^{-21} \times (900)^{2.87} \, e^{-1766/900}$

$$= 2.37 \times 10^{-13} \text{ cm}^3 \text{ molecule}^{-1} \text{s}^{-1}.$$

Theoretical rate constant $= 6.9 \times 10^{-20} \times (900)^{2.60} \, e^{-2454/900}$

$$= 2.17 \times 10^{-13} \text{ cm}^3 \text{ molecule}^{-1} \text{s}^{-1}.$$

9.31. $T_1 = 15°C = 288.15$ K; $1/T_1 = 3.4704 \times 10^{-3} \text{ K}^{-1}$

$T_2 = 25°C = 298.15$ K; $1/T_2 = 3.3540 \times 10^{-3} \text{ K}^{-1}$

Slope of a plot of $\ln k$ against $1/T$ is

$$-\frac{0.6931 \text{ K}}{0.1164 \times 10^{-3}} = 5.954 \times 10^3 \text{ K}$$

$$E = 8.3145 \times 5.954 \times 10^3 \text{ J mol}^{-1} = 49\ 500 \text{ J mol}^{-1}$$

$$= 49.5 \text{ kJ mol}^{-1}$$

9.32. ln 40 = 3.689

$T_1 = 4°C = 277.15$ K; $1/T_1 = 3.6082 \times 10^{-3}$ K^{-1}

$T_2 = 25°C = 298.15$ K; $1/T_2 = 3.3540 \times 10^{-3}$ K^{-1}

Slope of a plot of ln k against $1/T$ $= -\dfrac{3.689}{0.2542 \times 10^{-3}}$

$= -14\ 500$ K

$E = 120\ 700$ J mol^{-1} = 120.7 kJ mol^{-1}

9.33.

T/K	$\dfrac{k \times 10^{-9}}{cm^6\ mol^{-2}\ s^{-1}}$	$\log_{10}\left(\dfrac{T}{K}\right)$	$\log_{10}\left(k/\dfrac{10^9}{mol^2}\right)$
80.0	41.8	1.903	1.62
143.0	20.2	2.155	1.31
228.0	10.1	2.358	1.00
300.0	7.1	2.477	0.85
413.0	4.0	2.616	0.60
564.0	2.8	2.751	0.45

Slope of plot of $\log_{10}(k/10^9\ cm^6\ mol^{-2}\ s^{-1})$ against $\log_{10}(T/K)$, to the nearest half-integer, is −1.5. The plot is shown on the following page.

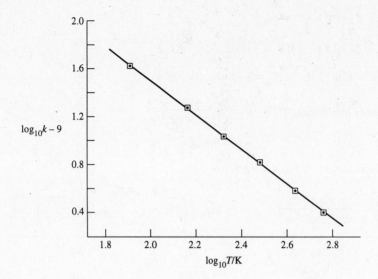

9.34. For the overall reaction,

$$\Delta H° = 139.5 - (218.0 - 74.8) = -3.7 \text{ kJ mol}^{-1}$$

The activation energy for the reverse reaction is thus

$$49.8 - (-3.7) = 53.5 \text{ kJ mol}^{-1}.$$

9.35. a. The rate equation is

$$\frac{dz}{dt} = k_1 ab - k_{-1} z.$$

Since at equilibrium, $\frac{dz}{dt} = 0$, we get , $k_1 a_e b_e - k_{-1} z_e$, or

$$k_1 a_e b_e = k_{-1} z_e = 0.$$

b. Since Z is formed as A and B are consumed, for a change in a_e and b_e by an amount $-x$, we get

$$\frac{d(z_e + x)}{dt} = \frac{dx}{dt} = k_1 (a_e - x)(b_e - x) - k_{-1}(z_e + x).$$

c. Multiplying out the results of part (b), we obtain

$$\frac{dx}{dt} = k_1 [a_e b_e - (a_e + b_e)x + x^2] - k_{-1}(z_e + x).$$

Since x is small, we drop x^2 from this expression. Now, using the result of part (a), we find

$$\frac{dx}{dt} = [k_1 (a_e + b_e) + k_{-1}]x.$$

d.   Since $x = x_0 e^{-t/t^*}$, we can differentiate both sides to obtain

$$\frac{dx}{dt} = \frac{-1}{t^*} \, x_0 e^{-t/t^*} = \frac{-x}{t^*}.$$

Comparing this to the result of part (c), we find

$$\frac{1}{t^*} = k_1(a_e + b_e) + k_{-1} = 2k_1 a_e + k_{-1} \quad \text{if } a_e = b_e.$$

e.   Now, for the equilibrium constant, we have $K = \dfrac{k_1}{k_{-1}} = \dfrac{z_e}{a_e b_e} = \dfrac{z_e}{a_e^2}$ for $a_e = b_e$. Therefore, $a_e = \sqrt{k_{-1} z_e / k_1}$. Substituting this result into the result of part (d) gives the desired result:

$$\frac{1}{t^*} = 2\sqrt{k_1 k_{-1} z_e} + k_{-1}.$$

9.36.   Based on the result of the previous problem, we plot $\sqrt{z_e}$ against $\dfrac{1}{t^*}$. This leads to a straight line with the equation (from linear regression) $1/(t^*/s) = 79.3 + 4.52 \times 10^3 \sqrt{z_e}$, from which we obtain $k_{-1} = 79.3 \ s^{-1}$, $k_1 = 28.5 \ M^{-1} \ s^{-1}$, and $K = 0.360 \ M^{-1}$.

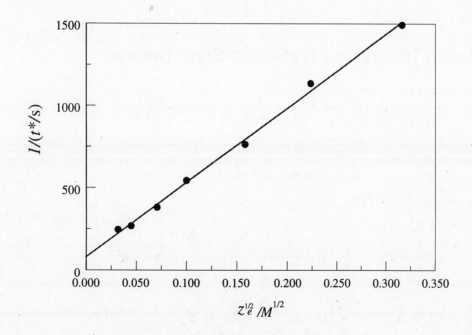

9.37. Since $-d[H_2O]/dt = k_1[H_2O] - k_{-1}[H^+][OH^-]$, from the equilibrium condition, we obtain

$$\frac{k_1}{k_{-1}} = \frac{[H^+]_e[OH^-]_e}{[H_2O]_e} = K = \frac{K_w}{[H_2O]_e} = 1.80 \times 10^{-16}\ M,$$

where we have used the fact that for 1.0 dm^3 of pure water, $[H_2O]_e \approx$ 1000 g/18.016 g mol^{-1} = 55.51 M. Now, since $k_1 = K\,k_{-1}$, the equation relating the relaxation time to concentration can be written as

$$\frac{1}{t^*} = k_{-1}[K + ([H^+]_e + [OH^-]_e)].$$

Using the fact that $[H^+]_e = [OH^-]_e = \sqrt{K_w}$, we get

$$\frac{1}{t^*} = k_{-1}\left(K + 2\sqrt{K_w}\right)$$

Therefore,

$$k_{-1} = (1/t^*)\left(K + 2\sqrt{K_w}\right)^{-1} = 1.25 \times 10^{12}\ (M\ s)^{-1},$$

and

$$k_1 = K\,k_{-1} = 2.25 \times 10^{-4}\ s^{-1}.$$

■ Collision Theory and Transition-State Theory

9.38.   Rate constant ratio = $e^{50/8.3145}$ = 409

9.39. 2.34×10^{-2} dm^3 mol^{-1} s^{-1} $= A\ e^{-150\ 000/8.3145 \times 673}$

$$= A \times 2.291 \times 10^{-12}$$

$$A = 1.02 \times 10^{10}\ dm^3\ mol^{-1}\ s^{-1}$$

For a second-order reaction

$$\Delta^{\ddagger}H^{\circ} = E_a - 2RT \tag{Eq. 9.100}$$

$$= 150\ 000 - 2 \times 8.3145 \times 673.15\ J\ mol^{-1}$$

$$= 138\ 806\ J\ mol^{-1} = 138.8\ kJ\ mol^{-1}$$

$$A = e^2 \frac{k_BT}{h}\ e^{\Delta^{\ddagger}S^{\circ}/R} \tag{Eq. 9.101}$$

$$1.02 \times 10^{10} = \frac{2.718^2 \times 1.381 \times 10^{-23} \times 673.15}{6.626 \times 10^{-34}}\ e^{\Delta^{\ddagger}S^{\circ}/R}$$

$$= 1.0365 \times 10^{14}\ e^{\Delta^{\ddagger}S^{\circ}/R}$$

$$9.84 \times 10^{-5} = e^{\Delta^{\ddagger}S^{\circ}/8.3145 \text{ J K}^{-1} \text{ mol}^{-1}}$$

$$\frac{\Delta^{\ddagger}S^{\circ}}{8.3145 \text{ J K}^{-1} \text{ mol}^{-1}} = -9.226$$

$$\Delta^{\ddagger}S^{\circ} = -76.71 \text{ J K}^{-1} \text{ mol}^{-1}$$

$$\Delta^{\ddagger}G^{\circ} = \Delta^{\ddagger}H^{\circ} - T\Delta^{\ddagger}S^{\circ} = 138\ 806 + 76.71 \times 673.15$$

$$= 190\ 400 \text{ J mol}^{-1} = 190.4 \text{ kJ mol}^{-1}$$

9.40. From the data:

$T/^{\circ}C$	$10^3/(T/\text{K})$	$\log_{10}(k/s^{-1})$
15.2	3.468	−5.380
20.0	3.4112	−5.118
25.0	3.3540	−4.863
30.0	3.2987	−4.618
37.0	3.2243	−4.288

A graph of $\log_{10}k$ against $10^3/T$ gives a slope equal to -4.393×10^3 K

$$E = 84\ 082 \text{ J mol}^{-1} = 84.1 \text{ kJ mol}^{-1}$$

$$\Delta^{\ddagger}H^{\circ} = 84\ 082 - 2479 = 81\ 603 \text{ J mol}^{-1}$$

$$= 81.6 \text{ kJ mol}^{-1}$$

$$\log_{10}k = \log_{10}\frac{k_B T}{h} - \frac{\Delta^{\ddagger}G^{\circ}}{19.14T}$$

At 25.0°C, $k_B T/h = 6.21 \times 10^{12}$ s^{-1}, and $\log_{10}(k/s^{-1}) = -4.86$.

Then $-4.86 = 12.79 - \dfrac{\Delta^{\ddagger}G^{\circ}}{5706.6 \text{ J mol}^{-1}}$

and therefore

$$\Delta^{\ddagger}G^{\circ} = 100\ 721 \text{ J mol}^{-1} = 100.7 \text{ kJ mol}^{-1}$$

The preexponential factor can be calculated from

At 298.15 K,

$$-4.86 = \log_{10}(A/s^{-1}) - \frac{84\ 082}{5706.6} \text{ or}$$

$$\log_{10}(A/s^{-1}) = 9.87$$

$$A = 7.48 \times 10^9 \text{ s}^{-1}.$$

$$\Delta^\ddagger S^\circ = \frac{81\ 603 - 100\ 721}{298.15} = -64.1 \text{ J K}^{-1} \text{ mol}^{-1}.$$

9.41.

T/K	k/s^{-1}	$10^3 (T/K)^{-1}$	$\log_{10}(k/s^{-1})$
313.05	4.67×10^{-6}	3.194	−5.331
316.95	7.22×10^{-6}	3.155	−5.141
320.25	10.0×10^{-6}	3.123	−5.000
323.35	13.9×10^{-6}	3.093	−4.857

Slope of the plot of $\log_{10}k$ against $1/T$ is -4.70×10^3. Therefore,

$$E = 89\ 958 \text{ J mol}^{-1} = 90.0 \text{ kJ mol}^{-1}.$$

$$\Delta^\ddagger H^\circ = 89\ 958 - 2479 = 87\ 479 \text{ J mol}^{-1}$$

$$= 87.5 \text{ kJ mol}^{-1}.$$

$$\log_{10}k = \log_{10}\frac{k_B T}{h} - \frac{\Delta^\ddagger G^\circ}{(19.14 \text{ J K}^{-1} \text{ mol}^{-1})\, T}$$

At $40.0\ °C = 313.15$ K, $\dfrac{k_B T}{h} = 6.527 \times 10^{12} \text{ s}^{-1}$

$$\log_{10}\left[\frac{k_B T}{h}/s^{-1}\right] = 12.81$$

From the graph, $\log_{10}(k/s^{-1})$ at $40.0° = -5.335$.

$$-5.335 = 12.81 - \frac{\Delta^\ddagger G^\circ/\text{J mol}^{-1}}{19.14 \times 313.15}$$

$$\Delta^\ddagger G^\circ = 108\ 783 \text{ J mol}^{-1} = 108.8 \text{ kJ mol}^{-1}$$

$$\Delta^\ddagger S^\circ = \frac{87\ 479 - 108\ 783}{313.15} = -68.03 \text{ J K}^{-1} \text{ mol}^{-1}$$

$$A = e \frac{k_B T}{h}\ e^{\,\Delta^\ddagger S^\circ/R}$$

$$\log_{10}(A/s^{-1}) = \log_{10}e + 12.81 - \frac{68.03}{19.14}$$

$$= 0.434 + 12.81 - 3.55 = 9.67$$

$$A = 4.94 \times 10^9 \text{ s}^{-1}$$

9.42. $\ln \dfrac{3460}{530} = 1.876$

$T_1 = 333.15$ K $1/T_1 = 3.001\ 65 \times 10^{-3}$ K^{-1}

$T_2 = 338.15$ K $1/T_2 = 2.957\ 27 \times 10^{-3}$ K^{-1}

Slope of the plot of $\ln(k_1/k_2)$ against $1/T$

$$= -\frac{1.876}{4.438 \times 10^{-5} \text{ K}^{-1}} = -42\ 270 \text{ K}$$

$E = 351\ 403$ J mol^{-1} = 351.4 kJ mol^{-1}

$\Delta^{\ddagger}H^{\circ} = 351\ 403 - 2770 = 348\ 633$ J mol^{-1}

$\qquad = 348.6$ kJ mol^{-1}

At 60.0°C, $k = (\ln 2)/3460$ s $= 2.003 \times 10^{-4}$ s^{-1}

$$\frac{k_{\text{B}}T}{h} = 6.94 \times 10^{12} \text{ s}^{-1}$$

$2.003 \times 10^{-4} = 6.94 \times 10^{12} e^{-\Delta^{\ddagger}G/(8.3145 \times 333.15)}$

$\Delta^{\ddagger}G^{\circ} = 105\ 470$ J mol^{-1} = 105.5 kJ mol^{-1}

$\Delta^{\ddagger}S^{\circ} = \dfrac{\Delta^{\ddagger}H^{\circ} - \Delta^{\ddagger}G^{\circ}}{T} = \dfrac{348\ 630 - 105\ 500}{333.15}$ J K^{-1} mol^{-1}

$\qquad = 730$ J K^{-1} mol^{-1}

9.43.   a.   Mass of HI molecule $= \dfrac{127.91 \times 10^{-3} \text{ kg mol}^{-1}}{6.022 \times 10^{23} \text{ mol}^{-1}}$

$\qquad = 2.124 \times 10^{-25}$ kg

Then, from Eq. 9.73,

$Z_{\text{AA}} = 2 \times (0.35 \times 10^{-9} \text{ m})^2 \times (6.022 \times 10^{23} \text{ m}^{-3})^2 \times$

$$\left(\frac{\pi \times 1.381 \times 10^{-23} \text{ J K}^{-1} \times 573.15 \text{ K}}{2.124 \times 10^{-25} \text{ kg}} \right)^{1/2}$$

$\qquad = 3.040 \times 10^{31}$ m^{-3} s^{-1}

b.   Rate of reaction under these conditions with $E = 184$ kJ mol^{-1} is

$3.040 \times 10^{31} e^{-184\ 000/8.3145 \times 573.15}$ m^{-3} s^{-1}

$\qquad = 5.179 \times 10^{14}$ m^{-3} s^{-1} = 8.68×10^{-10} mol m^{-3} s^{-1}

The rate constant is this rate divided by the square of the concentration, which is 1 mol m^{-3}.

$$k = \frac{8.60 \times 10^{-10} \text{ mol m}^{-3} \text{ s}^{-1}}{(1 \text{ mol m}^{-3})^2}$$

$$= 8.60 \times 10^{-10} \text{ m}^3 \text{ mol}^{-1} \text{ s}^{-1}$$

$$= 8.60 \times 10^{-7} \text{ dm}^3 \text{ mol}^{-1} \text{ s}^{-1}$$

The collision frequency factor (the preexponential factor) is k divided by

$e^{-184\,000/8.3145 \times 573.15}$ and is $5.048 \times 10^{-10} \text{ dm}^3 \text{ mol}^{-1} \text{ s}^{-1}$

To obtain the entropy of activation we use the relationship

$$A = e^2 \frac{k_B T}{h} e^{\Delta^{\ddagger}S^{\circ}/R} \qquad\qquad \text{(Eq. 9.101)}$$

With $A = 5.048 \times 10^{10} \text{ dm}^3 \text{ mol}^{-1} \text{ s}^{-1}$ this gives

$$5.048 \times 10^{10} = e^2 \frac{k_B T}{h} e^{\Delta^{\ddagger}S^{\circ}/R}$$

$$= \frac{2.718^2 \times 1.381 \times 10^{-23} \times 573.15}{6.626 \times 10^{-34}} e^{\Delta^{\ddagger}S/R}$$

$$e^{\Delta^{\ddagger}S^{\circ}/R} = 5.720 \times 10^{-4}$$

$$\frac{\Delta^{\ddagger}S^{\circ}}{R} = -7.47$$

$$\Delta^{\ddagger}S^{\circ} = -62.1 \text{ J K}^{-1} \text{ mol}^{-1} \text{ (standard state: 1 mol dm}^{-3})$$

9.44.
$$k = (k_B T/h)e^{-\Delta^{\ddagger}G^{\circ}/RT}$$

$$7.40 \times 10^{-9} = 6.213 \times 10^{12} e^{-\Delta^{\ddagger}G^{\circ}/RT}$$

$$= 6.213 \times 10^{12} e^{-\Delta^{\ddagger}G^{\circ}/(8.3145 \times 298.15)}$$

whence $\quad \Delta^{\ddagger}G^{\circ} = 119\,436 \text{ J mol}^{-1} = 119.4 \text{ kJ mol}^{-1}$

For a first-order reaction $E_a = \Delta^{\ddagger}H^{\circ} + RT$

$$\Delta^{\ddagger}H^{\circ} = 112\,000 - 8.3145 \times 298.15$$

$$= 109\,521 \text{ J mol}^{-1} = 109.5 \text{ kJ mol}^{-1}$$

$$\Delta^{\ddagger}S^{\circ} = \frac{\Delta^{\ddagger}H^{\circ} - \Delta^{\ddagger}G^{\circ}}{T}$$

$$= (109\,521 - 119\,436)/298.15$$

$$= -33.3 \text{ J K}^{-1} \text{ mol}^{-1}$$

$$A = e \frac{k_B T}{h} e^{\Delta^{\ddagger}S^{\circ}/R}$$

$$= 3.08 \times 10^{11} \text{ s}^{-1}$$

(*A* could alternatively be calculated using the relationship $k = A\,e^{-E_a/RT}$. $\Delta^{\ddagger}S^{\circ}$ could then be calculated from *A*, and $\Delta^{\ddagger}G$ from $\Delta^{\ddagger}S^{\circ}$ and $\Delta^{\ddagger}H^{\circ}$.)

9.45.
$$k = \frac{k_B T}{h}\,e^{-\Delta^{\ddagger}G^{\circ}/RT}$$

$$3.95 \times 10^{-4} = 6.214 \times 10^{12}\,e^{-\Delta^{\ddagger}G^{\circ}/RT}$$

whence $\quad -\dfrac{\Delta^{\ddagger}G^{\circ}}{RT} = \ln 6.357 \times 10^{-17}$

$$\Delta^{\ddagger}G^{\circ} = 37.29 \times 8.3145 \times 298.15 = 92.44$$

$$= 92.4 \text{ kJ mol}^{-1}$$

$$E_a = \Delta^{\ddagger}H^{\circ} + RT$$

$$\Delta^{\ddagger}H^{\circ} = 120\,000 - 8.3145 \times 298.15$$

$$= 117\,520 \text{ J mol}^{-1} = 117.5 \text{ kJ mol}^{-1}$$

$$\Delta^{\ddagger}S^{\circ} = \frac{\Delta^{\ddagger}H^{\circ} - \Delta^{\ddagger}G^{\circ}}{T} = \frac{117\,520 - 92\,440}{298.15}$$

$$= 84.1 \text{ J K}^{-1} \text{ mol}^{-1}$$

$$k = A\,e^{-E_a/RT}$$

$$3.94 \times 10^{-4} = A\,e^{-120\,000/8.3145 \times 298.15}$$

$$A = \frac{3.94 \times 10^{-4}}{e^{-48.407}} = \frac{3.94 \times 10^{-4}}{9.484 \times 10^{-22}}$$

whence $\quad A = 4.15 \times 10^{17} \text{ dm}^3 \text{ mol}^{-1} \text{ s}^{-1}$

■ Ionic-Strength Effects

9.46.

$\dfrac{I}{10^{-3}\ \text{mol dm}^{-3}}$	$\dfrac{k}{\text{dm}^3\ \text{mol}^{-1}\ \text{s}^{-1}}$	$\sqrt{\dfrac{I}{\text{mol dm}^{-1}}}$	$\log_{10}\!\left(\dfrac{k}{\text{dm}^3\ \text{mol}^{-1}\ \text{s}^{-1}}\right)$
2.45	1.05	0.0495	0.0212
3.65	1.12	0.0604	0.0492
4.45	1.16	0.0667	0.0645
6.45	1.18	0.0815	0.0719
8.45	1.26	0.0941	0.1004
12.45	1.39	0.1116	0.1430

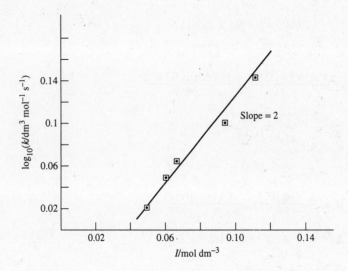

Slope of plot of $\log_{10}(k/\text{dm}^3\ \text{mol}^{-1}\ \text{s}^{-1})$ against $(\sqrt{I/\text{mol dm}^{-3}})$ is approximately 2:

$$z_A z_B = 2$$

9.47. Ionic strengths of the five mixtures are:

(1) $\dfrac{1}{2}\,[4 \times 5.0 \times 10^{-4} + 1.0 \times 10^{-3} + (2 \times 7.95 \times 10^{-4})] = 2.295 \times 10^{-3}\ M$

(2) $\dfrac{1}{2}\,[4 \times 5.96 \times 10^{-4} + 11.92 \times 10^{-4} + (2 \times 1.004 \times 10^{-3})] = 2.79 \times 10^{-3}\ M$

(3) $\frac{1}{2}[4 \times 6.0 \times 10^{-4} + 12.0 \times 10^{-4} + (2 \times 0.696 \times 10^{-3}) + 0.01] = 7.496 \times 10^{-3}\ M$

(4) $\frac{1}{2}(24.0 \times 10^{-4} + 12.0 \times 10^{-4} + 1.392 \times 10^{-3} + 0.04) = 22.50 \times 10^{-3}\ M$

(5) $\frac{1}{2}(24.0 \times 10^{-4} + 12.0 \times 10^{-4} + 1.392 \times 10^{-3} + 0.06) = 32.5 \times 10^{-3}\ M$

$\dfrac{I}{10^{-3}\ \text{mol cm}^{-3}}$	$\dfrac{k}{\text{dm}^3\ \text{mol}^{-1}\ \text{s}^{-1}}$	$\sqrt{\dfrac{I}{\text{mol dm}^{-3}}}$	$\log_{10}\!\left(\dfrac{k}{\text{dm}^3\ \text{mol}^{-1}\ \text{s}^{-1}}\right)$
2.295	1.52	0.0479	0.1818
2.790	1.45	0.0528	0.1614
7.496	1.23	0.0866	0.0899
22.50	0.97	0.1500	−0.0132
32.5	0.91	0.1803	−0.0409

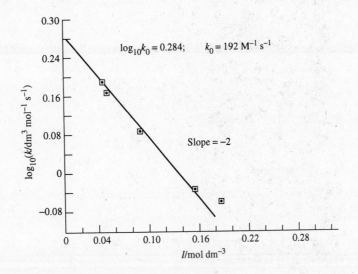

Plot of $\log_{10}(k/\text{dm}^3\ \text{mol}^{-1}\ \text{s}^{-1})$ against $\sqrt{I/(\text{mol dm}^{-3})}$ extrapolates to

$\log_{10}(k/\text{dm}^3\ \text{mol}^{-1}\ \text{s}^{-1}) = 0.284$

$k_0 = 1.92\ \text{dm}^3\ \text{mol}^{-1}\ \text{s}^{-1}$

The slope is −2, which is consistent with $z_A z_B = -2$

9.48. $$K^{\ddagger} = \frac{[X^{\ddagger}]}{[A][B]} \frac{\gamma^{\ddagger}}{\gamma_A \gamma_B}$$

Suppose that the rate $v = k [X^{\ddagger}] \gamma^{\ddagger}$; then

$$v = k^{\ddagger} k [A][B] \gamma_A \gamma_B$$

and

$$k = K^{\ddagger} k \gamma_A \gamma_B = k_0 \gamma_A \gamma_B$$

$$\log_{10}k = \log_{10}k_0 + \log_{10}\gamma_A\gamma_B$$

$$= \log_{10}k_0 - B[z_A^2 + z_B^2] \sqrt{I}$$

Plots of $\log_{10}k$ against $\sqrt{I}$ will always have negative slopes; this conclusion is inconsistent with the results in Fig. 9.22.

9.49. If there were a single negative charge on the muonium, $z_A z_B$ would be –2, and according to Eq. 9.124,

$$\log_{10}(k/k_0) = 2 B z_A z_B \sqrt{I}$$

$$= 2 \times 0.51 \times (-2) \times \sqrt{0.9}$$

or

$$\log_{10} (k/6.50 \times 10^9) = -1.935$$

whence

$$k = 7.54 \times 10^7 \text{ dm}^3 \text{ mol}^{-1} \text{ s}^{-1}$$

Since the reduction in rate constant was much less than this, we conclude that the charge on muonium is zero.

9.50. The slope of a plot of $\log_{10}k$ against $\sqrt{I}$ is 2.04. According to Eq. 9.125, this slope is equal to 1.02 $z_A z_B$. The value of $z_A z_B$ is therefore 2.

9.51. $$\log_{10}k = \log_{10}k_0 + 2z_A z_B I$$

$$-3.553 = \log_{10}k_0 - 4 \times 0.51 \sqrt{1.0 \times 10^{-3}}$$

$$\log_{10}k_0 = -3.553 + 0.0645 = -3.488$$

$$k_0 = 3.25 \times 10^{-4} \text{ dm}^3 \text{ mol}^{-1} \text{ s}^{-1}$$

9.52. $\ln \dfrac{k}{k_0} = \dfrac{\Delta^{\ddagger}V}{RT}P$ (Eq. 9.129)

1999 bar $= 1.999 \times 10^8$ Pa

$\ln 2 = \dfrac{(\Delta^{\ddagger}V/m^3) \times 1.999 \times 10^8}{8.3145 \times 573.15 \text{ mol}^{-1}}$

       $= -41\ 947 \times \Delta^{\ddagger}V/m^3 \text{ mol}^{-1})$

$\Delta^{\ddagger}V = -1.652 \times 10^{-5} \text{ m}^3 \text{ mol}^{-1} = -16.5 \text{ cm}^3 \text{ mol}^{-1}$

9.53.

Pressure/10^2 kPa	1.00	345	689	1033
$k/10^{-6}$ s^{-1}	7.18	9.58	12.2	15.8
$\ln(k/\text{s}^{-1})$	−11.84	−11.56	−11.31	−11.06

From plot of $\ln k$ against pressure, slope $= 7.55 \times 10^{-9}$ Pa^{-1}

$7.55 \times 10^{-9} = \dfrac{\Delta^{\ddagger}V/m^3 \text{ mol}^{-1}}{8.3145 \times 298.15}$

$\Delta^{\ddagger}V = -1.87 \times 10^{-5} \text{ m}^3 \text{ mol}^{-1}$

     $= -18.7 \text{ cm}^3 \text{ mol}^{-1}$

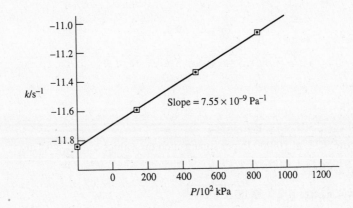

9.54.

P/kPa	$k/10^4 \text{ dm}^3 \text{ mol}^{-1} \text{ s}^{-1}$	$\ln(k/10^4 \text{ dm}^3 \text{ mol}^{-1} \text{ s}^{-1})$
101.3	9.30	2.23
2.76×10^4	11.13	2.41
5.51×10^4	13.1	2.57
8.27×10^4	15.3	2.73
11.02×10^4	17.9	2.88

Slope of plot of $\ln k$ against pressure = $5.75 \times 10^{-9} \text{ Pa}^{-1}$

$$5.75 \times 10^{-9} = -\frac{\Delta^{\ddagger}V/\text{m}^3 \text{ mol}^{-1}}{8.3145 \times 298.15}$$

$$\Delta^{\ddagger}V = -1.43 \times 10^{-5} \text{ m}^3 \text{ mol}^{-1}$$

$$= -14.3 \text{ cm}^3 \text{ mol}^{-1}$$

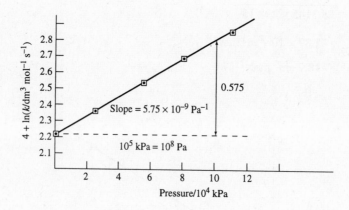

9.55. From Figure 9.23, at 10 000 lb per square inch the values of $\log_{10} k/k_0$ are

Ethyl acetate	0.105
Methyl acetate	0.11
Propionamide	0.20

The slope of a plot of $\ln k/k_0$ against P is therefore

for ethyl acetate,

$$\frac{0.125 \times 2.303}{10\ 000 \text{ psi} \times 6.89 \times 10^3 \text{ Pa/psi}} = 4.18 \times 10^{-9} \text{ Pa}^{-1}$$

The slope is $\Delta^{\ddagger}V^{\circ}/RT$ and therefore

$$\Delta^{\ddagger}V^{\circ} = 8.3145 \times 298.15 \text{ J mol}^{-1} \times 4.18 \times 10^{-9} \text{ Pa}^{-1}$$

$$= 1.04 \times 10^{-5} \text{ m}^3 \text{ mol}^{-1}$$

$$= 10.4 \text{ cm}^3 \text{ mol}^{-1}$$

Similarly for methyl acetate the slope is

$4.4 \times 10^{-9} \text{ Pa}^{-1}$, and thus

$$\Delta^{\ddagger}V^{\circ} = 8.3145 \times 298.15 \text{ J mol}^{-1} \times 4.4 \times 10^{-9} \text{ Pa}^{-1}$$

$$= 1.09 \times 10^{-5} \text{ m}^3 \text{ mol}^{-1}$$

$$= 10.9 \text{ cm}^3 \text{ mol}^{-1}$$

For propionamide the slope is

$8.0 \times 10^{-9} \text{ Pa}^{-1}$, and thus

$$\Delta^{\ddagger}V^{\circ} = 8.3145 \times 298.15 \text{ J mol}^{-1} \times 8.0 \times 10^{-9} \text{ Pa}^{-1}$$

$$= 2.00 \times 10^{-5} \text{ m}^3 \text{ mol}^{-1}$$

$$= 20.0 \text{ cm}^3 \text{ mol}^{-1}$$

10. CHEMICAL KINETICS II. COMPOSITE MECHANISMS

■ Composite Mechanisms and Rate Equations

10.1. $v = k_1[A][B]$

10.2. $v = k_2(k_1/k_{-1})^{1/2}[A]^{1/2}[B]$

10.3. $v = \dfrac{k_1 k_2[A][B]}{k_{-1} + k_2[B]}$

      a. If $k_2[B]$ is small compared with k_{-1}, it may be dropped in the denominator.

$$v = \frac{k_1 k_2}{k_{-1}}[A][B]$$

      b. If $k_2[B]$ is very large compared with k_{-1}, k_{-1} may be dropped in the denominator.

$$v = k_1[A]$$

10.4. $2A \rightarrow X$ (very slow)

        $X + 2B \rightarrow 2Y + 2Z$ (very fast)

10.5.   Two simultaneous reactions

10.6.   Two consecutive reactions

10.7.   The steady-state equation for NO_3 is

$$k_1[N_2O_5] - k_{-1}[NO_2][NO_3] - k_2[NO][NO_3] = 0$$

Thus

$$[NO_3] = \frac{k_1[N_2O_5]}{k_{-1}[NO_2] + k_2[NO]}$$

The rate of consumption of N_2O_5 is

$$v_{N_2O_5} = k_1[N_2O_5] - k_{-1}[NO_2][NO_3]$$

$$= k_1[N_2O_5] - \frac{k_1 k_{-1}[N_2O_5][NO_2]}{k_{-1}[NO_2] + k_2[NO]}$$

$$= \frac{k_1 k_2[N_2O_5][NO]}{k_{-1}[NO_2] + k_2[NO]}$$

Alternatively, we can note that $v_{N_2O_5} = v_{NO}$ from the stoichiometric equation, and the latter is $k_2[NO][NO_3]$.

10.8.   The steady-state equation is

$$k_1[NO]^2 - k_{-1}[N_2O_2] - k_2[N_2O_2][O_2] = 0$$

and therefore

$$[N_2O_2] = \frac{k_1[NO]^2}{k_{-1} + k_2[O_2]}$$

The rate is

$$v = v_{O_2} = k_2[N_2O_2][O_2] = \frac{k_1 k_2[NO]^2[O_2]}{k_{-1} + k_2[O_2]}$$

This reduces to

$$v = \frac{k_1 k_2}{k_{-1}} [NO]^2[O_2]$$

if $k_{-1} \gg k_2[O_2]$

10.9.   The mechanism is

$$Cl_2 \xrightarrow{k_1} 2Cl \qquad\qquad \text{initiation}$$

$$Cl + CH_4 \xrightarrow{k_2} HCl + CH_3$$

$$CH_3 + Cl_2 \xrightarrow{k_3} CH_3Cl + Cl \left.\right\} \text{chain propagation}$$

$$2Cl \xrightarrow{k_{-1}} Cl_2$$

The steady-state equations are

For Cl: $k_1[Cl_2] - k_2[Cl][CH_4] + k_3[CH_3][Cl_2] - k_{-1}[Cl_2]^2 = 0$

For CH$_3$: $k_2[Cl][CH_4] - k_3[CH_3][Cl_2] = 0$

Adding: $k_1[Cl_2] - k_{-1}[Cl]^2 = 0$

Therefore $[Cl] = \left(\dfrac{k_1}{k_{-1}}\right)^{1/2}[Cl_2]^{1/2}$

The rate of reaction is the rate of formation of HCl:

$$v = v_{HCl} = k_2[Cl][CH_4] = k_2\left(\frac{k_1}{k_{-1}}\right)^{1/2}[Cl_2]^{1/2}[CH_4]$$

10.10. The steady-state equation for O is

$$k_1[O_3]^2 - k_{-1}[O_3][O_2][O] - k_2[O][O_3] = 0$$

$$[O] = \frac{k_1[O_3]^2}{k_{-1}[O_3][O_2] + k_2[O_3]} = \frac{k_1[O_3]}{k_{-1}[O_2] + k_2}$$

$$v_{O_2} = k_1[O_3]^2 - k_{-1}[O_3][O_2][O] + 2k_2[O][O_3]$$

Subtraction of the steady-state equation gives

$$v_{O_2} = 3k_2[O][O_3]$$

$$= \frac{3k_1k_2[O_3]^2}{k_{-1}[O_2] + k_2}$$

The overall reaction is $2O_3 \rightarrow 3O_2$, and therefore

$$3v_{O_3} = 2v_{O_2}$$

Thus

$$v_{O_3} = \frac{2k_1k_2[O_3]^2}{k_{-1}[O_2] + k_2}$$

$$= 2k_1[O_3]^2 \text{ in the absence of } O_2$$

10.11. The steady-state equation is

$$k_1[A][B] - (k_{-1} + k_2)[X] = 0$$

and the rate is

$$v = \frac{k_1 k_2}{k_{-1} + k_2} [A][B]$$

and the rate constant is $k = k_1 k_2 / (k_{-1} + k_2)$.

$$\ln k = \ln k_1 + \ln k_2 - \ln (k_{-1} + k_2)$$

$$\frac{d \ln k}{dT} = \frac{d \ln k_1}{dT} + \frac{1}{k_2} \frac{dk_2}{dT} - \frac{1}{k_{-1} + k_2} \frac{d(k_{-1} + k_2)}{dT}$$

$$= \frac{d \ln k_1}{dT} + \frac{k_{-1}}{k_{-1} + k_2} \frac{d \ln k_2}{dT} - \frac{k_{-1}}{k_{-1} + k_2} \frac{d \ln k_1}{dT}$$

$$E \equiv RT^2 \frac{d \ln k}{dT}; \quad E_1 \equiv RT^2 \frac{d \ln k_1}{dT};$$

$$E_{-1} \equiv RT^2 \frac{d \ln k_{-1}}{dT}; \quad E_2 \equiv RT^2 \frac{d \ln k_2}{dT}$$

$$E = E_1 + \frac{k_{-1}}{k_{-1} + k_2} E_2 - \frac{k_{-1}}{k_{-1} + k_2} E_{-1}$$

$$= \frac{k_2 E_1 + k_{-1}(E_1 + E_2 - E_{-1})}{k_{-1} + k_2}$$

10.12. The steady-state equation is

$$k_1[A]^2 - k_{-1}[A][A^*] - k_2[A^*] = 0$$

Therefore

$$[A^*] = \frac{k_1[A]^2}{k_{-1}[A] + k_2}$$

The rate is

$$v = k_2[A^*] = \frac{k_1 k_2 [A]^2}{k_{-1}[A] + k_2}$$

At high pressures $k_{-1}[A] \gg k_2$ and thus

$$v = \frac{k_1 k_2}{k_{-1}} [A] \quad \text{first order}$$

At low pressures $k_{-1}[A] \ll k_2$ and thus

$$v = k_1[A]^2 \quad \text{second order}$$

10.13. With the boundary condition $t = 0$, $c = c_0$, the differential equation integrates as follows:

$$-\int \frac{dc}{c^2} = kdt$$

$$\frac{1}{c} = kt + I \quad \text{and} \quad I = \frac{1}{c_0}$$

Thus

$$\frac{1}{c} - \frac{1}{c_0} = kt = \frac{c_0 - c}{c\, c_0}$$

The fraction of functional groups remaining is

$$f = \frac{c_0 - c}{c_0}$$

Elimination of c gives

$$\frac{f}{1-f} = c_0 kt \quad \text{or} \quad f = \frac{c_0 kt}{1 + c_0 kt}$$

10.14. Since the first two reactions are fast we may write down the equilibrium equations for them:

$$[I]^2/[I_2] = k_1/k_{-1} \quad \text{so that} \quad [I] = (k_1/k_{-1})^{1/2}[I_2]^{1/2}$$

$$[H_2I]/[I][H_2] = k_2/k_{-2} \quad \text{so that}$$

$$[H_2I] = (k_2/k_{-2})[H_2][I] = (k_2/k_{-2})(k_1/k_{-1})^{1/2}[H_2][I_2]^{1/2}$$

The overall rate is the rate of the third (slow) reaction:

$$v = k_3[H_2I][I] = k_3(k_1k_2/k_{-1}k_{-2})[H_2][I_2]$$

$$= \frac{k_1k_2k_3}{k_{-1}k_{-2}}[H_2][I_2]$$

10.15. The steady-state equation for X is

$$k_1[A][B] - [X](k_{-1} + k_2[A] + k_3[B]) = 0$$

and therefore

$$[X] = k_1[A][B]/(k_{-1} + k_2[A] + k_3[B])$$

The rate of formation of Z is thus

$$v_Z = k_3[X][B] = k_1k_3[A][B]^2/(k_{-1} + k_2[A] + k_3[B])$$

   a. If A is in excess such that $k_2[A] \gg k_{-1} + k_3[B]$,

$$v_Z = (k_1k_3/k_2)[B]^2$$

b. If B is in excess such that $k_3[B] \gg k_{-1} + k_2[A]$,

$$v_z = k_1[A][B]$$

(In this case, since the form is the same as that derived from the first reaction, the first reaction is rate controlling.)

■ Photochemistry and Radiation Chemistry

10.16. The dissociation energy per molecule is

$$390.4 \times 10^3 \text{ J mol}^{-1}/6.022 \times 10^{23} \text{ mol}^{-1}$$

$$= 6.48 \times 10^{-19} \text{ J}$$

This corresponds to a frequency of

$$6.48 \times 10^{-19} \text{ J}/6.626 \times 10^{-34} \text{ J s} = 9.78 \times 10^{14} \text{ s}^{-1}$$

and to a wavelength of

$$\lambda = \frac{2.998 \times 10^8 \text{ m s}^{-1}}{9.78 \times 10^{14} \text{ s}^{-1}} = 3.06 \times 10^{-7} \text{ m}$$

$$= 306 \text{ nm}$$

This is the maximum wavelength that will cause dissociation.

10.17. 440 μg of HI is $\dfrac{440 \times 10^{-6} \text{ g}}{127.9 \text{ g mol}} = 3.44 \times 10^{-6} \text{ mol}$

$$= 2.07 \times 10^{18} \text{ molecules}$$

207 nm corresponds to a frequency of

$$\frac{2.998 \times 10^8 \text{ m s}^{-1}}{207 \times 10^{-9} \text{ m}} = 1.448 \times 10^{15} \text{ s}^{-1}$$

and to an energy of

$$6.626 \times 10^{-34} \text{ J s} \times 1.448 \times 10^{15} \text{ s}^{-1} = 9.60 \times 10^{-19} \text{ J}$$

Therefore 1 J of radiation corresponds to

$$1/9.60 \times 10^{-19} = 1.04 \times 10^{18} \text{ photons}$$

One photon therefore decomposes 2.07/1.04

$$= 2 \text{ molecules}$$

Possible mechanism:

$$HI + h\nu \rightarrow HI^*$$

$$HI^* + HI \rightarrow H_2 + I_2$$

or

$$HI + h\nu \rightarrow H + I$$

(This is a quantum yield of 2. In other words, 2 molecules of reactant are decomposed for each photon absorbed.)

Then

$$H + HI \rightarrow H_2 + I$$

$$I + I \rightarrow I_2$$

The net reaction is

$$2HI + h\nu \rightarrow H_2 + I_2$$

10.18. The frequency of the radiation is

$$2.998 \times 10^8 \text{ m s}^{-1}/253.7 \times 10^{-9} \text{ m} = 1.18 \times 10^{15} \text{ s}^{-1}$$

and its energy is

$$6.626 \times 10^{-34} \text{ J s} \times 1.18 \times 10^{15} \text{ s}^{-1} = 7.83 \times 10^{-19} \text{ J}$$

A 100-W lamp emits 100 J per second and therefore

$$\frac{100 \text{ J s}^{-1}}{7.83 \times 10^{-19} \text{ J}} = 1.28 \times 10^{20} \text{ photons per second}$$

This is

$$\frac{1.28 \times 10^{20} \text{ s}^{-1}}{6.022 \times 10^{23} \text{ mol}^{-1}} = 2.12 \times 10^{-4} \text{ moles of photons per second}$$

To decompose 0.01 mol requires

$$0.01/2.12 \times 10^{-4} = 47 \text{ seconds}$$

10.19. The lamp emits 2.12×10^{-4} moles of photons per second (see previous problem). In one hour it therefore emits

$$2.12 \times 10^{-4} \text{ mol s}^{-1} \times 3600 \text{ s} = 0.76 \text{ mol}$$

This is the amount of ethyne produced.

10.20. The energy of a photon of radiation of wavelength 253.7 nm is 7.83×10^{-19} J. (See solution to Problem 10.18.) A 1000-W lamp emits 1000 J per second, and therefore it emits

$$1000 \text{ J s}^{-1}/7.83 \times 10^{-19} \text{ J} = 1.277 \times 10^{21} \text{ photons s}^{-1}$$

In one microsecond it emits 1.277×10^{15} photons, and this is the number of excited mercury atoms formed.

10.21. The steady-state equations are

For Cl: $2I_a - k_2[\text{Cl}][\text{CHCl}_3] + k_3[\text{CCl}_3][\text{Cl}_2] - k_4[\text{Cl}]^2 = 0$

For CCl$_3$: $k_2[\text{Cl}][\text{CHCl}_3] - k_3[\text{CCl}_3][\text{Cl}_2] = 0$

Adding: $2I_a - k_4[\text{Cl}]^2 = 0$

$$[\text{Cl}] = \left(\frac{2I_a}{k_4}\right)^{1/2}$$

$$v = v_{\text{HCl}} = k_2[\text{Cl}][\text{CHCl}_3] = k_2\left(\frac{2}{k_4}\right)^{1/2} I_a^{1/2}[\text{CHCl}_3]$$

10.22. H$^+$ ions are most easily formed by the process

$$e^- + \text{H}_2\text{O} \rightarrow \text{H}^+ + \text{OH} + 2e^-$$

This is the sum of $\text{H}_2\text{O} \rightarrow \text{H} + \text{OH}$ and $e^- + \text{H} \rightarrow \text{H}^+ + 2e^-$ and its $\Delta H°$ is therefore

$$498.7 + 1312.2 \quad = 1810.9 \text{ kJ mol}^{-1}$$
$$= 18.1 \text{ eV}$$

This is lower than the observed value of 19.5 eV, indicating that the system passes through a state of higher energy. The OH radical probably dissipates energy in the form of translational, vibrational, and rotational energy.

O$^-$ ions are most easily formed by

$$e^- + \text{H}_2\text{O} \rightarrow 2\text{H} + \text{O}^-$$

which is the sum of $\text{H}_2\text{O} \rightarrow \text{H} + \text{OH}$, $\text{OH} \rightarrow \text{H} + \text{O}$, and $\text{O} + e^- \rightarrow \text{O}^-$.

Its $\Delta H°$ is therefore

$$498.7 - 482.2 - 213.4 = 713.5 \text{ kJ mol}^{-1} = 7.4 \text{ eV}$$

This is close to the observed appearance potential of 7.5 eV.

10.23. From the suggested mechanism, the rate of formation of ethane is

$$\frac{d[\text{C}_2\text{H}_6]}{dt} = k_3[\text{C}_2\text{H}_5][\text{H}_2].$$

Applying the steady state approximation for [C$_2$H$_5$], we obtain

$$\frac{d[\text{C}_2\text{H}_5]}{dt} = 0 = k_2[\text{C}_2\text{H}_4][\text{H}] - k_3[\text{C}_2\text{H}_5][\text{H}_2]. \tag{1}$$

Applying the steady state approximation for [H], we obtain

$$\frac{d[\text{H}]}{dt} = 0 = 2k_1[\text{Hg*}][\text{H}_2] - k_2[\text{C}_2\text{H}_4][\text{H}] + k_3[\text{C}_2\text{H}_5][\text{H}_2] - k_4[\text{H}]^2. \tag{2}$$

Now, using Eq. (1) in (2), we obtain

$$2k_1[Hg^*][H_2] - k_4[H]^2 = 0.$$

Therefore,

$$[H] = \left(\frac{2k_1[Hg^*][H_2]}{k_4} \right)^{1/2}.$$

Substituting this expression into Eq. (1) yields

$$[C_2H_5] = \frac{k_2[C_2H_4]}{k_3[H_2]} \left(\frac{2k_1[Hg^*][H_2]}{k_4} \right)^{1/2}.$$

Therefore, the rate of formation of ethane is given by

$$\frac{d[C_2H_6]}{dt} = k_2[C_2H_4] \left(\frac{2k_1[Hg^*][H_2]}{k_4} \right)^{1/2},$$

which is first order with respect to ethylene and half-order with respect to Hg* and H_2. Since the number of moles of Hg* produced is directly proportional to the intensity of the light absorbed (number of quanta), the rate is also proportional to the square root of the intensity of the light.

The observed rate constant is given by

$$k_{obs} = k_2 \left(\frac{2k_1}{k_4} \right)^{1/2}.$$

■ Catalysis

10.24. The concentration of H^+ ions in the solution is

$$\frac{3.2 \times 10^{-5}}{4.7 \times 10^{-2}} = 6.81 \times 10^{-4} \text{ mol dm}^{-3}$$

The dissociation constant is therefore

$$K_a = \frac{(6.81 \times 10^{-4})^2}{10^{-3}} = 4.64 \times 10^{-4} \text{ mol dm}^{-3}$$

10.25. The steady-state equations are:

For I: $k_1[I_2] - k_2[CH_3CHO][I] + k_4[CH_3][HI] - k_{-1}[I]^2 = 0$ (1)

For CH_3CO: $k_2[CH_3CHO][I] - k_3[CH_3CO] = 0$ (2)

For CH_3: $k_3[CH_3CO] - k_4[CH_3][HI] = 0$ (3)

(1) + (2) + (3) = $k_1[I_2] - k_{-1}[I]^2 = 0$

Then

$$[I] = \left(\frac{k_1}{k_{-1}}\right)^{1/2} [I_2]^{1/2}$$

$$v = v_{CO} = k_3[CH_3CO] = k_2[CH_3CHO][I]$$

$$= k_2 \left(\frac{k_1}{k_{-1}}\right)^{1/2} [I_2]^{1/2}[CH_3CHO]$$

10.26. The equations can be written as

$$\frac{1}{\rho}\log_{10}k = \frac{1}{\rho}\log_{10}k_0 + \sigma$$

and

$$\frac{1}{\rho'}\log_{10}K = \frac{1}{\rho'}\log_{10}K_0 + \sigma$$

Subtraction leads to

$$\frac{1}{\rho}\log_{10}k = \frac{1}{\rho'}\log_{10}K = \text{const}$$

and therefore

$$\log_{10}\frac{k^{1/\rho}}{K^{1/\rho'}} = \text{const}$$

or

$$\log_{10}\frac{k}{K^{\rho/\rho'}} = \text{const}$$

Thus

$$k = G\,K^{\rho/\rho'} = GK^\alpha$$

where G and α are constants, the latter being equal to the ratio ρ/ρ'.

10.27. The HCl concentration remains unchanged throughout the experiment and the reaction is pseudo-first order with a rate constant of

$$2.80 \times 10^{-5} \text{ dm}^3 \text{ mol}^{-1} \text{ s}^{-1} \times 0.01 \text{ mol dm}^{-3}$$

$$= 2.80 \times 10^{-7} \text{ s}^{-1}$$

The half-life (Eq. 9.40) is

$$t_{1/2} = \frac{\ln 2}{2.80 \times 10^{-7} \text{ s}^{-1}} = 2.48 \times 10^6 \text{ s } (\approx 29 \text{ days})$$

10.28. Put $\quad$ S = $Co(NH_3)_5Cl^{2+}$

$$X = Co(NH_3)_4(NH_2)Cl^+$$

$$Y = Co(NH_3)_4(NH_2)^{2+}$$

and the reactions are rewritten as

$$S + OH^- \xrightarrow{k_1} X + H_2O$$

$$X \xrightarrow{k_2} Y + Cl^-$$

$$Y \xrightarrow{k_3} Co(NH_3)_5(OH)^{2+}$$

Steady-state equations:

For X: $\quad k_1[S][OH^-] - (k_{-1} + k_2)[X] = 0$

For Y: $\quad k_2[X] - k_3[Y] = 0$

$$[X] = \frac{k_1[S][OH^-]}{k_{-1} + k_2}$$

$$[Y] = \frac{k_1(k_2/k_3)[S][OH^-]}{k_{-1} + k_2}$$

$$v = k_3[Y] = \frac{k_1 k_2[S][OH^-]}{k_{-1} + k_2}$$

The dependence of v on [S] and [OH$^-$] is independent of the relative magnitudes of the rate constants.

10.29. Differentiation of the expression for k with respect to [H$^+$] gives

$$dk/d[H^+] = k_{H^+} - k_{OH^-} K_w/[H^+]^2$$

If this is set equal to zero, the required equation is obtained.

10.30. The acid dissociation constants may be converted into base constants K_b by division into 10^{-14} $mol^2\ dm^{-6}$. The following table can then be constructed:

	K_b	k	$\ln K_b$	$\ln k$
$ClCH_2COO^-$	7.19×10^{-12}	1.41×10^{-3}	-25.65	-6.56
CH_3COO^-	5.56×10^{-10}	1.34×10^{-2}	-21.3	-4.31
HPO_4^{2-}	1.60×10^{-7}	0.26	-15.6	-1.35

A plot of Eq. 10.75 is shown where $\ln k$ is plotted against $\ln K_b$ in the figure; its slope, 0.53, is the value of the Brønsted coefficient β.

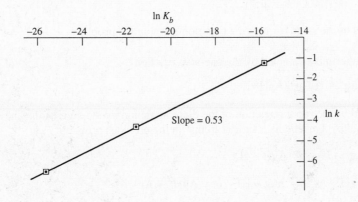

10.31. The hydroxide ion is assumed to abstract a proton from the acetone molecule, giving the ion $CH_3COCH_2^-$, which exists in a resonant state with the species

$$
\begin{array}{c}
O^- \\
| \\
CH_3 - C = CH_2
\end{array}
$$

If this species reacts rapidly with a bromine molecule, the rate is independent of the bromine concentration.

10.32. The inhibition by hydrogen ions suggests the mechanism

$$Cu^{2+} + H_2 \underset{k_{-1}}{\overset{k_1}{\rightleftharpoons}} CuH^+ + H^+$$

$$CuH^+ + Cu^{2+} \overset{k_2}{\longrightarrow} 2Cu^+ + H^+$$

This may be followed by rapid reaction with $Cr_2O_7{}^{2-}$. Application of the steady-state treatment to this mechanism leads to

$$v = \frac{k_1 k_2 [H_2][Cu^{2+}]}{k_{-1}[H^+] + k_2[Cu^{2+}]}$$

which is of the same form as the equation given in the problem.

10.33. The first term is explained by the one-step reaction

$$2Ag^+ + H_2 \rightarrow 2AgH^+$$

followed by the rapid reduction of the dichromate ion by the AgH^+ ion. The second term is explained by a mechanism similar to that in the previous problem:

$$Ag^+ + H_2 \rightleftharpoons AgH + H^+$$

$$AgH + Ag^+ \rightarrow 2Ag + H^+ \quad or \quad AgH^+ + Ag$$

There can then be rapid reaction of Ag or AgH^+ with the dichromate ion. Application of the steady-state treatment to AgH leads to an equation of the form of the second term in the problem.

10.34. A simple explanation is that there is an initial fast equilibrium:

$$Ag^+ + Ce^{4+} \rightleftharpoons Ce^{3+} + Ag^{2+}$$

followed by a slow reaction of Ag^{2+} with Tl^+:

$$Ag^{2+} + Tl^+ \rightarrow Tl^{2+} + Ag^+$$

This may be followed by a fast reaction of Tl^{2+} with Ce^{4+}:

$$Tl^{2+} + Ce^{4+} \rightarrow Tl^{3+} + Ce^{3+}$$

Since the first reaction is fast we can write the equilibrium equation:

$$\frac{[Ce^{3+}][Ag^{2+}]}{[Ce^{4+}][Ag^+]} = K_1$$

The overall rate is

$$v = k_2[Ag^{2+}][Tl^+]$$

$$= k_2 K_1 [Ce^{4+}][Ag^+][Tl^+]/[Ce^{3+}]$$

■ Enzyme-Catalyzed Reactions

10.35.

[S]/(10⁻³ mol dm⁻³)	v	$1/v$	$1/([S]/10^{-3}$ mol dm⁻³)	$v/([S]/10^{-3}$ mol dm⁻³)
0.4	2.41	0.415	2.50	6.025
0.6	3.33	0.300	1.667	5.55
1.0	4.78	0.209	1.000	4.78
1.5	6.17	0.162	0.667	4.11
2.0	7.41	0.135	0.500	3.705
3.0	8.70	0.115	0.333	2.90
4.0	9.52	0.105	0.250	2.38
5.0	10.5	0.095	0.200	2.10
10.0	12.5	0.080	0.100	1.25

From plots, $K_m = 2.0$ mmol dm⁻³.

Curve 1

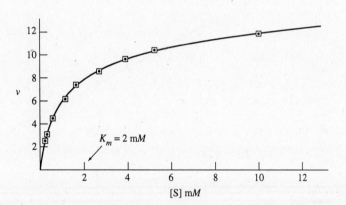

Curve 2

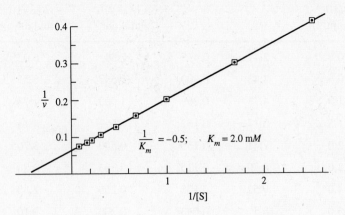

$$\frac{1}{K_m} = -0.5; \quad K_m = 2.0 \text{ m}M$$

1/[S]

Curve 3

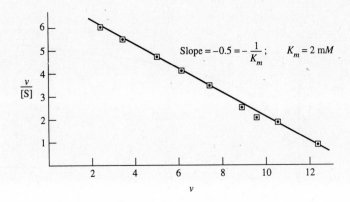

$$\text{Slope} = -0.5 = -\frac{1}{K_m}; \quad K_m = 2 \text{ m}M$$

Curve 1 is nonlinear. Curve 2 is too sensitive to high $\frac{1}{S}$ values. Curve 3 gives evenly weighted data. Curve 3 of $v/[S]$ against v gives the best statistical results since it spreads out the data most evenly in a linear plot.

10.36.

$[S]/(10^{-6}$ $dm^{-3})$	$v/(10^{-6}$ mol dm^{-3} $s^{-1})$	$1/v(10^{-6}$ mol dm^{-3} $s^{-1})$	$1/([S]/10^{-6}$ mol $dm^{-3})$	$\dfrac{v/[S]}{10^{-3}\ s^{-1}}$
7.5	0.067	14.9	0.133	8.933
12.5	0.095	10.5	0.080	7.60
20.0	0.119	8.40	0.050	5.95
32.5	0.149	6.71	0.0308	4.58
62.5	0.185	5.41	0.0160	2.96
155.0	0.191	5.24	0.00645	1.232
320.0	0.195	5.13	0.00313	0.609

m graphs; $K_m = 16\ \mu mol\ dm^{-3}$

$V = 0.22\ \mu mol\ dm^{-3}$

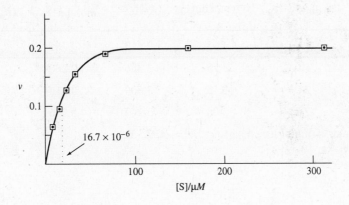

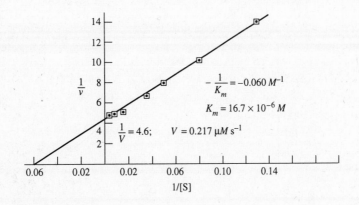

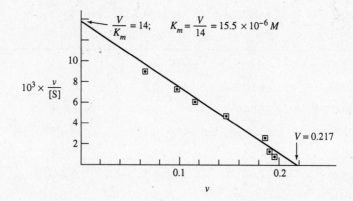

$$\frac{V}{K_m} = 14; \qquad K_m = \frac{V}{14} = 15.5 \times 10^{-6}\,M$$

$V = 0.217$

$10^3 \times \dfrac{v}{[S]}$

v

10.37.

T/K	$V/10^{-6}$ mol dm^{-3} s^{-1}	$\dfrac{K_m}{10^{-4}\,\text{mol dm}^{-3}}$	$\dfrac{10^3}{(T/K)}$	$\log_{10}(V/10^{-6}$ mol dm^{-3} s$^{-1})$	$\log_{10}$ $(K_m/10^{-4}\,\text{mol dm}^{-3})$
293	1.84	4.03	3.413	0.265	0.605
298	1.93	3.75	3.356	0.286	0.574
303	2.04	3.35	3.300	0.310	0.525
308	2.17	3.05	3.247	0.336	0.484

a. Slope of plot of $\log_{10}V$ against $1/T = -427.5$ K.

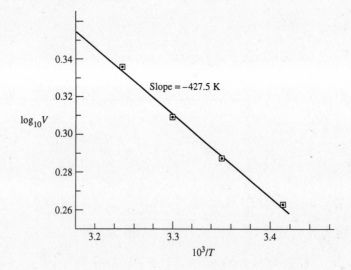

Slope = −427.5 K

$\log_{10}V$

$10^3/T$

$$E = 19.14 \times 427.5 = 8180 \text{ J mol}^{-1} = 8.18 \text{ kJ mol}^{-1}$$

$$RT \text{ at } 25°C = 2479 \text{ J mol}^{-1} = 2.48 \text{ kJ mol}^{-1}$$

$$\Delta^{\ddagger}H = 8.18 - 2.48 = 5.7 \text{ kJ mol}^{-1}$$

At 25.0°C, $V = 1.93 \times 10^{-6} \text{ dm}^3 \text{ mol}^{-1} \text{ s}^{-1}$

$$[E]_0 = 1.0 \times 10^{-11} \text{ mol dm}^3$$

$$k_c = V/[E]_0 = 1.93 \times 10^5 \text{ s}^{-1}$$

$$k_c = \frac{k_B T}{h} e^{-\Delta^{\ddagger}G/RT}; \quad \frac{k_B T}{h} \text{ at } 25°C = 6.21 \times 10^{12} \text{ s}^{-1}$$

$$e^{-\Delta^{\ddagger}G/RT} = 1.93 \times 10^5/6.21 \times 10^{12} = 3.108 \times 10^{-8}$$

$$= 10^{-7.51}$$

$$\Delta^{\ddagger}G = 19.14 \times 298.15 \times 7.51 = 42\,856 \text{ J mol}^{-1}$$

$$= 42.9 \text{ kJ mol}^{-1}$$

$$\Delta^{\ddagger}S = \frac{5700 - 42\,856}{298.15} = -124 \text{ J K}^{-1} \text{ mol}^{-1}$$

b. Slope of $\log_{10} K_m$ against $1/T$ plot = 740 K

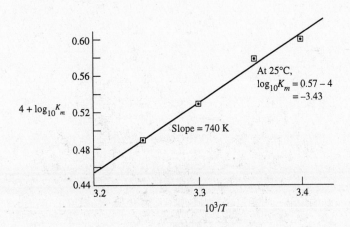

$$\Delta H° \text{ (for dissociation)} = -14\,164 \text{ J mol}^{-1}$$

$$= -14.2 \text{ kJ mol}^{-1}$$

At 25.0°C, $\log_{10} (K_m/10^{-4} \text{ dm}^{-3}) = 0.57 - 4 = -3.43$

$$\Delta G° \text{ (for dissociation)} = 19.14 \times 298.15 \times 3.43$$

$$= 19\ 570\ \text{J mol}^{-1}$$

For association, $\Delta G° = -19.6\ \text{kJ mol}^{-1}$;

$$\Delta H° = 14.2\ \text{kJ mol}^{-1}$$

$$\Delta S° = \frac{14\ 164 + 19\ 570}{298.15} = 113\ \text{J K}^{-1}\ \text{mol}^{-1}$$

c.

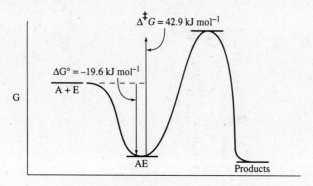

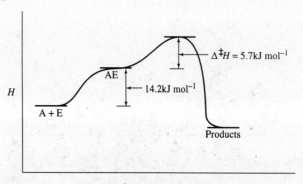

10.38.

$[S]/10^{-3}$ mol dm^{-3}	$V/10^{-5}$ mol dm^{-3} s^{-1}	$\dfrac{1}{[S]/\text{mol dm}^{-3}}$	$1/v$
2.0	13	500	7692
4.0	20	250	5000
8.0	29	125	3448
12.0	33	83	3030
16.0	36	62.5	2778
20.0	38	50.0	2630

From a plot $1/v$ against $1/[S]$,

$$V = 5.0 \times 10^{-4} \text{ mol dm}^{-3} \text{ s}^{-1}$$

$$K_m = 5.8 \times 10^{-3} \text{ mol dm}^{-3}$$

$$[E]_0 = \frac{2 \text{ g dm}^{-3}}{50\,000 \text{ g mol}^{-1}} = 4.0 \times 10^{-5} \text{ mol dm}^{-3}$$

$$k_c = \frac{V}{[E]_0} = \frac{5.0 \times 10^{-4}}{4.0 \times 10^{-5}} = 12.5 \text{ s}^{-1}$$

10.39.

Temperature				
(°C)	T/K	$k_c/10^{-6}$ s^{-1}	$10^3/(T/K)$	$\log_{10}(k_c/10^{-6}$ s$^{-1})$
39.9	312.9	4.67	3.196	0.669
43.8	316.8	7.22	3.157	0.858
47.1	320.1	10.0	3.124	1.00
50.2	323.2	13.9	3.094	1.143

Slope of plot of $\ln k_c$ against $1/T$

$$= -10\,890 \; K$$

$$E = 8.3145 \times 10\,890 = 90\,540 \text{ J mol}^{-1} = 90.5 \text{ kJ mol}^{-1}$$

RT at 40 °C $= 8.3145 \times 313.15 = 2604 \text{ J mol}^{-1}$

$$\Delta^{\ddagger}H° = 90\,500 - 2604 = 87\,900 \text{ mol}^{-1} = 87.9 \text{ kJ mol}^{-1}$$

$\dfrac{k_B T}{h}$ at 40 °C $= 6.52 \times 10^{12}$ s^{-1}; k_c at 40 °C $= 4.72 \times 10^{-6}$ s^{-1}

$$e^{-\Delta^{\ddagger}G°/RT} = \frac{4.72 \times 10^{-6}}{6.52 \times 10^{12}} = 7.24 \times 10^{-19} = 10^{-18.14}$$

$$\Delta^{\ddagger}G° = 19.14 \times 313.15 \times 18.14 \quad = 108\,700 \text{ J mol}^{-1}$$

$$= 108.7 \text{ kJ mol}^{-1}$$

$$\Delta^\ddagger S^\circ = \frac{87\,900 - 108\,700}{313.15} = -66.5 \text{ J K}^{-1} \text{ mol}^{-1}$$

10.40. Steady-state equation for EA:

$$k_1[A][E] - k_{-1}[EA] - k_2[EA][B] = 0$$

Steady-state equation for EY:

$$k_2[EA][B] - k_3[EY] = 0$$

$$[E]_0 = [E] + [EA] + [EY]$$

$$= [EA] \left\{ \frac{k_{-1} + k_2[B]}{k_1[A]} + 1 + \frac{k_2[B]}{k_3} \right\}$$

$$v = k_2[EA][B]$$

$$= \frac{k_2[B][E]_0}{\dfrac{k_{-1} + k_2[B]}{k_1[A]} + 1 + \dfrac{k_2[B]}{k_3}}$$

$$= \frac{k_1 k_2 k_3 [E]_0 [A][B]}{k_{-1}k_3 + k_1 k_3[A] + k_2 k_3[B] + k_1 k_2[A][B]}$$

10.41.   **a.** Steady-state condition of [ES]:

$$k_1[E][S] - (k_{-1} + k_2)[ES] = 0 \tag{1}$$

Steady-state condition of [EI]:

$$k_i[E][I] - k_{-i}[EI] = 0 \tag{2}$$

Find [ES] from (1) and [EI] from (2). Use (1) to find [E] and eliminate [E] from the [ES] and [EI] terms.

$$[E]_0 = [E] + [ES] + [EI]$$

$$[E]_0 = [ES] \left\{ \frac{k_{-1} + k_2}{k_1[S]} + 1 + \frac{k_i[I](k_{-1} + k_2)}{k_{-i}k_1[S]} \right\}$$

$$v = k_2[ES] = \frac{k_2[E]_0}{\dfrac{k_{-1} + k_2}{k_1[S]} + 1 + \dfrac{k_i[I](k_{-1} + k_2)}{k_{-i}k_1[S]}}$$

$$= \frac{k_2[E]_0[S]}{\dfrac{k_{-1} + k_2}{k_1} + [S] + \dfrac{k_i(k_{-1} + k_2)}{k_{-i}k_1}[I]} = \frac{k_2[E]_0[S]}{K_m\left(1 + \dfrac{[I]}{K_i}\right) + [S]}$$

b. $\varepsilon = \dfrac{v_0 - v}{v_0} = 1 - \dfrac{v}{v_0}$

$$= 1 - \frac{k_2[E]_0[S]}{K_m\left(1 + \frac{[I]}{K_i}\right) + [S]} \cdot \frac{K_m + [S]}{k_2[E]_0[S]}$$

$$= 1 - \frac{K_m + [S]}{K_m\left(1 + \frac{[I]}{K_i}\right) + [S]} = \frac{\dfrac{K_m}{K_i}[I]}{K_m\left(1 + \frac{[I]}{K_i}\right) + [S]}$$

10.42.
$$E + S \underset{k_{-1}}{\overset{k_1}{\rightleftarrows}} ES \overset{k_2}{\underset{Y}{\searrow}} ES'$$

$$k_1[E][S] - (k_{-1} + k_2)[ES] = 0$$

$$k_2[ES] - k_3[ES'] = 0$$

$$[E]_0 = [E] + [ES] + [ES']$$

$$= [ES]\left(\frac{k_{-1} + k_2}{k_1[S]} + 1 + \frac{k_2}{k_3}\right)$$

$$v = k_2[ES] = \frac{k_2[E]_0}{\dfrac{k_{-1} + k_2}{k_1[S]} + 1 + \dfrac{k_2}{k_3}}$$

$$= \frac{\dfrac{k_2 k_3}{k_2 + k_3}[E]_0[S]}{\dfrac{k_{-1} + k_2}{k_1}\dfrac{k_3}{k_2 + k_3} + [S]}$$

When [S] is large

$$v = \frac{k_1 k_2}{k_{-1} + k_2}[E]_0$$

and the catalytic constant is therefore

$$k_c = \frac{k_1 k_2}{k_{-1} + k_2}$$

K_m is the first term in the denominator of the rate equation:

$$K_m = \frac{k_{-1} + k_2}{k_1}\frac{k_3}{k_2 + k_3}$$

10.43. Suppose that ΔH_m is positive: K_m will then increase with increasing temperature, and at sufficiently low temperatures it will be much smaller than [S]. The effective activation energy is thus E_c, and the rate will raise with increasing temperature.

At high enough temperatures K_m will be much larger than [S], and the effective activation energy is $E_c - \Delta H_m$. If ΔH_m is larger that E_c, the effective activation energy is negative, and the rate has gone through a maximum as the temperature is increased.

If ΔH_m is not larger than E_c, the rate will not go through a maximum, but the observed activation energy will be lower at higher temperatures. An Arrhenius plot will therefore show curvature, bending out from the axes.

10.44. Steady-state equation for EA:

$$k_1[E][A] - k_{-1}[EA] - k_2[EA][B] = 0$$

Steady-state equation for EAB:

$$k_2[EA][B] - (k_{-2} + k_3)[EAB] = 0$$

$$[E]_0 = [E] + [EA] + [EAB]$$

$$= [EAB]\left(\frac{k_{-1} + k_2[B]}{k_1[A]} \cdot \frac{k_{-2} + k_3}{k_2[B]} + \frac{k_{-2} + k_3}{k_2[B]} + 1\right)$$

$$v = k_2[EAB] = \frac{k_2[E]_0}{\dfrac{k_{-1} + k_2[B]}{k_1[A]} \cdot \dfrac{k_{-2} + k_3}{k_2[B]} + \dfrac{k_{-2} + k_3}{k_2[B]} + 1}$$

$$= \frac{k_1 k_2 k_3 [E]_0 [A][B]}{k_{-1}(k_{-2} + k_3) + k_1(k_{-2} + k_3)[A] + k_2(k_{-2} + k_3)[B] + k_1 k_2[A][B]}$$

10.45. $[E]° = [E] + [EA] + [EA'] + [EA'B]$

The steady-state equations are:

for EA: $\quad k_1[E][A] = (k_{-1} + k_2)\,[EA]$

for EA': $\quad k_2[EA] = k_3[EA'][B]$

for EA'B: $\quad k_3[EA'][B] = k_4[EA'B]$

Then, by use of the steady-state equations, $[E]°$ can be expressed in terms of $[EA']$:

$$[E]° = [EA']\left\{\frac{(k_{-1} + k_2)}{k_1[A]}\frac{k_3[B]}{k_2} + \frac{k_3[B]}{k_2} + \frac{k_3[B]}{k_2} + 1 + \frac{k_3[B]}{k_4}\right\}$$

$$v = k_3[EA'][B]$$

$$= \frac{k_3[E]°[B]}{\dfrac{(k_{-1} + k_2)}{k_1[A]} \cdot \dfrac{k_3[B]}{k_2} + \dfrac{k_3[B]}{k_2} + 1 + \dfrac{k_3[B]}{k_4}}$$

$$= \frac{k_1 k_2 k_3 k_4 [E]°[A][B]}{k_1 k_2 k_4[A] + k_3 k_4(k_{-1} + k_2)[B] + k_1 k_3(k_2 + k_4)[A][B]}.$$

Polymerization

10.46. From Eq. 10.97, the polymerization rate is

$$-\frac{d[M]}{dt} = k_p\left(\frac{k_i}{k_t}\right)^{1/2}[M]^{3/2}[C]^{1/2}$$

and the rate of initiation is $k_i[M][C]$ (Eq. 10.96). The chain length is therefore

$$\frac{k_p[M]^{1/2}}{(k_i k_t)^{1/2}[C]^{1/2}}$$

10.47. The rate of formation of CH_3 is $2I$. The steady-state equations are

$$2I - k_p[CH_3][M] - k_t[CH_3]\Sigma[R_n] = 0$$

From Eq. 20.2,

$$k_p[CH_3][M] - k_p[CH_3CH_2CH_2{-}][M] - k_t[CH_3CH_2CH_2{-}]\Sigma[R_n] = 0$$

and so on.

The sum of all the equations is

$$2I - k_t(\Sigma[R_n])^2 = 0$$

so that

$$\Sigma[R_n] = \left(\frac{2I}{k_t}\right)^{1/2}$$

The rate of removal of monomer is

$$v = k_p[M]\Sigma[R_n] = k_p\left(\frac{2I}{k_t}\right)^{1/2}[M].$$

11. QUANTUM MECHANICS AND ATOMIC STRUCTURE

■ Electromagnetic Radiation and Wave Motion

11.1. a. The frequency is calculated from $c = \lambda v$ where c is the speed of light ($c = 2.998 \times 10^8$ m s^{-1}):

$$v = \frac{c}{\lambda} = \frac{2.998 \times 10^8 \text{ m s}^{-1}}{325 \times 10^{-9} \text{ m}} = 9.22 \times 10^{14} \text{ s}^{-1}$$

b. $\lambda = 325 \times 10^{-9}$ m $= 3250 \times 10^{-8}$ cm

$$\bar{v} = \frac{1}{\lambda} = 3.08 \times 10^4 \text{ cm}^{-1}$$

c. $\varepsilon = hv = 6.626 \times 10^{-34}$ (J s) 9.22×10^{14} s^{-1}

$$= 6.11 \times 10^{-19} \text{ J}$$

In molar units:

$$6.022 \times 10^{23} \text{ mol}^{-1} \times 6.11 \times 10^{-19} \text{ J}$$

$$= 3.68 \times 10^5 \text{ J mol}^{-1} = 368 \text{ kJ mol}^{-1}$$

$$E = hv = 6.11 \times 10^{-19} \text{ J} \times 6.2420 \times 10^{18} \text{ eV/J}$$

$$= 3.81 \text{ eV}$$

d. The momentum is calculated by the use of Eq. 11.56:

$$p = \frac{h}{\lambda} = \frac{6.626 \times 10^{-34} \text{ J s}}{3.25 \times 10^{-9} \text{ m}} = 2.04 \times 10^{-25} \text{ kg m s}^{-1}$$

Handwritten notes (left margin):
$m \tilde{=} 0.$
$E^2 = p^2 c^2 + m^2 c^4 \approx 0.$
$E = pc$
$E = p(c = p\lambda v)$
$p = \frac{E}{\lambda v} = \frac{hv}{\lambda v}$

Handwritten notes (right margin):
For EM radiation:
$\lambda = \frac{h}{p} = \frac{h}{mu}$ ← velocity, mass of particle.
$\Rightarrow p = \frac{h}{\lambda}$

11.2. a. $\lambda = c/v = \dfrac{2.998 \times 10^8 \text{ m s}^{-1}}{1.965 \times 10^8 \text{ s}^{-1}} = 1.526$ m

b. $E = hv = 6.626 \times 10^{-34}$ J s $\times 196.5 \times 10^6$ s^{-1}

$$= 1.302 \times 10^{-25} \text{ J}$$

$$= 8.127 \times 10^{-7} \text{ eV} = 0.07841 \text{ J mol}^{-1}$$

c. $p = \dfrac{h}{\lambda} = \dfrac{6.626 \times 10^{-34} \text{ J s}}{1.526 \text{ m}} = 4.342 \times 10^{-34} \text{ kg m s}^{-1}$

11.3. $v_1 = \dfrac{c}{\lambda_1} = \dfrac{2.998 \times 10^8 \text{ m s}^{-1}}{766.494 \times 10^{-9} \text{ m}} = 3.9113 \times 10^{14} \text{ s}^{-1}$

$v_2 = \dfrac{c}{\lambda_2} = \dfrac{2.998 \times 10^8 \text{ m s}^{-1}}{769.901 \times 10^{-9} \text{ m}} = 3.8940 \times 10^{14} \text{ s}^{-1}$

$v_1 - v_2 = 0.0173 \times 10^{14} \text{ s}^{-1} = 1.73 \times 10^{16} \text{ s}^{-1}$

11.4. The form of the equation is (compare Eq. 11.6)

$y = A \sin(\omega t + \delta)$

The angular frequency ω is $(3\pi/5)$ rad s^{-1}

From Eq. 11.4, $\omega = (2\pi \text{ rad}) v$ and therefore

$v = \dfrac{w}{2\pi rad} = \dfrac{(3\pi/5) \text{ rad s}^{-1}}{2\pi \text{ rad}} = 0.3 \text{ s}^{-1}$

11.5. a. The period, $\tau = 1/v$, is related to the force constant of the spring by Eq. 11.15.

$$\text{Period} = \frac{1}{v} = 2\pi \sqrt{\frac{m}{k_h}}$$

Solving for k_h:

$$k_h = \frac{4\pi^2 m}{\tau^2} = \frac{4\pi^2 (0.2 \text{ kg})}{(3.0 \text{ s})^2} = 0.88 \text{ N m}^{-1}$$

b. To obtain the velocity, differentiate the equation of motion for the harmonic oscillator (Eq. 11.6). Thus

$$v = \frac{dy}{dx} = \omega A \cos(\omega \tau + \delta)$$

Since $|\cos \theta| \leq 1$ for all θ, it follows that the maximum velocity is

$$v_{max} = \omega A = \sqrt{\frac{k_h}{m}} \, A = \sqrt{\frac{0.88 \text{ N/m}}{0.2 \text{ kg}}} \times 0.01 \text{ m}$$

$$= 0.021 \text{ m s}^{-1}$$

11.6. The low-frequency limit can be obtained by use of the series expression

$$e^x = 1 + x + \frac{x^2}{2!} + \dots$$

When x is small we may approximate e^x by $1 + x$. Thus, when $h v \ll k_B T$

$$\bar{E} = \frac{h v}{1 + \dfrac{h v}{k_B T} - 1} = \boxed{k_B T} \quad \text{divided btw the two degrees of freedom of EM waves.}$$

✳ ✳ ✳   Electromagnetic waves are transverse waves having two degrees of freedom. This value k_BT is divided between these two degrees of freedom.

■ Particles and Waves

11.7.   The energy of each photon is

$$h\nu = \frac{hc}{\lambda} = \frac{6.626 \times 10^{-34} \text{ J s} \times 2.998 \times 10^8 \text{ m s}^{-1}}{550 \times 10^{-9} \text{ m}} = 3.61 \times 10^{-19} \text{ J}$$

The lamp emits 50 J per second and the number of photons is therefore

$$\frac{50}{3.61 \times 10^{-19}} = 1.39 \times 10^{20}$$

The momentum of each is

$$p = \frac{h}{\lambda} = \frac{6.626 \times 10^{-34} \text{ J s}}{550 \times 10^{-9} \text{ m}} = 1.204 \times 10^{-27} \text{ kg m s}^{-1}$$

11.8.   Eq. 11.37, $h\nu = \frac{1}{2}mu^2 + w$, gives $h\nu_0 = w$ when the kinetic energy $\frac{1}{2}mu^2$ is zero; ν_0 is the threshold frequency. Thus,

$$h\nu_0 = w \quad = 6.626 \times 10^{-34} \text{ (J s)} \, 43.9 \times 10^{13} \text{ s}^{-1}$$

$$= 2.91 \times 10^{-19} \text{ J}$$

or, in eV,

$$w = 2.91 \times 10^{-19} \text{ J}/1.602 \times 10^{-19} \text{ J/eV} = 1.82 \text{ eV}$$

With the more modern value,

$$w = 6.626 \times 10^{-34} \text{ J s} \times 5.5 \times 10^{13} \text{ s}^{-1}$$

$$= 3.6 \times 10^{-20} \text{ J} = 0.23 \text{ eV}$$

11.9.   a. Using Eq. 11.56, $\lambda = \frac{h}{mu}$, we have

$$\lambda = \frac{6.626 \times 10^{-34} \text{ J s}}{9.11 \times 10^{-31} \text{ kg} \times 6.0 \times 10^7 \text{ m s}^{-1}}$$

$$= 1.21 \times 10^{-11} \text{ m} = 12.1 \text{ pm}$$

  b. $\lambda = \frac{h}{mu} = \dfrac{6.626 \times 10^{-34} \text{ J s}}{\dfrac{32.0 \text{ g mol}^{-1}}{6.022 \times 10^{23} \text{ mol}^{-1}} \dfrac{1 \text{ kg}}{1000 \text{ g}} 425 \text{ m s}^{-1}}$

$$= 2.93 \times 10^{-11} \text{ m} = 29.3 \text{ pm}$$

c. $\lambda = \dfrac{h}{mu} = \dfrac{6.626 \times 10^{-34} \text{ J s}}{\dfrac{4.0 \text{ g mol}^{-1}}{6.022 \times 10^{23} \text{ mol}^{-1}} \dfrac{1 \text{ kg}}{1000 \text{ g}} 1.5 \times 10^7 \text{ m s}^{-1}}$

$\qquad = 6.65 \times 10^{-15} \text{ m} = 6.65 \text{ fm}$

d. $\lambda = \dfrac{h}{mu} = \dfrac{6.626 \times 10^{-34}}{9.11 \times 10^{-31}(2.818 \times 10^8)} = 2.58 \times 10^{-12} \text{ m} = 2.58 \text{ pm}$

11.10. From Eq. 11.61, the product of the uncertainties in position and velocity is

$$\Delta q \Delta u \approx \dfrac{h}{4\pi m} = \dfrac{6.626 \times 10^{-34} \text{ J s } (= \text{kg m}^2 \text{ s}^{-1})}{4 \times 3.14 \times 6 \times 10^{-16} \text{ kg}}$$

$$= 10^{-19} \text{ m}^2 \text{ s}^{-1}$$

Therefore, since $\Delta q = 10^{-9}$ m,

$$\Delta u = 10^{-10} \text{ m s}^{-1}$$

With this uncertainty in velocity, the position of the particle one second later would be uncertain to within 2×10^{-10} m, or 0.2 nm. This is only 0.2% the diameter of the particle, and the uncertainty principle therefore does not present a serious problem for particles of this magnitude. For particles of molecular sizes, the uncertainty is much greater.

11.11.   a. The kinetic energy of the electron is

$$E_k = 10 \times 1.602 \times 10^{-19} \text{ V C} = 1.602 \times 10^{-18} \text{ J}$$

This energy is $\frac{1}{2} mu^2$ and therefore

$$u = \sqrt{\dfrac{2 \times 1.602 \times 10^{-18} \text{ J}}{9.110 \times 10^{-31} \text{ kg}}} = 1.875 \times 10^6 \text{ m s}^{-1}$$

Then, from Eq. 11.56,

$$\lambda = \dfrac{6.626 \times 10^{-34} \text{ J s}}{9.110 \times 10^{-31} \text{ kg} \times 1.875 \times 10^6 \text{ m s}^{-1}}$$

$$= 3.88 \times 10^{-10} \text{ m} = 388 \text{ pm}$$

b. $u = \sqrt{\dfrac{2 \times 10^3 \times 1.602 \times 10^{-19} \text{ J}}{9.110 \times 10^{-31} \text{ kg}}} = 1.875 \times 10^7 \text{ m s}^{-1}$

$\lambda = \dfrac{6.626 \times 10^{-34}}{9.110 \times 10^{-31} \times 1.875 \times 10^7} = 3.879 \times 10^{-11}$

$\qquad = 38.8 \text{ pm}$

c. $u = \sqrt{\dfrac{2 \times 10^6 \times 1.602 \times 10^{-19} \text{ J}}{9.110 \times 10^{-31} \text{ kg}}} = 5.930 \times 10^8 \text{ m s}^{-1}$

$\lambda = \dfrac{6.626 \times 10^{-34}}{9.110 \times 10^{-31} \times 5.930 \times 10^8} = 1.226 \times 10^{-12} \text{ m}$

$$= 1.23 \text{ pm}$$

11.12.   A particle of mass m has a frequency of

$$\nu = \frac{1}{2} \, mu^2/h$$

and a de Broglie wavelength (Eq. 11.56) of

$$\lambda = \frac{h}{mu}$$

Elimination of u between these equations gives

$$\nu = \frac{h}{2m}\left(\frac{1}{\lambda^2}\right)$$

The group velocity is thus

$$v_g = \frac{d\nu}{d(1/\lambda)} = 2 \cdot \frac{h}{2m} \cdot \frac{1}{\lambda} = \frac{h}{m\lambda} = u$$

11.13.   a.   The minimum frequency ν_{min} is such that

$$h\nu_{min} = 5 \text{ eV} = 8.01 \times 10^{-19} \text{ J}$$

$$\nu_{min} = 8.01 \times 10^{-19} \text{ J}/6.626 \times 10^{-34} \text{ J s}$$

$$= 1.209 \times 10^{15} \text{ s}^{-1}$$

The corresponding wavelength is

$$2.998 \times 10^8 \text{ m s}^{-1}/1.209 \times 10^{15} \text{ s}^{-1}$$

$$= 2.480 \times 10^{-7} \text{ m} = 248 \text{ nm}$$

b.   The wavelength of 150 nm corresponds to a frequency of

$$2.998 \times 10^8 \text{ m s}^{-1}/1.50 \times 10^{-7} \text{ s}^{-1} = 2.00 \times 10^{15} \text{ s}^{-1}$$

and to an energy of 6.626×10^{-34} J s $\times 2.00 \times 10^{15}$ s^{-1}

$$= 1.33 \times 10^{-18} \text{ J}$$

The excess energy is therefore

$$(1.33 \times 10^{-18} \text{ J}) - (8.01 \times 10^{-19}) = 5.2 \times 10^{-19} \text{ J}$$

This is $\frac{1}{2} \, mu^2$ where m is the mass of the electron and u is its velocity. The velocity is therefore

$$(2 \times 5.29 \times 10^{-19} \text{ J}/9.110 \times 10^{-31} \text{ kg})^{1/2}$$

$$= 1.078 \times 10^6 \text{ m s}^{-1}$$

11.14.   According to the de Broglie equation, Eq. 11.56, $p = h/\lambda$, and the kinetic energy E_k is $p^2/2m$. Thus

$$E_k - \frac{h^2}{2m\lambda^2} = \frac{(6.626 \times 10^{-24} \text{ J s})^2}{2(9.110 \times 10^{-31} \text{ kg}) \lambda^2}$$

$$= 2.410 \times 10^{-37} \text{ kg m}^4 \text{ s}^{-2}/\lambda^2$$

a. With $\lambda = 1.0 \times 10^{-8}$ m,

$$E_k = 2.410 \times 10^{-21} \text{ J}$$

b. With $\lambda = 1.00 \times 10^{-7}$ m,

$$E_k = 2.410 \times 10^{-23} \text{ J}$$

11.15.   a. The energy of the α-particle is $100 \text{ V} \times 2 \times 1.602 \times 10^{-19} \text{ C} = 3.204 \times 10^{-17}$ J. Therefore, the velocity is

$$u = \sqrt{\frac{2 \times 3.204 \times 10^{-17} \text{ J}}{6.64 \times 10^{-27} \text{ kg}}} = 9.824 \times 10^4 \text{ m s}^{-1}.$$

The de Broglie wavelength, then, is

$$\lambda = \frac{h}{mu} = \frac{6.626 \times 10^{-34} \text{ J s}}{6.64 \times 10^{-27} \text{ kg} \times 9.824 \times 10^4 \text{ m s}^{-1}} = 1.016 \times 10^{-12} \text{ m}.$$

Thus, the wavelength is about 1000 times larger than the diameter of the particle.

b. For a tennis ball, we get

$$\lambda = \frac{6.626 \times 10^{-34} \text{ J s}}{55.4 \times 10^{-3} \text{ kg} \times 2.20 \times 10^5 \text{ m h}^{-1}} \times 3600 \text{ s h}^{-1} = 1.96 \times 10^{-34} \text{ m},$$

which is about 3×10^{32} times smaller than the diameter of a tennis ball!

■ Quantum-Mechanical Principles

11.16.   a. $N^2 \int (\psi_1 + \psi_2)^2 \, d\tau = 1$

$N^2 [\int \psi_1{}^2 d\tau + \int \psi_2{}^2 d\tau + 2 \int \psi_1 \psi_2 d\tau] = 1$

The first two integrals are equal to unity because ψ_1 and ψ_2 are normalized, and $2 \int \psi_1 \psi_2 d\tau = 0$ because the wave functions are orthogonal. The result is $2N^2 = 1$ or

$N = \frac{1}{\sqrt{2}}$. The normalized wave function is thus $\frac{1}{\sqrt{2}}(\psi_1 + \psi_2)$.

b. $N^2 \int (\psi_1 - \psi_2)^2 \, d\tau = 1$

$N^2 [\int \psi_1{}^2 d\tau + \int \psi_2{}^2 d\tau + 2 \int \psi_1 \psi_2 d\tau] = 1$

Because of normalization, the first two integrals are equal to unity. The integral $\int \psi_1 \psi_2 d\tau = 0$ by the orthogonality condition. Thus, $N^2(2) = 1$ or $N = \dfrac{1}{\sqrt{2}}$ and the normalized wave function is $\dfrac{1}{\sqrt{2}}(\psi_1 - \psi_2)$.

c. $N^2 \int (\psi_1 + \psi_2 + \psi_3)^2 \, d\tau = 1$

$N^2 (\int \psi_1{}^2 d\tau + 2\int \psi_1 \psi_2 d\tau + 2\int \psi_1 \psi_3 d\tau + \int \psi_2{}^2 d\tau + 2\int \psi_2 \psi_3 d\tau + \int \psi_3{}^2) d\tau = 1$

By the normalization and orthogonality conditions,

$$N^2[1 + 2\,(0) + 2\,(0) + 1 + 2\,(0) + 1] = 1$$

$$3\,N^2 = 1$$

$$N = \frac{1}{\sqrt{3}}$$

and the wave function is $\dfrac{1}{\sqrt{3}} \, (\psi_1 + \psi_2 + \psi_3)$.

d. $N^2 \int \left(\psi_1 - \dfrac{1}{\sqrt{2}} \, \psi_2 + \dfrac{\sqrt{3}}{\sqrt{2}} \, \psi_3 \right)^2 \, d\tau = 1$

$N^2 [\int \psi_1^2 d\tau + \dfrac{1}{2} \int \psi_2^2 d\tau + \dfrac{3}{2} \int \psi_3^2 d\tau - \dfrac{2}{\sqrt{2}} \int \psi_1 \psi_2 d\tau - \dfrac{2\sqrt{3}}{\sqrt{2}} \int \psi_1 \psi_3 d\tau - \dfrac{2\sqrt{3}}{2} \int \psi_2 \psi_3 d\tau]$

By the normalization and orthogonality conditions

$$N^2 \left(1 + \frac{1}{2} + \frac{3}{2} - 0 - 0 - 0 \right) = 1.$$

$$3\,N^2 = 1$$

$$N = \frac{1}{\sqrt{3}}$$

and the wave function is

$$\frac{1}{\sqrt{3}} \left(\psi_1 - \frac{1}{\sqrt{2}} \psi_2 + \frac{\sqrt{3}}{\sqrt{2}} \psi_3 \right) = \frac{1}{\sqrt{3}} \, \psi_1 - \frac{1}{\sqrt{6}} \, \psi_2 + \frac{1}{\sqrt{2}} \, \psi_3$$

11.17. If Ae^{-ax} is an eigenfunction of $\dfrac{d^2}{dx^2}$, operation on Ae^{-ax} twice by d/dx will give the original function multiplied by a constant:

$$\frac{d}{dx}(Ae^{-ax}) = -Aae^{-ax}$$

$$\frac{d}{dx}(-Aae^{-ax}) = Aa^2 e^{-ax}.$$

Therefore, it is an eigenfunction, with eigenvalue a^2.

11.18.   For the wave function to be single valued,

$$\sin[m_l(\phi + 2\pi)] \text{ must equal } \sin m_l\phi.$$

The former function is

$$\sin[m_l\phi + 2\pi m_l] = \sin m_l\phi \cos 2\pi m_l + \sin 2\pi m_l \cos m_l\phi$$

For this to equal $\sin m_l\phi$, $\cos 2\pi m_l$ must be 1 and $\sin 2\pi m_l$ must be 0.

This can only occur if m_l is an integer.

11.19. $\Psi(x, y, z, t) = \psi(x, y, z)e^{-2\pi iEt/h}$

so that

$$\Psi^* = \psi^*(x, y, z)\, e^{2\pi iEt/h}$$

and

$$\Psi^*\Psi = \psi^*\psi\, e^0 = \psi^*\psi$$

which is independent of time.

11.20.   The operator for p_x (Table 11.1) is

$$\frac{h}{2\pi i}\frac{\partial}{\partial x}$$

If $\phi(x)$ and $\psi(x)$ are any two functions, the Hermitian condition is

$$\int_{-\infty}^{\infty}\phi^*\frac{h}{2\pi i}\frac{d\psi}{dx}\,dx = \int_{-\infty}^{\infty}\psi\left(\frac{h}{2\pi i}\right)^*\frac{d\phi^*}{dx}\,dx \qquad (1)$$

Integration by parts of the left-hand side gives

$$\int_{-\infty}^{\infty}\phi^*\frac{h}{2\pi i}\frac{d\psi}{dx}\,dx = \frac{h}{2\pi i}\left[\phi^*\psi\right]\Big|_{-\infty}^{\infty} - \int_{-\infty}^{\infty}\psi\left(\frac{h}{2\pi i}\cdot\frac{d\phi^*}{dx}\right)dx \qquad (2)$$

The first term is zero since all wave functions must asymptotically go to zero at $\pm\infty$. The second term is equal to

$$\int_{-\infty}^{\infty}\psi\left(\frac{h}{2\pi i}\right)^*\frac{d\phi^*}{dx}dx$$

which is the right-hand side of Eq. 1. The operator is therefore Hermitian.

11.21.   a. $\dfrac{dk}{dx} = 0 = 0 \times k$; k is an eigenfunction with the eigenvalue 0.

   b. $\dfrac{dkx^2}{dx} = 2kx$; kx^2 is not an eigenfunction.

   c. $\dfrac{d\sin kx}{dx} = k\cos kx$; $\sin kx$ is not an eigenfunction.

   d. $\dfrac{de^{kx}}{dx} = ke^{kx}$; e^{kx} is an eigenfunction with the eigenvalue k.

e. $\dfrac{de^{kx^2}}{dx} = 2kxe^{kx^2}$; e^{kx^2} is not an eigenfunction.

f. $\dfrac{de^{ikx}}{dx} = ike^{ikx}$; e^{ikx} is an eigenfunction with the eigenvalue ik.

11.22. The length of the vector in each case is given by
$$L = \sqrt{2(2+1)}\, \hbar = \sqrt{6}\, \hbar.$$

The Z component in each case is given by $m_l \hbar$. The cosine of the angle is, therefore, given by

$$\cos \theta = \frac{m_l \hbar}{\sqrt{6}\hbar} = \frac{m_l}{\sqrt{6}}.$$

We then find the angles to be as follows:

m_l	2	1	0	−1	−2
θ/deg.	35.3	65.9	90.0	114.1	144.7

11.23. According to Eq. (11.203), $\hat{L}_x^2 + \hat{L}_y^2 = \hat{L}^2 - \hat{L}_z^2$. Therefore,
$$\left(\hat{L}_x^2 + \hat{L}_y^2\right)\psi_{nlm} = \left(\hat{L}^2 - \hat{L}_z^2\right)\psi_{nlm} = [l(l+1) - m^2]\hbar^2 \psi_{nlm}.$$

Therefore, the wavefunctions ψ_{nlm} are eigenfunctions of the operator $\left(\hat{L}_x^2 + \hat{L}_y^2\right)$.

The operator corresponds to $x^2 + y^2$, and the equation
$$x^2 + y^2 = a^2$$

is the equation for a circle of radius a. The physical property corresponding to the operator is thus the square of the radius of the base of the angular momentum vector as it rotates about the Z axis (See Fig 11.20).

11.24. If the harmonic oscillator ground state energy were zero, the implication is that the kinetic and potential energies are both zero (since neither can take on negative values). Therefore, the momentum is exactly zero. Also, from the nature of the potential-energy function, the total energy can only be zero at $x = 0$. Therefore, this leads to a situation where we know the values of the momentum and position simultaneously and exactly. This is a violation of the Uncertainty Principle.

■ Particle in a Box

11.25. The lowest energy is given by Eq. 11.151, with
$$n_1 = n_2 = n_3 = 1.$$

a. If $a = 1.0 \times 10^{-11}$ m

$$E = \frac{3(6.626 \times 10^{-34} \text{ J s})^2}{8 \times 9.11 \times 10^{-31} \text{ kg} \times (10^{-11} \text{ m})^2}$$

$$= 1.81 \times 10^{-15} \text{ J}$$

$$= 1.13 \times 10^4 \text{ eV}$$

b. If $a = 10^{-15}$ m

$$E = \frac{3(6.626 \times 10^{-34} \text{ J s})^2}{8 \times 9.11 \times 10^{-31} \text{ kg} \times (10^{-15} \text{ m})^2}$$

$$= 1.81 \times 10^{-7} \text{ J}$$

$$= 1.13 \times 10^{12} \text{ eV}$$

The latter energy is so large that the electron would not remain in the nucleus, but would be emitted as a β particle.

11.26.   a.   The normalization condition is

$$\int_a^b \psi \psi^* dx = 1 = \int_a^b \frac{A^2}{x^2} dx$$

$$= A^2 \left[-\frac{1}{x} \right]_a^b = A^2 \left[-\frac{1}{b} + \frac{1}{a} \right]$$

Therefore

$$A^2 = \frac{ab}{b-a} \quad \text{and} \quad A = \sqrt{\frac{ab}{b-a}}$$

b.   The average value of x is

$$<x> = \int_a^b \psi^* x \psi \, dx$$

$$= A^2 \int_a^b \frac{1}{x} \, dx = A^2 \left[\ln x \right] \Big|_a^b$$

$$= \frac{ab}{b-a} \ln \frac{b}{a}$$

11.27.   The energy of the nth level is, from Eq. 11.149,

$$E_n = \frac{n^2 h^2}{8ma^2} = \frac{n^2 (6.626 \times 10^{-34} \text{ J s})^2}{8 \times 9.110 \times 10^{-31} \text{ kg} \times (10^{-9} \text{ m})^2}$$

$$= 6.024 \times 10^{-20} n^2 \text{ J}$$

$$= 0.376 n^2 \text{ eV}$$

At 10 eV, $n = \sqrt{\frac{10}{0.376}} = 5.17$

Thus, levels 1 through 5 have energies less than 10 eV.

$$\text{At } 100 \text{ eV}, \quad n = \sqrt{\frac{100}{0.376}} = 16.31$$

Therefore, levels 6 through 16, i.e., 11 levels, have energies between 10 and 100 eV.

11.28. The solution for a particle in a one-dimensional box (Eq. 11.148) is

$$\psi_n = \sqrt{\frac{2}{a}} \sin \frac{n\pi x}{a}$$

The momentum operator is $\boxed{\dfrac{\hbar}{2\pi i} \dfrac{d}{dx}}$:

$$\frac{\hbar}{2\pi i} \frac{d\psi_n}{dx} = \frac{\hbar}{2\pi i} \frac{n\pi}{a} \cdot \sqrt{\frac{2}{a}} \cos \frac{n\pi x}{a}$$

Since the result is not a constant multiplied by ψ_n, ψ_n is not an eigenfunction of the momentum operator. This conclusion is related to the Heisenberg uncertainty principle; the position and momentum operators do not commute, there are no common eigenfunctions, and the two properties cannot be measured simultaneously and precisely. However, the eigenfunction ψ_n, like any other function, can be expressed as a linear combination of the set of momentum eigenfunctions (compare Eq. 11.117 to 11.120). The physical significance of this is that the function ψ_n corresponds to a wave train of particular momentum being reflected at the walls of the box and giving rise to a wave train in the opposite direction.

11.29. a. The basic equation is

$$-\frac{h^2}{8\pi^2 m} \nabla^2 \psi + E_p(x, y, z)\psi = E\psi$$

b. The potential energy, E_p, can be set equal to zero inside the box, and therefore

$$\frac{\partial^2 \psi}{\partial x^2} + \frac{\partial^2 \psi}{\partial y^2} + \frac{\partial^2 \psi}{\partial z^2} = -\frac{8\pi^2 mE}{h^2}\psi$$

The energy, E, is separated into its component parts: $E = E_x + E_y + E_z$, and ψ is factored into the functions $X(x)$, $Y(y)$, and $Z(z)$. These values are substituted and the whole expression is divided by XYZ:

$$\frac{1}{XYZ} \frac{\partial^2 XYZ}{\partial x^2} + \frac{1}{XYZ} \frac{\partial^2 XYZ}{\partial y^2} + \frac{1}{XYZ} \frac{\partial^2 XYZ}{\partial z^2} = -\frac{8\pi^2 mE}{h^2}$$

In the first term, Y and Z do not depend on x and can be brought outside the derivative. We thus obtain three equations: one is

$$-\frac{1}{X} \frac{\partial^2 X}{\partial x^2} = \frac{8\pi^2 mE_x}{h^2}$$

and there are equivalent equations for Y and Z.

c. The preceding equations are identical with Eq. 11.128. The solutions are of the form of Eq. 11.144:

$$\psi_n = C \sin \frac{n\pi x}{a}$$

and normalization gives $C = \sqrt{\dfrac{2}{a}}$

d. The total energy is

$$E = E_x + E_y + E_z = \frac{n_x^2 h^2}{8ma^2} + \frac{n_y^2 h^2}{8mb^2} + \frac{n_z^2 h^2}{8mc^2}$$

If $a = b = c$

$$E = (n_x^2 + n_y^2 + n_z^2)\frac{h^2}{8ma^2}$$

11.30. The quantum mechanical probability is given by

$$P_{QM} = \left(\frac{2}{a}\right)\int_{a/3}^{2a/3} \sin^2\left(\frac{n\pi x}{a}\right) dx.$$

Using the fact that $\sin^2(bx)dx = \frac{1}{2}[1 - \cos(2bx)]$, we can evaluate the integral to obtain

$$P_{QM} = \frac{1}{3} - \frac{1}{n\pi}\left[\sin\left(\frac{4n\pi}{3}\right) - \sin\left(\frac{2n\pi}{3}\right)\right].$$

11.31. a. The classical probability is

$$P_{Cl} = \int_{a/3}^{2a/3} \frac{dx}{a} = \frac{1}{3}.$$

b. Since n can only be an integer, the quantity in the square brackets in the solution to the previous problem can only have three values:

$$\left[\sin\left(\frac{4n\pi}{3}\right) - \sin\left(\frac{2n\pi}{3}\right)\right] = 0, \text{ if } n = 3,6,9,...$$

$$= -\sqrt{3}, \text{ if } n = 1,4,7,10,...$$

$$= \sqrt{3}, \text{ if } n = 2,5,8,11,....$$

Therefore, we determine that as $n \to \infty$, the second term will vanish, yielding a result identical to the classical probability above.

11.32. $\Delta q\Delta p = \dfrac{h}{4p}$ and $\dfrac{p^2}{2m} = E$

$\Delta p = \sqrt{2m\Delta E}$

$\Delta E = \dfrac{(\Delta p)^2}{2m} = \dfrac{h^2}{32\pi^2 m(\Delta q)^2}$

a. For a cube of sides 10 pm $= 1.0 \times 10^{-11}$ m

$$\Delta E = \frac{(6.626 \times 10^{-34} \text{ J s})^2}{32\pi^2 \times 9.110 \times 10^{-31} \text{ kg } (1.0 \times 10^{-11} \text{ m})^2}$$

$$= 1.53 \times 10^{-17} \text{ J} = 95.3 \text{ eV}$$

b. For a cube of sides 1 fm $= 1.0 \times 10^{-15}$ m

$$\Delta E = \frac{(6.626 \times 10^{-34} \text{ J s})^2}{32\pi^2 \times 9.110 \times 10^{-31} \text{ kg } (1.0 \times 10^{-15} \text{ m})^2}$$

$$= 1.53 \times 10^{-9} \text{ J} = 9.53 \times 10^9 \text{ eV}$$

These uncertainties are considerably smaller than the energies calculated for the particle in a box. They are in fact smaller by the factor

$$\frac{3/8}{1/32\pi^2} = 12\pi^2 = 118.4$$

11.33. Wave functions for levels m and n are given by (Eq. 11.148),

$$\psi_m = \sqrt{\frac{2}{a}} \, \sin \frac{m\pi x}{a} \quad \text{and} \quad \psi_n = \sqrt{\frac{2}{a}} \, \sin \frac{n\pi x}{a}$$

Then

$$\int_0^a \psi_m \psi_n \, dx = \int_0^a \frac{2}{a} \, \sin \frac{m\pi x}{a} \sin \frac{n\pi x}{a} dx$$

Put $y = \pi x/a$; the limits are now π and 0

Then $dy = (\pi/a)dx$ and the integral becomes

$$\frac{2}{\pi} \int_0^\pi \sin my \sin ny \, dy$$

This is a standard integral and has the value

$$\frac{2}{\pi} \left[\frac{\sin(m-n)y}{2(m-n)} - \frac{\sin(m+n)y}{2(m+n)} \right]_0^\pi$$

At the lower limit ($y = 0$), both terms are zero since $\sin 0 = 0$. At the upper limit, both terms are zero since if m and n are integers, $\sin(m-n)$ and $\sin(m+n)$ are zero. The integral is thus zero and the wave functions are orthogonal.

$$E = \frac{-\dfrac{\hbar^2}{2m} \displaystyle\int_0^a x(a-x) \dfrac{d^2}{dx^2} x(a-x)}{\displaystyle\int_0^a x^2(a-x)^2 \, dx}.$$

11.34. Since $\hat{H} = -\dfrac{\hbar^2}{2m} \dfrac{d^2}{dx^2}$, we need to calculate

The numerator is easily found to be equal to $\hbar a^3/(6m)$, and the denominator is $a^5/30$. Therefore,

$$E = \frac{5\hbar^2}{ma^2} = \frac{5h^2}{4\pi^2 ma^2} = 1.0132 \frac{h^2}{8ma^2}.$$

The exact energy is $h^2/(8ma^2)$ so that the error is only 1.32%.

11.35. a. For $n = 2$, there is one node; for $n = 3$, there are two. (Note that we do not count the points $x = 0$ and $x = a$, where the wavefunction becomes zero but does not pass *through* it.)

b. From the expression in the Table, we need to find the values of r for which the equation

$$6 - \frac{4Zr}{a_0} + \frac{4Z^2 r^2}{9a_0^2} = 0$$

is satisfied. Since this is a quadratic in r, the solutions are

$$r_{\pm} = \frac{1}{2 \times \left(\frac{4Z^2}{9a_0^2} \right)} \left\{ \frac{4Z}{a_0} \pm \sqrt{ \left(\frac{4Z}{a_0} \right)^2 - 4 \times \frac{4Z^2}{9a_0^2} \times 6 } \right\}$$

$$= \frac{1}{2} \left(\frac{9a_0^2}{4Z^2} \right) \left\{ \frac{4Z}{a_0} \left[1 \pm \frac{\sqrt{3}}{3} \right] \right\}$$

or, upon simplification, $r_{\pm} = \frac{3a_0}{2Z} \left(3 \pm \sqrt{3} \right)$. Therefore, the two radial nodes are located at

distances of $r = \frac{3a_0}{2Z} \left(3 - \sqrt{3} \right)$ and $r = \frac{3a_0}{2Z} \left(3 + \sqrt{3} \right)$ from the nucleus.

■ Vibration and Rotation

11.36. Frequency of vibration,

$$\nu_0 = 2.998 \times 10^{10} \text{ cm s}^{-1} \times 2360 \text{ cm}^{-1}$$

$$= 7.075 \times 10^{13} \text{ s}^{-1}$$

$$h\nu_0 = 7.075 \times 10^{13} \times 6.626 \times 10^{-34} = 4.69 \times 10^{-20} \text{ J}$$

Zero-point energy $= \frac{1}{2} h\nu_0 = 2.34 \times 10^{-20} \text{ J}$

Energy at $v = 1$ is $\frac{3}{2} h\nu_0 = 7.03 \times 10^{-20} \text{ J}$

11.37. The energy is related to the angular momentum by Eq. 11.215:

$$E = \frac{L^2}{2I}$$

The Hamiltonian operator is therefore

$$\hat{H} = \frac{1}{2I} \left(\frac{h}{2\pi i} \right)^2 \frac{\partial^2}{\partial \phi^2}$$

$$= -\frac{h^2}{8\pi^2 I} \frac{\partial^2}{\partial \phi^2}.$$

■ The Atom

11.38. The energy required to remove the electron from the lowest energy level in hydrogen ($n_1 = 1$) to infinity ($n_2 = \infty$) is the ionization potential. Using Eq. 11.50 and the value of the Rydberg constant, we have

$$\bar{v} = 1.0968 \times 10^7 \ (\text{m}^{-1})\left(\frac{1}{1^2} - \frac{1}{\infty^2}\right) = 1.0968 \times 10^7 \ \text{m}^{-1}$$

The energy required is

$$E = hc\bar{v} = 6.626 \times 10^{-34} \ (\text{J s}) \ 2.998 \times 10^8 \ (\text{m s}^{-1}) \cdot 1.0968 \times 10^{17} \ \text{m}^{-1}$$

$$= 2.179 \times 10^{-18} \ \text{J}$$

$$= \frac{2.179 \times 10^{-18} \ \text{J}}{1.602 \times 10^{-19} \ \text{J/eV}} = 13.60 \ \text{eV}$$

11.39. From Eq. 11.39, with $n = 1$,

$$u = \frac{h}{2\pi mr}$$

and by Eq. 11.44, $r = a_0 = 52.92$ pm.

Therefore

$$u = \frac{6.626 \times 10^{-34} \ \text{J s}}{2\pi \times 9.110 \times 10^{-31} \ \text{kg} \times 5.29 \times 10^{-11} \ \text{m}}$$

$$= 2.19 \times 10^6 \ \text{m s}^{-1}$$

By Eq. 11.56,

$$\lambda = \frac{h}{mu} = \frac{6.626 \times 10^{-34} \ \text{J s}}{9.110 \times 10^{-31} \ \text{kg} \times 2.1 \ \text{g} \times 10^6 \ \text{m s}^{-1}}$$

$$= 3.32 \times 10^{-10} \ \text{m} = 332 \ \text{pm}$$

From Eq. 11.39,

$$u = \frac{nh}{2\pi mr}$$

and therefore, since for $Z = 1$, $r = n^2 a_0$

$$\lambda = \frac{h}{mu} = \frac{2\pi r}{n} = 2\pi a_0 n$$

The circumference of the orbit is $2\pi n^2 a_0$, and the ratio of the circumference to λ is therefore n.

11.40. The radius of the sphere of maximum probability corresponds to the maximum in the radial probability function $4\pi r^2 |\psi_{1s}|^2$. This is found by differentiating the radial probability function with respect to r, setting it equal to zero, and solving for the value of r that satisfies the resulting equation. In other words, we set

$$\frac{d}{dr}\left[4\pi r^2 \left(\frac{Z}{a_0}\right)^3 e^{-2Zr/a_0} \right] = 4\pi \left(\frac{Z}{a_0}\right)^3 \left(2r - \frac{2Z^2 r^2}{a_0}\right) e^{-2Zr/a_0} = 0.$$

This means that the value of r that satisfies the equation is

$$r = a_0/Z,$$

which, for the case of the 1s electron ($n = 1$), is identical to the expression of Eq. (11.43).

11.41. From Eq. 11.148,

$$\mu_H = \frac{m_e m_H}{m_e + m_H} = \frac{9.1095 \times 1.6727 \times 10^{-58}}{9.1095 \times 10^{-31} + 1.6727 \times 10^{-27}}$$

$$= \frac{15.2374 \times 10^{-58}}{1.6736 \times 10^{-27}} = 9.1046 \times 10^{-31} \text{ kg}$$

$$\mu_D = \frac{m_e m_D}{m_e + m_D} = \frac{9.1095 \times 3.3434 \times 10^{-58}}{9.1095 \times 10^{-31} + 3.3434 \times 10^{-27}}$$

$$= \frac{30.4567 \times 10^{-58}}{3.3434 \times 10^{-27}} = 9.1070 \times 10^{-31} \text{ kg}$$

a. By Eq. 11.44, the Bohr radius is inversely proportional to μ, and is therefore slightly smaller for D than for H.

   The energies are inversely proportional to the Bohr radii (Eq. 11.44) and are therefore slightly greater for D than for H. The frequencies of the transitions are therefore slightly greater for D, and the wavelengths are slightly shorter.

b. The wavelengths are in the inverse ratio of the reduced masses, and are therefore in the ratio

$$\frac{\lambda(D)}{\lambda(H)} = \frac{9.1046}{9.1070}$$

   The wavelength of the line in the spectrum of D is therefore

$$\frac{9.1046}{9.1070} \times 656.47 = 656.30 \text{ nm}$$

11.42. $\bar{v} = \frac{1}{\lambda} = 1.0968 \times 10^7 \text{ m}^{-1} \left(\frac{1}{4^2} - \frac{1}{5^2}\right)$

$$= 2.468 \times 10^5 \text{ m}^{-1}$$

$$\lambda = \frac{1}{\bar{v}} = 4.052 \times 10^{-6} \text{ m}$$

This is in the infrared region of the electromagnetic spectrum.

$$E = hc\bar{v} = 6.626 \times 10^{-34} \text{ (J s)} \ 2.998 \times 10^8 \text{ (s m}^{-1}) \ 2.468 \times 10^5 \text{ m}^{-1} = 4.903 \times 10^{-20} \text{ J}$$

11.43. From Eq. 11.48, with $Z = 1$ and $n = 2$,

$$E_p = -\frac{e^2}{4\pi\varepsilon_0 4 a_0}$$

$$= -\frac{(1.602 \times 10^{-19} \ C)^2}{16\pi(8.854 \times 10^{-12} \ C^2 \ N^{-1} \ m^{-2})(5.292 \times 10^{-11} \ m)}$$

$$= -10.9 \times 10^{-18} \ J$$

In atomic units of $e^2/4\pi\varepsilon_0 a_0$,

$$E_p = -0.250 \ au$$

11.44. Problems 11.44 and 11.45 are conveniently worked out with reference to Problem 11.38, where the ionization energy of H was calculated to be 13.6 eV. According to Eq. 11.49, the first ionization energy is proportional to Z_{eff}^2/n^2, where n is the quantum number of the most easily removed electron. For H, $Z_{eff} = 1$ and $n = 1$, so that the ionization energy is

$$I/eV = 13.6 \frac{Z_{eff}^2}{n^2}$$

For Li, $n = 2$ and therefore

$$5.39 = 13.6 \frac{Z_{eff}^2}{4}$$

Thus

$$Z_{eff} = 1.26$$

11.45. Similarly, for Na, where $n = 3$,

$$5.14 = 13.6 \frac{Z_{eff}^2}{9}$$

and

$$Z_{eff} = 1.84$$

11.46. a. The nuclear charge of Cl is 17. We subtract

   0.30 for the other 3s electron,

   5×0.35 for the five 3p electrons,

   8×0.85 for the eight 2s and 2p electrons,

   2×1.00 for the two 1s electrons,

Thus, $\sigma = 10.85$

and $Z_{eff} = 17 - 10.85 = 6.15$

   b. The nuclear charge of P is 15. We subtract

   2×0.30 for the 3s electron,

   2×0.35 for the other 3p electrons,

   8×0.85 for the eight 2s and 2p electrons,

2×1.00 for the two 1s electrons.

Thus, $\sigma = 10.1$

and $Z_{eff} = 15 - 10.1 = 4.9$

c. The nuclear charge of K is 19. We subtract

8×0.85 for the 3s and 3p electrons,

10×1.00 for the 1s, 2s, and 2p electrons.

Thus, $\sigma = 16.8$

and $Z_{eff} = 19 - 16.8 = 2.2$

11.47. We minimize E with respect to Z_{eff}:

$$\frac{\partial E}{\partial Z_{eff}} = 0 = \frac{e^2}{a_0}\left(2Z_{eff} - \frac{27}{8}\right),$$

from which we obtain $Z_{eff} = 27/16 = 1.6875$. The energy corresponding to this is

$$E = \frac{e^2}{a_0}\left(1.6875^2 - \frac{27 \times 1.6875}{8}\right) = \frac{-2.848\,e^2}{a_0}.$$

The effective charge is less than the true charge because of screening; each electron tends to screen the other from the nucleus.

11.48. From the expression for ψ in Table 11.5, p. 543,

$$\psi^*\psi = \frac{1}{\pi a_0^3}\,e^{-2r/a_0}$$

Multiplying by the volume element in spherical coordinates we have

$$\frac{1}{\pi a_0^3}\,e^{-2r/a_0}\,r^2dr\,\sin\theta\,d\theta\,d\phi$$

Integration of θ from 0 to π and ϕ from 0 to 2π gives

$$\int_0^\pi \sin\theta\,d\theta = -\cos\theta\,\Big|_0^\pi = -[(-1) - (1)] = 2$$

and

$$\int_0^\pi d\phi = 2\pi$$

so that

$$\int\int \frac{1}{\pi a_0^3}\,e^{-2r/a_0}\,r^2dr\,\sin\theta\,d\theta\,d\phi$$

$$= \frac{4\pi}{\pi a_0^3}\,r^2\,e^{-2r/a_0}\,dr$$

$$= \frac{4}{a_0^3} \, r^2 \, e^{-2r/a_0} \, dr$$

11.49.   From Tables 11.2 and 11.3, the functions we require are

$$\Theta_{10}\Phi_0 = \frac{\sqrt{6}}{2} \, \cos\theta \, \frac{1}{\sqrt{2\pi}}$$

$$\Theta_{11}\Phi_1 = \frac{\sqrt{3}}{2} \, \sin\theta \, \frac{1}{\sqrt{\pi}} \, \cos\phi$$

$$\Theta_{1-1}\Phi_{-1} = \frac{\sqrt{3}}{2} \, \sin \, \frac{1}{\sqrt{\pi}} \, \sin\phi$$

The sum of their squares is

$$\frac{3}{4\pi} \cos^2\theta + \frac{3}{4\pi}\sin^2\theta \cos^2\phi + \frac{3}{4\pi}\sin^2\theta \sin^2\phi$$

$$= \frac{3}{4\pi}\cos^2\theta + \frac{3}{4\pi}\sin^2\theta \, (\cos^2\phi + \sin^2\phi)$$

$$= \frac{3}{4\pi} \, (\text{since } \cos^2\theta + \sin^2\theta = 1)$$

This is independent of θ and ϕ.

12. THE CHEMICAL BOND

■ Bond Energies, Shapes of Molecules, and Dipole Moments

12.1. The resultant energy is

$$E_p/\text{kJ mol}^{-1} = -\frac{137.2}{r/\text{nm}} + \frac{0.0975}{(r/\text{nm})^6}$$

$$\frac{d(E_p/\text{kJ mol}^{-1})}{d(r/\text{nm})} = \frac{137.2}{(r/\text{nm})^2} - \frac{6 \times 0.0975}{(r/\text{nm})^7}$$

This is zero when $r = r_0$; thus

$$(r_0/\text{nm})^5 = \frac{6 \times 0.0975}{137.2} = 4.264 \times 10^{-3}$$

$$r_0 = 0.336 \text{ nm} = 336 \text{ pm}$$

When $r = 0.336$ nm,

$$E_p = \frac{-137.2}{0.336} + \frac{0.0975}{0.336^6} = -341 \text{ kJ mol}^{-1}$$

12.2. If the bond were completely ionic, the dipole moment would be

$$\mu = (1.602 \times 10^{-19} \text{ C}) (239 \times 10^{-12} \text{ m})$$

$$= 3.83 \times 10^{-29} \text{ C m}$$

The percentage ionic character is thus

$$\frac{2.09}{3.83} \times 100 = 55\%$$

12.3.   a. $E_\text{ionic} = D(\text{LiH}) - [D(\text{Li}_2) D(\text{H}_2)]^{1/2}$

$$= 243 - (113 \times 435)^{1/2} = 21 \text{ kJ mol}^{-1}$$

$$\therefore \left| \chi_{Li} - \chi_{H} \right| = \frac{\sqrt{21}}{10} = 0.46$$

Thus, the estimated dipole moment is 0.46 D.

    b.   If the molecule were completely ionic, the dipole moment would be

$$4.8 \times (1.26 + 0.36) = 7.78 \text{ D}$$

$$\% \text{ ionic character} = \frac{0.46}{7.78} \times 100 = 5.9\%$$

12.4.

$BeCl_2$ (no lone pairs)	Linear
SF_6 (no lone pairs)	Octahedral
H_3O^+ (one lone pair)	Triangular-pyramid
NH_4^+ (no lone pair)	Tetrahedral
PCl_6^- (no lone pair)	Octahedral
AlF_6^{3-} (no lone pair)	Octahedral
PO_4^{3-} (no lone pair)	Tetrahedral
CO_2 (no lone pair)	Linear
SO_2 (one lone pair)	Bent
NH_3^{2+} (no lone pair)	Trigonal Planar
CO_3^{2-} (no lone pair)	Trigonal Planar
NO_3^- (no lone pair)	Trigonal Planar

12.5.   HCl: μ_{ionic} $= 1.602 \times 10^{-19}$ C $\times 127 \times 10^{-12}$ m

$$= 2.03 \times 10^{-29} \text{ C m}$$

$$\% \text{ ionic character} = \frac{3.60 \times 10^{-30} \times 100}{2.03 \times 10^{-29}} = 17.7\%$$

HBr: μ_{ionic} $= 1.602 \times 10^{-19}$ C $\times 141 \times 10^{-12}$ m

$$= 2.26 \times 10^{-29} \text{ C m}$$

$$\% \text{ ionic character} = \frac{2.67 \times 10^{-30} \times 100}{2.26 \times 10^{-29}} = 11.8\%$$

HI: μ_{ionic} $= 1.602 \times 10^{-19}$ C $\times 160 \times 10^{-12}$ m

$$= 2.56 \times 10^{-29} \text{ C m}$$

$$\% \text{ ionic character} = \frac{1.40 \times 10^{-30} \times 100}{2.56 \times 10^{-29}} = 5.5\%$$

$$CO: \mu_{\text{ionic}} = 1.602 \times 10^{-19} \text{ C} \times 113 \times 10^{-12} \text{ m}$$

$$= 1.81 \times 10^{-29} \text{ C m}$$

$$\% \text{ ionic character} = \frac{0.33 \times 10^{-30} \times 100}{1.81 \times 10^{-29}} = 1.8\%$$

■ Molecular Orbitals

12.6.   The MO diagrams are based on Figures 12.21(a) and 12.23. Then:

No. of electrons		Configuration	Bond Order	Number of Unpaired Electrons
B_2	10	$(1s\sigma_g)^2(1s\sigma_u^*)^2(2s\sigma_g)^2(2s\sigma_u^*)^2(2p\sigma_g)^2$	1	0
CO	14	$(1s\sigma)^2(1s\sigma^*)^2(2s\sigma)^2(2s\sigma^*)^2(2p\sigma)^2(2p\pi)^4$	3	0
BN	12	$(1s\sigma)^2(1s\sigma^*)^2(2s\sigma)^2(2s\sigma^*)^2(2p\sigma)^2(2p\pi)^2$	2	2
BN^{2-}	14	Same as CO	3	0
BO	13	$(1s\sigma)^2(1s\sigma^*)^2(2s\sigma)^2(2s\sigma^*)^2(2p\sigma)^2(2p\pi)^3$	2.5	1
BF	14	Same as CO	3	0
OF	17	$(1s\sigma)^2(1s\sigma^*)^2(2s\sigma)^2(2s\sigma^*)^2(2p\sigma)^2(2p\pi)^4(2p\pi^*)^3$	1.5	1
OF^-	18	$(1s\sigma)^2(1s\sigma^*)^2(2s\sigma)^2(2s\sigma^*)^2(2p\sigma)^2(2p\pi)^4(2p\pi^*)^4$	1	0
OF^+	16	$(1s\sigma)^2(1s\sigma^*)^2(2s\sigma)^2(2s\sigma^*)^2(2p\sigma)^2(2p\pi)^4(2p\pi^*)^2$	2	2

Species with no unpaired electrons are diamagnetic; others are paramagnetic.

12.7. N_2: See Fig. 12.21(a). An added electron goes into the antibonding $2p\pi_g^*$ level; the bond is weakened. An electron removed leaves from the $2p\sigma_g$ level; the bond is weakened.

O_2: See Fig. 12.21(b). Added electron goes into antibonding $2p\pi_g^*$ level; the bond is weakened. An electron removed leaves from the $2p\pi_g^*$ level; the bond is strengthened.

C_2: The electronic configuration is

$$(1s\sigma_g)^2(1s\sigma_u^*)^2(2s\sigma_g)^2(2s\sigma_u^*)^2(2p\sigma_g)^2(2p\pi_u)^2$$

Removal of a $2p\pi_u$ electron will weaken the bond; addition will strengthen it.

F_2: The configuration is

$$(1s\sigma_g)^2(1s\sigma_u^*)^2(2s\sigma_g)^2(2s\sigma_u^*)^2(2p\sigma_g)^2(2p\pi_u)^4(2p\pi_g^*)^4$$

Removal of a $2p\pi_g^*$ electron will strengthen the bond; an added electron will go into the $2p\sigma_u^*$ level and weaken it.

CN: The configuration is

$$(1s\sigma)^2(1s\sigma^*)^2(2s\sigma)^2(2s\sigma^*)^2(2p\sigma)^2(2p\pi)^3$$

Removal of a $2p\pi$ electron will weaken the bond; an added electron will go into the $2p\pi$ level and strengthen it.

NO: The configuration is

$$(1s\sigma)^2(1s\sigma^*)^2(2s\sigma)^2(2s\sigma^*)^2(2p\sigma)^2(2p\pi)^4\ 2p\pi^*$$

Removal of a $2p\pi^*$ electron will strengthen the bond; an added electron will go into the $2p\pi^*$ level and weaken it.

Thus,

a. Addition of an electron will strengthen C_2 and CN, and will weaken N_2, O_2, F_2, and NO.

b. Removal of an electron will strengthen O_2, F_2, and NO, and will weaken N_2, C_2, and CN.

12.8. We are given that

$$\int (1s_A)^2\ d\tau = 1 \quad \text{and} \quad \int (1s_B)^2\ d\tau = 1$$

We are required to prove that

$$\int (1s_A + 1s_B)\ (1s_A - 1s_B)\ d\tau = 0$$

(The 1s wave functions are real so that they are the same as their complex conjugates). The integral is

$$\int (1s_A)^2\ d\tau - \int (1s_B)^2\ d\tau$$

which is zero since each term is unity.

(Note that another approach is to recognize that the integral of the product of wave functions is a product of even and odd functions about zero. Since the resultant function is odd, the integral must go to zero.)

12.9. a. $\int t_3 t_3^*\ d\tau = \int t_3^2\ d\tau$ (since the functions are real)

$$= \frac{1}{4}\int (s - p_x + p_y - p_z)^2\ d\tau$$

$$= \frac{1}{4}\int (s^2 + p_x^2 + p_y^2 + p_z^2 - 2sp_x + 2sp_y + \ldots)\ d\tau$$

All cross terms are zero, and the remainder are unity; the integral is thus

$$\frac{1}{4}(1 + 1 + 1 + 1) = 1$$

b. $\int t_2 t_4 {}^* \, d\tau = \int t_2 t_4 \, d\tau$ (since the functions are real)

$$= \frac{1}{4} \int (s + p_x - p_y - p_z)(s - p_x - p_y + p_z) \, d\tau$$

$$= \frac{1}{4} \int (s^2 - p_x^2 + p_y^2 - p_z^2 + \text{cross terms, which are zero}) \, d\tau$$

$$= \frac{1}{4}(1 - 1 + 1 - 1) = 0$$

12.10. $\int \psi_1^2 \, d\tau = \int (as + bp_x)^2 \, d\tau = \int (a^2 s^2 + b^2 p_x^2 + 2ab s p_x) \, d\tau = a^2 + b^2 = 1$ (1)

$\int \psi_2^2 \, d\tau = \int (as - \frac{1}{2} b p_x + \frac{\sqrt{3}}{2} c p_y)^2 \, d\tau$

$$= a^2 + \frac{b^2}{4} + \frac{3}{4} c^2 = 1 \qquad\qquad (2)$$

$\int \psi_3^2 \, d\tau = \int (as - \frac{1}{2} b p_x - \frac{\sqrt{3}}{2} c p_y)^2 \, d\tau$

$$= a^2 + \frac{b^2}{4} + \frac{3}{4} c^2 = 1 \qquad\qquad (3)$$

$\int \psi_1 \psi_2 \, d\tau = \int (as + bp_x)(as - \frac{1}{2} b p_x + \frac{\sqrt{3}}{2} c p_y) \, d\tau$

$$= a^2 + \frac{1}{2} b^2 = 0 \qquad\qquad (4)$$

From (1) and (4) $\frac{3}{2} b^2 = 1$; $b = \sqrt{\frac{2}{3}}$

$$a^2 = \frac{1}{3}; \quad a = \frac{1}{\sqrt{3}}$$

From (3) $\frac{1}{3} + \frac{1}{6} + \frac{3}{4} c^2 = 1$; $c^2 = \frac{2}{3}$; $c = \sqrt{\frac{2}{3}}$

(This has only used the fact that ψ_1 and ψ_2 are orthogonal; it is easily confirmed that ψ_3 is also orthogonal to ψ_1 and ψ_2.) The normalized wave functions are thus

$$\psi_1 = \frac{1}{\sqrt{3}} s + \frac{2}{\sqrt{3}} p_x$$

$$\psi_2 = \frac{1}{\sqrt{3}} s - \frac{1}{\sqrt{6}} p_x + \frac{1}{\sqrt{2}} p_y$$

$$\psi_3 = \frac{1}{\sqrt{3}} s - \frac{1}{\sqrt{6}} p_x - \frac{1}{\sqrt{2}} p_y$$

This process, called Schmidt orthogonalization, is a common and useful method for resolving orthonormal functions in quantum chemistry.

12.11.   In the main text, the diagram in problem 12.10 looks down the Z axis and shows the contributions along the X and Y axes due to the s and p_z orbitals. The wave functions are thus

$$\psi_1 = as + bp_z$$

$$\psi_2 = as - bp_z$$

The normalization condition for ψ_1 is

$$\int \psi_1^2 \, d\tau = (as + bp_z)^2 \, d\tau = a^2 + b^2 = 1$$

The orthogonality condition is

$$\int \psi_1 \psi_2 \, d\tau = (as + bp_z)(as - bp_z) \, d\tau$$

$$= a^2 - b^2 = 0$$

Thus

$$2b^2 = 1; \quad b = \frac{1}{\sqrt{2}}; \quad a = \frac{1}{\sqrt{2}}$$

The wave functions are therefore

$$\psi_1 = \frac{1}{\sqrt{2}}(s + p_z)$$

$$\psi_2 = \frac{1}{\sqrt{2}}(s - p_z) \, .$$

12.12.   We will use a short-hand notation for the integrals by writing

$$\left\langle L^2 \right\rangle_i = \int_{-\infty}^{\infty} \psi_i \hat{L}^2 \psi_i d\tau = \left\langle \psi_i \left| \hat{L}^2 \right| \psi_i \right\rangle; i = 1,2,3.$$

Now, recalling that $\hat{L}^2 \psi_{nlm} = l(l+1)\hbar^2 \, \psi_{nlm}$, and the orthonormal property of the orbitals belonging to the same atom (see Prob. 12.9), we find

$$\left\langle \psi_1 \left| \hat{L}^2 \right| \psi_1 \right\rangle = \frac{1}{3}[0(0+1)]\hbar^2 \left\langle 2s \left| 2s \right\rangle + \frac{2}{3}[1(1+1)]\hbar^2 \left\langle 2p_x \left| 2p_x \right\rangle = \frac{4}{3}\hbar^2.$$

$$\left\langle \psi_2 \left| \hat{L}^2 \right| \psi_2 \right\rangle = 0 + \frac{1}{6}[1(1+1)]\hbar^2 \left\langle 2p_x \left| 2p_x \right\rangle + \frac{1}{2}[1(1+1)]\hbar^2 \left\langle 2p_y \left| 2p_y \right\rangle = \frac{4}{3}\hbar^2.$$

$$\left\langle \psi_3 \left| \hat{L}^2 \right| \psi_3 \right\rangle = 0 + \frac{1}{6}[1(1+1)]\hbar^2 \left\langle 2p_x \left| 2p_x \right\rangle + \frac{1}{2}[1(1+1)]\hbar^2 \left\langle 2p_y \left| 2p_y \right\rangle = \frac{4}{3}\hbar^2.$$

12.13.   A paramagnetic molecule is one with unpaired electrons. The C_2 molecule has 12 electrons. Let us write the molecular electronic configuration based on (a) the energy level ordering of Fig. 12.21(a) and (b) the energy level ordering of Fig. 12.21(b).

a. C_2: $(1s\sigma_g)^2(1s\sigma_u^*)^2(2s\sigma_g)^2(2s\sigma_u^*)^2(2p\pi_u)^4$. All electrons are paired; the molecule is diamagnetic.

b. C_2: $(1s\sigma_g)^2(1s\sigma_u^*)^2(2s\sigma_g)^2(2s\sigma_u^*)^2(2p\sigma_g)^2(2p\pi_u)^2$. The electrons in the $2p\pi_u$ orbital are unpaired (Hund's rule) and, therefore, the molecule will be paramagnetic.

12.14. First determine the energy of the molecular orbital for each geometric arrangement. Then place electrons in the lowest energy orbital.

The determinant for the linear arrangement is

$$\begin{vmatrix} \alpha - E & \beta & 0 \\ \beta & \alpha - E & \beta \\ 0 & \beta & \alpha - E \end{vmatrix} = 0,$$

where we have assumed that $H_{11} = H_{22} = H_{33} = \alpha$, $H_{12} = H_{23} = \beta$ and $H_{13} = 0$. Dividing throughout by b and setting $(\alpha - E)/\beta = x$, the determinant reduces to

$$\begin{vmatrix} x & 1 & 0 \\ 1 & x & 1 \\ 0 & 1 & x \end{vmatrix} = 0,$$

which leads to

$$x^3 - 2x = 0, \text{ or } x = 0, \pm\sqrt{2}.$$

Then, $E = \alpha + \beta\sqrt{2}$, α, and $\alpha - \beta\sqrt{2}$.

For H_3^+ in the triangular arrangement, we get the determinant

$$\begin{vmatrix} \alpha - E & \beta & \beta \\ \beta & \alpha - E & \beta \\ \beta & \beta & \alpha - E \end{vmatrix} = 0.$$

Making the substitutions (see above) gives

$$\begin{vmatrix} x & 1 & 1 \\ 1 & x & 1 \\ 1 & 1 & x \end{vmatrix} = 0,$$

which gives

$$x^3 - 3x + 2 = 0, \text{ or } x = +1, +1, -2.$$

Then, $E = \alpha + 2\beta$, $\alpha - \beta$, and $\alpha - \beta$.

When two electrons occupy the same energy level, $E = 2\alpha + 4\beta$. This energy is lower than the energy $2(\alpha + \beta\sqrt{2})$ because β is a negative quantity. Therefore, the triangular arrangement is predicted to be lower in energy.

12.15. The c_n are obtained as the quotient of c_n/c_1, divided by $\sqrt{\Sigma(c_n/c_1)^2}$. For $x = -1.6180$, we set up the table:

n	c_n/c_1	$(c_n/c_1)^2$	c_n
1	1.0000	1.0000	0.3717
2	1.6180	2.6180	0.6015
3	1.6180	2.6180	0.6015
4	1.0000	1.0000	0.3717
	$\Sigma(c_n/c_1)^2) =$	7.23598	$(2.6900)^2$

■ Group Theory

12.16.

Shape		Symmetry Elements	Point Group
a.	Equilateral triangle	$E, 2C_3, 3C_2, \sigma_h, 3\sigma_v, 2S_3$	D_{3h}
b.	Isosceles triangle	$E, C_2, 2\sigma_v$	C_{2v}
c.	Cylinder	$E, C_\infty, \infty C_2, \infty\sigma_v, \sigma_h, i$	$D_{\infty h}$

12.17.

Molecule	Symmetry Elements	Point Group
$CHCl_3$	$E, C_3, 3\sigma_v$	C_{3v}
CH_2Cl_2	$E, C_2, 2\sigma_v$	C_{2v}
Naphthalene	$E, C_2, 2C_2, 3\sigma, i$	D_{2h}
Chlorobenzene	$E, C_2, 2\sigma_v$	C_{2v}
NO_2	$E, C_2, 2\sigma_v$	C_{2v}
Cyclopropane	$E, C_3, 3C_2, 3\sigma_v, \sigma_h$	D_{3h}
CO_3^{2-}	$E, C_3, 3C_2, 3\sigma_v, \sigma_h$	D_{3h}
C_2H_2	$E, C_\infty, \infty C_2, \infty\sigma_v, \sigma_h, i$	$D_{\infty h}$

12.18. H_2O_2 belongs to the C_2 point group and has neither a plane of symmetry nor a center of inversion. It therefore can exist in two enantiomeric forms. However, optical activity has not been detected because of the rapid interconversion of the two forms.

12.19.   a. H_2; point group $D_{\infty h}$; σ_g^+

b. H_2O; point group C_{2v}; a_1, b_2, a_1

c. CO_2; point group $D_{\infty h}$; σ_g^+, σ_u^+

d. BF_3; point group D_{3h}; a_1'

e. NH_3; point group C_{3v}; a_1

f. HCN; point group $C_{\infty v}$; σ^+

12.20.   Cyclopropane:

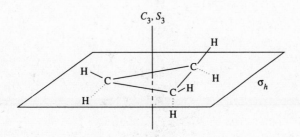

The C_3 axis is also the S_3 axis:

$$S_3 = C_3\sigma_h$$

Cyclopentane:

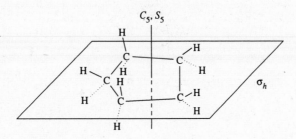

The C_5 axis is the S_5 axis:

$$S_5 = C_5\sigma_h$$

The point groups C_{2h}, D_{2h}, $D_{\infty h}$, and O_h have centers of symmetry. The groups D_{3h} and T_d have axes of improper rotation. The groups in Table 12.3 for which there can be a dipole moment are the remaining ones: C_1, C_2, C_{2v}, C_{3v}, and $C_{\infty v}$.

12.21. It is clear from Fig. 11.16 that the d_{z^2} and $d_{x^2-y^2}$ orbitals have lobes pointing along the cartesian axes while the other three have lobes pointing between cartesian axes. Therefore, if Fe lies at the origin and the ligands lie on the cartesian axes, the d_{z^2} and $d_{x^2-y^2}$ orbitals will experience most of the electron-electron repulsion. The other three, since their probability density is highest between the cartesian axes, experience the repulsive forces to a lesser extent. Therefore, in the presence of the approaching ligands, the set with E_g symmetry (d_{z^2} and $d_{x^2-y^2}$) will have higher energy, and the set with T_{2g} symmetry (d_{xy}, d_{xz}, and d_{yz}) will be lower in energy.

13. FOUNDATIONS OF CHEMICAL SPECTROSCOPY

■ Absorption of Radiation

13.1. $I/I_0 = 0.767$ $I_0/I = 1.304$

$$\log_{10} 1.304 = 0.1152 = A = \varepsilon cl$$

$$0.1152 = 532 \times c \times 1 \text{ cm}$$

$$c = 2.166 \times 10^{-4} \text{ mol dm}^{-3}$$

$$c = 13.86 \text{ g dm}^{-3}$$

13.2. $I_0/I = 78/55$

Absorbance $= \log_{10} (78/55) = \log_{10} 1.418 = 0.152$

Transmittance $= I/I_0 = 55/78 = 0.705$

Molar absorption coefficient.

$$\varepsilon = \frac{0.152}{0.1 \text{ mol dm}^{-3} \times 0.5 \text{ cm}} = 3.04 \text{ dm}^3 \text{ cm}^{-1} \text{ mol}^{-1}$$

13.3. From Eq. 13.45,

$$\log_{10} I/I_0 = -\varepsilon cl/2.303$$

a plot of $\log_{10} I/I_0$ (log T) against concentration should give a straight line, the slope of which is $-\varepsilon l/2.303$ where ε is the absorption coefficient. The plot is shown.

Some regular calculators and any programmable calculator can do a least squares fit of the data provided and give the slope of the line. Alternatively, from the plot of the data, the slope of the curve can be determined. The slope is

$$\frac{-0.450 - (-0.85)}{(0.96 - 1.785) \times 10^{-4}} = -\frac{0.400}{0.825 \times 10^{-4}} = -4.85 \times 10^3 \text{ dm}^3 \text{ mol}^{-1}$$

$$\frac{-0.450 - (-0.85)}{(0.96 - 1.785) \times 10^{-4}} = -\frac{0.400}{0.825 \times 10^{-4}} = -4.85 \times 10^3 \text{ dm}^3 \text{ mol}^{-1}$$

From the equation above, the slope is $-\varepsilon l/2.303$. From the graph, the slope is -4.85×10^3 dm^3 mol^{-1}. Equating, we have

$$\varepsilon l = 2.303 \times 4.85 \times 10^3 \text{ dm}^3 \text{ mol}^{-1}$$

Since $l = 1$ cm $= 0.1$ dm,

$$\varepsilon = 11\,170 \text{ dm}^3 \text{ mol}^{-1} \text{ cm}^{-1} = 1117 \text{ m}^2 \text{ mol}^{-1}$$

where ε is the molar (decadic) absorbance coefficient. For a 10 M solution

$$A = \varepsilon cl = 11\,170 \text{ (dm}^3 \text{ mol}^{-1} \text{ cm}^{-1}) \times 10 \times 10^{-6} \text{ (mol dm}^{-3}) \times 1 \text{ (cm)}$$

$$= 0.1117 \quad \text{rounded to} \quad 0.112$$

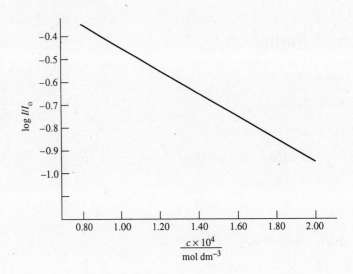

13.4.   Absorbance,

$$A = \log_{10} \frac{1}{0.87} = 0.0605$$

Since 1 dm^3 = (10 cm)^3 = 1000 cm^3, then $\overset{100}{\cancel{1000}}$ cm^3 is 1/10 of 1 dm^3.

Concentration,

$$c = \frac{9.5 \text{ g dm}^{-3}}{18\,800} = 5.05 \times 10^{-4} \text{ mol dm}^{-3}$$

$$\varepsilon = \frac{0.0605}{5.05 \times 10^{-4} \times 10.0} = 12.0 \text{ dm}^3 \text{ cm}^{-1} \text{ mol}^{-1}$$

13.5. $A = \log_{10}(1/0.28) = 0.55$

$=\varepsilon = 276 \ dm^3 \ mol^{-1} \ cm^{-1}$

In a cell 1 cm thick,

$A = 276 \times 1 \times 0.01 = 2.76$

From Eq. 13.43,

$\log_{10} T\% = 2 - A = -0.76$

$T\% = 0.17\%$

13.6. $c = \dfrac{0.155}{532 \times 1.0} = 2.91 \times 10^{-4} \ mol \ dm^{-3}$

13.7.   Remember that measured values are always assumed to have the required number of zeros to give three significant figures.

$A = 0.1028 = \varepsilon c l$

With $c = 10 \times 10^{-6} \ M$ and $l = 1$ cm,

$\varepsilon = \dfrac{0.1028}{10^{-5}} = 1.028 \times 10^4 \ dm^3 \ mol^{-1} \ cm^{-1}$

With $c = 10^{-6}$ and $l = 1$ cm,

$A = 1.028 \times 10^4 \times 10^{-6} = 1.1028 \times 10^{-2}$

$\log_{10} T\% = 2 - A = 1.9897$

$T\% = 97.7\%$

13.8.   The absorbance at 340 gives [NADH]:

$[NADH] = \dfrac{0.215}{6.22 \times 10^3 \times 1} = 3.46 \times 10^{-5} \ mol \ dm^{-3}$

At 260 nm this concentration will give an absorbance of

$1.8 \times 10^4 \times 3.46 \times 10^{-5} = 0.623.$

The remaining absorbance, $0.850 - 0.632 = 0.227$, is due to the NAD^+:

$[NAD^+] = \dfrac{0.227}{1.80 \times 10^4} = 1.04 \times 10^{-5} \ mol \ dm^{-3}$

13.9. $A = \log_{10} \dfrac{1}{0.280} = 0.553$

$\varepsilon = \dfrac{0.553}{0.01 \ mol \ dm^{-3} \times 0.20 \ cm} = 276 \ dm^3 \ mol^{-1} \ cm^{-1}$

In a cell 1 cm thick,

$A = 276 \times 0.01 \times 1 = 2.76$ dm³·mol⁻·cm⁻

$$\log T\% = 2 - A = -0.76$$

$$T\% = 0.17\%$$

13.10. $T = 0.0147 = I/I_0$

$A = \log_{10}(I_0/I) = 1.8327 = 458 \times c \times l$

$c = 4.002 \times 10^{-3}\ M$

$$K_c = \frac{(4.002 \times 10^{-3})^2}{0.1 - 4.002 \times 10^{-3}}$$

$$= 1.668 \times 10^{-4}\ M$$

$\Delta G° = -RT \ln 1.668 \times 10^{-4}$

$$= -8.3145 \times 298.15 \times (-8.699)$$

$$= 21\ 563\ \text{J mol}^{-1} = 21.6\ \text{kJ mol}^{-1}$$

■ Atomic Spectra

13.11.   From Eq. 11.50 we have, after substituting $n = 3$ for the first emission line,

$$1/\lambda = 1/656.3 \times 10^{-9}\ \text{m} = R(1/2^2 - 1/3^2)$$

$R = 1/(0.13889 \times 656.3 \times 10^{-9}\ \text{m})$

$$= 1.097 \times 10^7\ \text{m}^{-1}$$

The emitted light quanta has energy of

$$\varepsilon = hc/\lambda = \frac{6.626 \times 10^{-34}\ (\text{J s})(2.998 \times 10^8)(\text{m s}^{-1})}{656.3 \times 10^{-9}\ \text{m}}$$

$\varepsilon = 3.027 \times 10^{-19}\ \text{J}$

For 1 mol, the energy is

$L\varepsilon = 6.022 \times 10^{23}\ \text{mol}^{-1} \times 3.027 \times 10^{-19}\ \text{J}$

$$= 182.3\ \text{kJ mol}^{-1}$$

13.12.   The $1s^2$ electrons form a closed shell and need not be considered.

2s: $l = 0, s = \frac{1}{2}$;   therefore, $j = \frac{1}{2}$; $^2S_{1/2}$

2p: $l = 1, s = \frac{1}{2}$;   therefore, $j = \frac{3}{2}$ and $\frac{1}{2}$; $^2P_{3/2}, {}^2P_{1/2}$

13.13.   The closed shells need not be considered.

$l_1 = l_2 = 1$ so that $L = 0, 1, 2$

$$s_1 = s_2 = \frac{1}{2} \quad \text{so that} \quad S = 0, 1$$

When $S = 0$ we therefore have ^1S, ^1P, and ^1D.

When $S = 1$ we have ^3S, ^3P, ^3D.

For the singlet terms, $J = L$ and therefore only $^1S_0$, $^1P_1$, and $^1D_2$ occur.

For ^3D, since $L = 2$ and $S = 1$, we have $J = 3, 2, 1$; therefore $^3D_3$, $^3D_2$, $^3D_1$.

For ^3P, $J = 2, 1, 0$; therefore $^3P_2$, $^3P_1$, $^3P_0$.

For ^3S, $J = 1$ only; therefore $^3S_1$.

13.14. a. For the 3p electron, $l = 1$ and $s = \frac{1}{2}$; we therefore have the term ^2P. The j values are

$$1 + \frac{1}{2} = \frac{3}{2} \text{ and } 1 - \frac{1}{2} = \frac{1}{2}; \text{ therefore the terms are}$$

$$^2P_{3/2} \quad \text{and} \quad ^2P_{1/2}$$

b. For $3d^1$, $l = 2$ and $s = \frac{1}{2}$; therefore the terms are

$$^2D_{5/2} \quad \text{and} \quad ^2D_{3/2}$$

13.15. ^1P: $S = 0, L = 1; J = 1$

^3P: $S = 1, L = 1; J = 2, 1, 0$

^4P: $S = \frac{3}{2}, L = 1; J = \frac{5}{2}, \frac{3}{2}, \frac{1}{2}$

^1D: $S = 0, L = 2; J = 2$

^2D: $S = \frac{1}{2}, L = 2; J = \frac{5}{2}, \frac{3}{2}$

^3D: $S = 1, L = 2; J = 3, 2, 1$

^4D: $S = \frac{3}{2}, L = 2; J = \frac{7}{2}, \frac{5}{2}, \frac{3}{2}, \frac{1}{2}$

13.16. The Landé g factor is

$$g_{1/2} = 1 + \frac{\left(\frac{1}{2} \times \frac{3}{2}\right) + \left(\frac{1}{2} \times \frac{3}{2}\right) - (1 \times 2)}{2\left(\frac{1}{2} \times \frac{3}{2}\right)} = \frac{2}{3}$$

The $^2P_{1/2}$ level splits into two levels with $M_J = \frac{1}{2}$ and $M_J = -\frac{1}{2}$. The energy is given by Eq. 13.88:

$$E = g_J \mu_B B M_J$$

The separation is therefore

$$\Delta E = g_J \mu_B B$$

$$= \frac{2}{3} \times 9.273 \times 10^{-24} \times 4.0 \text{ J}$$

$$= 2.47 \times 10^{-23} \text{ J}$$

To convert to cm^{-1} we divide by hc with $c = 2.998 \times 10^{10}$ cm s^{-1}

$$\Delta E = \frac{2.47 \times 10^{-23} \text{ J}}{6.626 \times 10^{-34} \text{ J s} \times 2.998 \times 10^{10} \text{ cm s}^{-1}}$$

$$= 1.24 \text{ cm}^{-1}$$

13.17.   For the $^3P_0$ level, $g_J = 0$. The splitting of the line is therefore entirely due to the splitting of the $^3D_1$ level. For this level,

$$g_J = 1 + \frac{(1 \times 2) + (1 \times 2) - (2 \times 3)}{(2 \times 2)} = \frac{1}{2}$$

It will be split into three levels with $M_J = 1, 0, -1$, and the separation between the levels is

$$\Delta E = g_J \mu_B B$$

$$= \frac{1}{2} \times 9.273 \times 10^{-24} \times 4.0 = 1.85 \times 10^{-23} \text{ J}$$

$$= \frac{1.85 \times 10^{-23} \text{ J}}{6.626 \times 10^{-34} \text{ J} \times 2.998 \times 10^{10} \text{ cm s}^{-1}}$$

$$= 0.93 \text{ cm}^{-1}$$

■ Rotational and Microwave Spectra

13.18.   The separation is $2B$ and therefore,

$$B = 0.5115 \text{ cm}^{-1}$$

This is equal to $h/8\pi^2 Ic$, and the moment of inertia is therefore,

$$I = \frac{6.626 \times 10^{-34} \text{ J s}}{0.5115 \text{ cm}^{-1} (8 \pi^2) (2.998 \times 10^{10} \text{ cm s}^{-1})}$$

$$= 5.472 \times 10^{-46} \text{ kg m}^2$$

The reduced mass (Eq. 13.94) is

$$\mu = \frac{35 \times 19 \text{ g mol}^{-1}}{54 \times 6.022 \times 10^{23} \text{ mol}^{-1}} = 2.045 \times 10^{-26} \text{ kg}$$

The moment of inertia is μr^2 and therefore,

$$r = \sqrt{\frac{5.472 \times 10^{-46}}{2.045 \times 10^{-26}}} = 1.64 \times 10^{-10} \text{ m}$$

$$= 164 \text{ pm}$$

13.19. $B = 20.95 \text{ cm}^{-1}$ and the moment of inertia is

$$I = \frac{6.626 \times 10^{-34}}{20.95 \times 8\pi^2 \times 2.998 \times 10^{10}} = 1.336 \times 10^{-47} \text{ kg m}^2$$

The reduced mass is

$$\mu = \frac{1.008 \times 19.0 \text{ g}}{20.008 \times 6.022 \times 10^{23}} = 1.590 \times 10^{-23} \text{ kg}$$

The interatomic distance is thus,

$$r = \sqrt{\frac{1.343 \times 10^{-47}}{1.590 \times 10^{-27}}} = 9.19 \times 10^{-11} \text{ m}$$

$$= 92 \text{ pm}$$

The separation is inversely proportional to I and therefore to the reduced mass; the interatomic separations are assumed to be the same. The reduced masses for HF, DF, and TF are in the ratio:

$$: \quad \frac{2 \times 19}{21} \quad : \quad \frac{3 \times 19}{22}$$

$$= \quad 1 \quad : \quad 1.90 \quad : \quad 2.72$$

The predicted separations are therefore,

DF: 22.1 cm^{-1}

TF: 15.4 cm^{-1}

13.20. The separation is $2B$; therefore,

$$B = 5.7635 \times 10^{10} \text{ s}^{-1}$$

The moment of inertia is therefore,

$$I = \frac{6.626 \times 10^{-34} \text{ J s}}{8\pi^2 \times 5.7635 \times 10^{10} \text{ s}^{-1}} = 1.456 \times 10^{-46} \text{ kg m}^2$$

The reduced mass is

$$\mu = \frac{12.000 \times 15.995 \text{ g}}{27.995 \times 6.022 \times 10^{23}} = 1.1385 \times 10^{-26} \text{ kg}$$

The interatomic distance is

$$r = \sqrt{\frac{1.456 \times 10^{-46}}{1.1385 \times 10^{-26}}} = 1.131 \times 10^{-10} \text{ m}$$

$$= 113 \text{ pm}$$

13.21. a. For a rigid diatomic molecule, the rotational energy levels are given by Eq. 13.95, which relates the energy to the moment of inertia. The frequency associated with the transition is then given by Eq. 13.97 or in terms of wave members by Eq. 13.98. For the transition $J = 0 \rightarrow J = 1$, $\Delta \tilde{v} = 2\tilde{B}$, and Eq. 13.99 and Eq. 13.100 give

$$\Delta \tilde{v}_j = 2(J + 1) \frac{h}{8\pi^2 I c}$$

$$= 3.842\ 35 \times 10^2 = \frac{2 \times 6.625\ 608 \times 10^{-34}\ \text{J s}}{8\pi^2 (I/\text{kg m}^2) \times 2.997\ 92 \times 10^8\ \text{s}^{-1}}$$

$$I \quad = 1.456\ 96 \times 10^{-46}\ \text{kg m}^2$$

The isotopic mass of ^{16}O is 15.994 91 from Table 13.2. For the rigid molecule, $I = \mu r^2$ where the reduced mass is

$$I \quad = \frac{m_1 m_2}{m_1 + m_2} = \frac{12.0000 \times 15.994\ 91}{12.0000 + 15.994\ 91} \times \frac{10^{-3}}{6.022\ 14 \times 10^{23}}$$

$$= 1.13850 \times 10^{-26}\ \text{kg}$$

$$r^2 = I/\mu = 1.456\ 96 \times 10^{-46}\ \text{kg m}^2 / 1.138\ 50 \times 10^{-26}\ \text{kg}$$

$$= 1.279\ 72 \times 10^{-20}$$

$$r \quad = 1.1312 \times 10^{-10}\ \text{m}$$

$$= 0.1131\ \text{nm}$$

b. The rotational constant, B, is inversely proportional to the reduced mass. Letting the subscripts 1 refer to $^{12}C^{16}O$ and 2 refer to $^{13}C^{16}O$ we have, where M_r is the relative mass of ^{13}C:

$$\frac{B_1}{B_2} = \frac{\mu_1}{\mu_2} = \frac{3.842\ 35 \times 10^2}{3.673\ 37 \times 10^2}$$

$$= \frac{15.994\ 91 \times M_r}{15.994\ 91 + M_r} \times \frac{(12.0000 + 15.994\ 91)}{(12.0000 \times 15.994\ 91)}$$

$$1.046\ 00 \times (15.994\ 91 + M_r) \quad = 2.332\ 909\ M_r$$

$$M_r \quad = 13.0007$$

The value given in Table 13.2, 13.003 35, may be accounted for by the fact that the bond is not exactly rigid.

c. The bond length in $^{13}C^{16}O$ is found using the value of M_r in Table 13.2 and the value of $\tilde{v}$,

$$3.673\ 37 \times 10^2 = \frac{2 \times 6.625\ 608 \times 10^{-34}\ \text{J s}}{8\pi^2 (I/\text{kg m}) \times 2.997\ 92 \times 10^8\ \text{m s}^{-1}}$$

$$I = 1.523\ 99 \times 10^{-46}$$

Then from $I = \mu r^2$,

$$r^2 = \frac{1.523\ 99 \times 10^{-46} \times 6.022\ 14 \times 10^{23}}{\dfrac{13.003\ 35 \times 15.994\ 91}{13.003\ 35 + 15.994\ 91} \times 10^{-3}}$$

$$= 1.279\ 5812 \times 10^{-20}$$

$$r = 1.1312 \times 10^{-10} = 0.113\ 12 \text{ nm}$$

There is essentially no difference in the bond lengths with the ^{13}O isotope.

13.22. The B values are half the separations:

$$B(^{16}O^{12}C^{32}S) = 6.0815 \times 10^9 \text{ s}^{-1}$$

$$B(^{16}O^{12}C^{34}S) = 5.9325 \times 10^9 \text{ s}^{-1}$$

The moments of inertia are

$$I(^{16}O^{12}C^{32}S) = \frac{6.626 \times 10^{-34} \text{ J s}}{8\pi^2 \times 6.0815 \times 10^9 \text{ s}^{-1}}$$

$$= 1.3799 \times 10^{-45} \text{ kg m}^2$$

$$I(^{16}O^{12}C^{34}S) = \frac{6.626 \times 10^{-34} \text{ J s}}{8\pi^2 \times 5.9325 \times 10^9 \text{ s}^{-1}}$$

$$= 1.4146 \times 10^{-45} \text{ kg m}^2$$

The moment of inertia of a linear triatomic molecule is given by Eq. 13.103, and we write x for r_{12} and y for r_{23}, and M_1, M_2, and M_3 for the molar masses

$$\frac{I(^{16}O^{12}C^{32}S)}{L} = M_1 x^2 + M_3 y^2 - \frac{(M_1 x - M_3 y)^2}{M}$$

$$= 1.3799 \times 10^{-42} \text{ g m}^2$$

$$\frac{I(^{16}O^{12}C^{34}S)}{L} = M_1 x^2 + M_3' y^2 - \frac{(M_1 x - M_3' y)^2}{M}$$

$$= 1.4146 \times 10^{-42} \text{ g m}^2$$

Insertion of $L = 6.022 \times 10^{23} \text{ mol}^{-1}$, $M_1 = 16 \text{ g mol}^{-1}$, $M_3 = 32 \text{ g mol}^{-1}$, $M_3' = 34 \text{ g mol}^{-1}$, $M = 60$ g mol^{-1}, $x = 116$ pm, and $y = 156$ pm into the left-hand sides of these equations gives approximately the values on the right-hand sides. For example, for the first equation,

$$\text{L.H.S.} = 16(116)^2 + 32(156)^2 = \frac{[(16 \times 116) - (32 \times 156)]^2}{60}$$

$$= 830\ 140 \text{ g pm}^2 \text{ mol}^{-1} = 1.38 \times 10^{-42} \text{ g m}^2$$

■ Vibrational-Rotational and Raman Spectra

13.23. Set 1/2 kx^2 equal to the vibrational energy derived from Eqs. 13.123 and 13.85.

$$\frac{1}{2}\,kx^2 = \frac{h}{4\pi}\sqrt{\frac{k}{\mu}}$$

Substituting values where $\mu = 1.1385 \times 10^{-26}$ kg, we have

$$x^2 = \frac{h}{4\pi k}\sqrt{\frac{k}{\mu}} = \frac{h}{2\pi}\sqrt{\frac{1}{k\mu}}$$

$$= \frac{6.626 \times 10^{-34}}{2\pi}\sqrt{\frac{1}{(1.86 \times 10^3 \text{ N/m})(1.1385 \times 10^{-26} \text{ kg})}}$$

$$= (1.055 \times 10^{-34} \text{ kg m}^2 \text{ s}^{-1})\, 2.173 \times 10^{11} \text{ s kg}^{-1}$$

$$= 2.293 \times 10^{-23} \text{ m}^2$$

$$x = 4.788 \times 10^{-12} \text{ m} = 4.788 \times 10^{-3} \text{ nm} = 0.0479 \text{ Å}$$

The bond length is 0.1131 nm for the CO molecule found in Problem 13.21.

$$\frac{4.788 \times 10^{-3} \text{ nm}}{0.1131 \text{ nm}} \times 100 = 4.2\%$$

This extension represents about a 4% change.

13.24.

	a. Pure rotational spectrum	b. Vibrational-rotational spectrum	c. Rotational Raman spectrum	d. Vibrational Raman Spectrum
H_2			X	X
HCl	X	X	X	X
CO_2		X	X	X
CH_4		X		X
H_2O	X	X	X	X
CH_3Cl	X	X	X	X
CH_2Cl_2	X	X	X	X
H_2O_2	X	X	X	X
NH_3	X	X	X	X
SF_6		X		X

13.25. The force constant is related to the fundamental frequency by Eq. 13.122:

$$k = 4\pi^2 v_0^2 \mu$$

The reduced mass (see Example 13.9 on p. 668 of the text) is

$$\mu = 1.614 \times 10^{-27} \text{ kg}$$

The frequency v is

$$2988.9 \text{ cm}^{-1} \times 2.998 \times 10^{10} \text{ cm s}^{-1}$$

$$= 8.961 \times 10^{13} \text{ s}^{-1}$$

The force constant is thus,

$$k = 4\pi^2 (8.961 \times 10^{13} \text{ s}^{-1})^2 \, 1.614 \times 10^{-27} \text{ kg}$$

$$= 511.6 \text{ kg s}^{-2}$$

13.26. Equations 13.135 and 13.136 both show that a plot of the observed frequencies $\tilde{v}$ as a function of J'' will be a straight line with slope $= \pm 2\tilde{B}$ and intercept $\tilde{v}_0$. A linear regression analysis on the P branch data using Eq. 13.136 yields

$$\tilde{v} = 2888.70 - 22.614J'',$$

from which we obtain $\tilde{v}_0 = 2888.7 \text{ cm}^{-1}$ and $\tilde{B} = 11.307 \text{ cm}^{-1}$.

A similar treatment of the R branch data yields

$$\tilde{v} = 2907.18 - 18.380J'',$$

from which we calculate $\tilde{v}_0 = 2907.18 \text{ cm}^{-1}$ and $\tilde{B} = 9.190 \text{ cm}^{-1}$.

13.27. For $v = 0$ and 1, we write

$$T_{0, J''} = \tfrac{1}{2}\tilde{v}_0 - \tfrac{1}{4}\tilde{v}_0 x_e + J''(J'' + 1)\tilde{B}$$

$$T_{1, J'} = \tfrac{3}{2}\tilde{v}_0 - \tfrac{9}{4}\tilde{v}_0 x_e + J'(J' + 1)\tilde{B}$$

which gives us

$$\Delta T_{0, J'' \to 1, J'} = \tilde{v}_0 - 2\tilde{v}_0 x_e + \tilde{B}[J'(J' + 1) - J''(J'' + 1)].$$

For $\Delta J = J' - J'' = -1$ (P branch), this reduces to

$$\tilde{v} = \tilde{v}_0 - 2\tilde{v}_0 x_e - 2\tilde{B}J'',$$

and for $\Delta J = J' - J'' = +1$ (R branch), we get

$$\tilde{v} = \tilde{v}_0 - 2\tilde{v}_0 x_e + 2\tilde{B}J''.$$

13.28. The average of the two $\tilde{v}_0$ values calculated in Problem 13.26 is 2897.94 cm^{-1}. The more accurate treatment of Problem 13.27 shows that this is actually the value of $\tilde{v}_0 - 2\tilde{v}_0 x_e$. Therefore, equating $\tilde{v}_0$ to the experimental value of 2990 cm^{-1}, we obtain

$$\tilde{v}_0 x_e = \frac{1}{2}(2990 - 2897.94) \text{ cm}^{-1} = 46.03 \text{ cm}^{-1},$$

or

$$x_e = 46.03/2990 = 1.540 \times 10^{-2}.$$

13.29. The separation is $4B$ (Fig. 13.27) and B is therefore 0.2438 cm^{-1}. The moment of inertia is therefore,

$$I = \frac{6.626 \times 10^{-34} \text{ J s}}{0.2438 \text{ cm}^{-1} (8\pi^2) 2.998 \times 10^{10} \text{ cm s}^{-1}}$$

$$= 1.148 \times 10^{-45} \text{ kg m}^2$$

The reduced mass is

$$\mu = \frac{35 \times 35 \times 10^{-3}}{70 \times 6.022 \times 10^{23}} = 2.906 \times 10^{-26} \text{ kg}$$

The interatomic distance is thus,

$$r = \sqrt{\frac{1.148 \times 10^{-45}}{2.906 \times 10^{-26}}} = 1.99 \times 10^{-10} \text{ m} = 199 \text{ pm}$$

13.30. The zero-point energy of H_2 is

$$\frac{1}{2} h v_0 = \frac{1}{2}(6.626 \times 10^{-34} \text{ J s})(1.257 \times 10^{14} \text{ s}^{-1})$$

$$= 4.164 \times 10^{-20} \text{ J} = 25.1 \text{ kJ mol}^{-1}$$

Classical dissociation energy $= 432.0 + 25.1$

$$= 457.1 \text{ kJ mol}^{-1}$$

Reduced masses of H_2, HD, and D_2 are in the ratio

$$\frac{1}{2} \quad : \quad \frac{2}{3} \quad : \quad 1$$

$$= 1 \quad : \quad \frac{4}{3} \quad : \quad 2$$

Frequencies are inversely proportional to $\sqrt{\mu}$:

Estimated v_0(HD) $= 1.089 \times 10^{14}$ s^{-1}

$$\frac{1}{2} h v_0(\text{HD}) = 21.7 \text{ kJ mol}^{-1}$$

Estimated v_0(D_2) $= 8.89 \times 10^{13}$ s^{-1}

$$\frac{1}{2} h\nu_0(D_2) = 17.7 \text{ kJ mol}^{-1}$$

Estimated dissociation energies:

HD: $457.1 - 21.7 = 435.4 \text{ kJ mol}^{-1}$

D_2: $457.1 - 17.7 = 439.4 \text{ kJ mol}^{-1}$

13.31. The symmetric molecule has only two modes that are active in the infrared:

$$B \rightarrow\leftarrow A\!-\!B \rightarrow$$

and

$$\begin{array}{ccc} & \uparrow & \\ B & - \quad A \quad - & B \\ \downarrow & & \downarrow \end{array} \quad \text{(two degenerate modes)}$$

The symmetric stretch is inactive.

The unsymmetrical molecules have three modes active in the infrared:

$$\leftarrow A - B - B \rightarrow$$

$$A \rightarrow\leftarrow B - B$$

$$\begin{array}{ccc} & \uparrow & \\ A & - \quad B \quad - & B \\ \downarrow & & \downarrow \end{array} \quad \text{(two degenerate modes)}$$

The molecule is therefore A—B—B.

13.32. The reduced masses, $m_1 m_2/(m_1 + m_2)$, are

$$\text{O—H: } \mu_{OH} = \frac{1 \times 16}{(1 + 16)\,L} \text{ g}$$

$$\text{O—D: } \mu_{OD} = \frac{2 \times 16}{(2 + 16)\,L} \text{ g}$$

and are in the ratio

$$\frac{\mu_{OD}}{\mu_{OH}} = \frac{2 \times 17}{18} = 1.89$$

The force constants are the same, and the frequencies are inversely proportional to the square root of the reduced mass. Thus

$$\tilde{\nu}_{OD} = \frac{\tilde{\nu}_{OH}}{(1.89)^{1/2}} = \frac{3300}{1.37} \text{ cm}^{-1} = 2400 \text{ cm}^{-1}.$$

13.33. a. Using the formula derived in Problem 13.27, for P branch transitions, we write

$$\tilde{\nu} = \tilde{\nu}_0 - 2\tilde{\nu}_0 x_e - 2\tilde{B}J'' = 3737.76 - 2 \times 84.8813 - 2 \times 18.9108 \times J'',$$

from which we calculate

$$
\begin{array}{cccc}
J'' & 1 & 2 & 3 \\
\tilde{\nu}\ (\text{cm}^{-1}) & 3530.2 & 3492.4 & 3454.5
\end{array}
$$

b.   For the R branch transitions, we use

$$
\tilde{\nu} = \tilde{\nu}_0 - 2\tilde{\nu}_0 x_e + 2\tilde{B}J'' = 3737.76 - 2 \times 84.8813 + 2 \times 18.9108 \times J'',
$$

from which we calculate

$$
\begin{array}{cccc}
J'' & 0 & 1 & 2 \\
\tilde{\nu}\ (\text{cm}^{-1}) & 3568.0 & 3605.8 & 3643.6
\end{array}
$$

13.34.   The wavenumbers corresponding to the two wavelengths are

$$
435.83\ \text{nm:}\ \frac{1}{435.83 \times 10^{-9}}\ \text{m}^{-1} = 22\,994.7\ \text{cm}^{-1}
$$

$$
476.85\ \text{nm:}\ \frac{1}{476.85 \times 10^{-9}}\ \text{m}^{-1} = 20\,971.0\ \text{cm}^{-1}
$$

The difference, 1973.7 cm^{-1}, corresponds to a vibration in the C_2H_2 molecule (C—C symmetric stretch).

13.35.   The reduced mass of H^{127}I is

$$
\mu = \frac{1.008 \times 126.9}{127.908 \times 6.022 \times 10^{26}}\ \text{kg}
$$

$$
= 1.6607 \times 10^{-27}\ \text{kg}
$$

The frequency ν is

$$
2309.5\ \text{cm}^{-1} \times 2.998 \times 10^{10}\ \text{cm s}^{-1} = 6.924 \times 10^{13}\ \text{s}^{-1}
$$

The force constant k is

$$
\begin{aligned}
k &= 4\pi^2\nu^2\mu \quad \text{(Eq. 13.122)} \\
&= 4\pi^2(6.924 \times 10^{13}\ \text{s}^{-1})^2 \times 1.6607 \times 10^{-27}\ \text{kg} \\
&= 314.3\ \text{kg s}^{-1}
\end{aligned}
$$

13.36.   <u>cis–$C_2H_2Cl_2$</u> : The point group is C_{2v}:

a. a_1; active in infrared and Raman

b. a_2; inactive in infrared, active in Raman

c. b_1; active in infrared and Raman

d. a_1; active in infrared and Raman

<u>trans–$C_2H_2Cl_2$</u> : The point group is C_{2h}:

e. a_g; inactive in infrared, active in Raman

   f. a_u; active in infrared, inactive in Raman

   g. b_u; active in infrared, inactive in Raman

   h. a_g; inactive in infrared, active in Raman

   <u>Benzene</u> : The point group is D_{6h}:

   i. a_{1g}; inactive in infrared, active in Raman

   j. a_{2u}; active in infrared, inactive in Raman

   k. b_{1u}; inactive in infrared and Raman

13.37. At small x values the exponential in Eq. 13.146 may be expanded to $1 - ax$ and the potential energy is thus given by

$$E_p = D_e a^2 x^2$$

Then

$$\frac{dU}{dx} = 2D_e a^2 x$$

The restoring force, f, is $-dE_p/dx$:

$$f = -2D_e a^2 x = -kx$$

where k is the force constant. Thus

$$k = 2D_e a^2$$

The frequency ν_0 is

$$\nu_0 = \frac{1}{2\pi}\sqrt{\frac{2D_e a^2}{\mu}}$$

The reduced mass, μ, was calculated for $H^{35}Cl$ in the example on p. 668:

$$\mu = 1.627 \times 10^{-27} \text{ kg}$$

Then

$$\nu_0 = \frac{1}{2\pi}\sqrt{\frac{2 \times 4.67 \times 1.602 \times 10^{-19} \text{ J } (1.85 \times 10^{10} \text{ m}^{-1})^2}{1.627 \times 10^{-27} \text{ kg}}}$$

$$= 8.93 \times 10^{13} \text{ s}^{-1}.$$

13.38.   a.   From Equations 13.124 and 13.125, it is easily verified that for low amplitude vibrations (i.e., vibrations about $r = r_e$, or $x = 0$), the force constant $k = d^2 E_p/dx^2 \big|_{x=0}$. Differentiating the Extended Rydberg function twice with respect to x and setting $x = 0$, we obtain

$$\left.\frac{d^2 E_p}{dx^2}\right|_{x=0} = D_e(a_1^2 - 2a_2).$$

b.   Let us denote the exponential parameter as α. To find the minimum, we set $dE_p/dx = 0$:

$$\frac{dE_p}{dx} = 0 = -D_e[(a_1 + 2a_2 x + 3a_3 x^2) - \alpha(1 + a_1 x + a_2 x^2 + a_3 x^3)]e^{-\alpha x}.$$

In order for this minimum to occur at $r = r_e$, or $x = 0$, we require

$$-D_e(a_1 - \alpha) = 0,$$

which means that $a_1 = \alpha$.

13.39.   For $N = 2$, we get

$$E_p(x) = -D_e(c_1 e^{-\beta x} + c_2 e^{-2\beta x}).$$

The Morse potential of Equation 13.146 can be written as

$$E_p(x) = D_e(1 - 2e^{-\beta x} + e^{-2\beta x}) = -D_e(2e^{-\beta x} - e^{-2\beta x}) + D_e.$$

Comparing these two relationships, we see that the bond order model with $c_1 = 2$ and $c_2 = -1$ yields the Morse potential to within an additive constant D_e. [The absence of this constant in the bond order model leads to a potential with a minimum of $-D_e$ at $r = r_e$ that goes smoothly to zero as $r \to \infty$.]

13.40.   The force constant for the bond order potential is calculated as

$$k = \frac{d^2 E_p}{dx^2}\bigg|_{x=0} = -D_e \beta^2 \sum_{n=1}^{4} n^2 c_n = 0.302775 \text{ hartree}/a_0^2.$$

Converting this to SI units, we obtain

$$k = 0.302775 \frac{\text{hartree}}{a_0^2} \times \frac{4.3598 \times 10^{-18} \text{ J hartree}^{-1}}{(0.529177 \times 10^{-10} \text{m}/a_0)^2} = 471.40 \text{ J m}^{-2} \text{ (or N m}^{-1}).$$

Now, the vibrational frequency is calculated as

$$\nu_0 = \frac{1}{2\pi}\sqrt{\frac{k}{\mu}} = \frac{1}{2\pi}\sqrt{\frac{471.40 \text{ N m}^{-1}}{1.8234 \times 10^{-26} \text{ kg}}} = 2.5590 \times 10^{14} \text{ s}^{-1}$$

$$= 853.58 \text{ cm}^{-1}.$$

13.41.   Using the masses $m_1 = m_2 = 14.00 \times 10^{-3}$ kg mol$^{-1}/6.022 \times 10^{23}$ mol^{-1}, $m_3 = 16.00 \times 10^{-3}$ kg mol$^{-1}/6.022 \times 10^{23}$ mol^{-1}, we solve the quadratic equation 13.171

$$\lambda^2 - \left(\frac{m_1 + m_2}{m_1 m_2}k_{12} + \frac{m_2 + m_3}{m_2 m_3}k_{23}\right)\lambda + k_{12}k_{23}\frac{m_1 + m_2 + m_3}{m_1 m_2 m_3} = 0,$$

to obtain $\lambda_1 = 3.9060 \times 10^{28}$ s^{-2}, $\lambda_2 = 1.2678 \times 10^{29}$ s^{-2}, from which we calculate

$$\tilde{\nu}_1 = \frac{\sqrt{3.9060 \times 10^{28} \text{ s}^{-2}}}{2\pi \times 2.998 \times 10^{10} \text{ cm s}^{-1}}) = 1049.2 \text{ cm}^{-1}.$$

$$\tilde{v}_2 = \frac{\sqrt{1.2678 \times 10^{29} \text{ s}^{-2}}}{2\pi \times 2.998 \times 10^{10} \text{ cm s}^{-1}}) = 1890.2 \text{ cm}^{-1}.$$

13.42. a. From Equation 13.139, we write

$$\tilde{G}_{v'} = \tilde{v}_0[1 - x_e(v + \tfrac{1}{2})](v + \tfrac{1}{2}).$$

Therefore, for a transition $v'' \rightarrow v'$, we may write

$$\Delta\tilde{G}_{v'} = \tilde{v}_0(v' - v'') - \tilde{v}_0 x_e[(v'^2 - v''^2) + (v' - v'')],$$

or

$$\frac{\Delta\tilde{G}_{v'}}{v' - v''} = \tilde{v}_0 - \tilde{v}_0 x_e[v' + v'' + 1] = (\tilde{v}_0 - \tilde{v}_0 x_e) - \tilde{v}_0 x_e(v' + v'').$$

Therefore, plotting the left-hand side as a function of $(v' + v'')$ yields a straight line from which both $\tilde{v}_0$ and x_e can be calculated. If we consider the case where $v' - v'' = 1$, the left-hand side is simply the observed fundamental and overtone signals, and the analysis is particularly simple.

b. From the given data, we obtain

$$\Delta\tilde{G}_{v'} = 2239 - 58.64(v' + v''),$$

from which we calculate

$$\tilde{v}_0 = 2239 + 58.64 = 2298 \text{ cm}^{-1}, \text{ and } x_e = 58.64/2298 = 2.552 \times 10^{-2}.$$

■ Electronic Spectra

13.43.   a.   The curves are similar in both states:

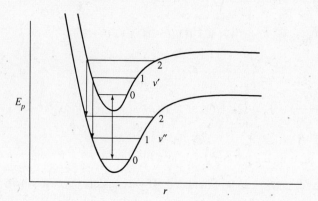

b.

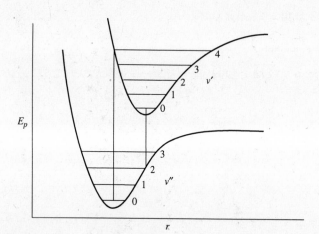

c.   Absorption from $v'' = 0$ gives the upper state with enough energy to dissociate at once. Emission occurs from lower vibrational states.

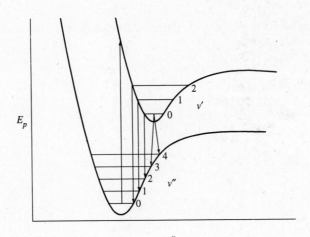

d. Predissociation:

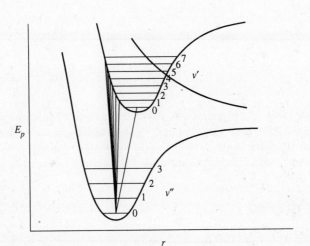

13.44. If a numerical integration program is not available, a graphical technique will always work. Plot $\tilde{\nu}$ against ν' using the data given. Use finely ruled graph paper and set off convenient bars to form layers of steps up to, but not beyond the curve. Calculate the value of each step and add each. Then estimate the value of the remaining area and add all areas together. The value obtained without a linear extrapolation, because of the large anharmonicity, is 56 858.6 cm^{-1}. The ν_{max} from the curve is 17. The values not listed are:

$\Delta\tilde{\nu}$ = 160, 100, and 9

The values you obtain may be slightly different depending on the proficiency of counting/estimating squares.

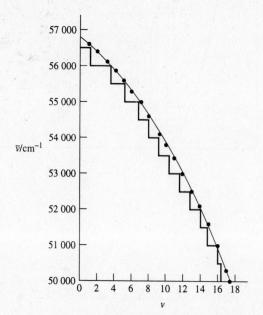

13.45.   As in Problem 13.44, plot ν against v as shown in the figure. The area under the linear extrapolation is 22 550 cm^{-1} (269.8 kJ mol^{-1}), and under the curve the value is 22 350 cm^{-1} (281.1 kJ mol^{-1}). Again, slight differences may be obtained depending on the curve drawn. A computer may be used to obtain an expression for the curve followed by integration.

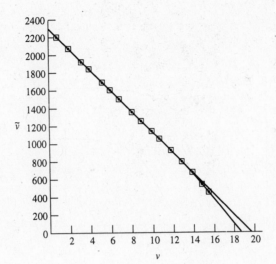

13.46. From Equation 13.138, the ground state vibrational energies are given by

$$\widetilde{G}_{v'} = (v'' + \tfrac{1}{2})\widetilde{v}_0'' - (v'' + \tfrac{1}{2})^2 \widetilde{v}_0'' x_e,$$

while the excited state energies are expressed as

$$\widetilde{G}_{v'} = \widetilde{T}_e + (v' + \tfrac{1}{2})\widetilde{v}_0' - (v' + \tfrac{1}{2})^2 \widetilde{v}_0' x_e.$$

Therefore, the energy difference for the $v'' \rightarrow v'$ transitions are given by

$$\widetilde{v} = \widetilde{G}_{v'} - \widetilde{G}_{v'} = \widetilde{T}_e + \left[(v' + \tfrac{1}{2})\widetilde{v}_0' - (v' + \tfrac{1}{2})^2 \widetilde{v}_0' x_e \right] - \left[(v'' + \tfrac{1}{2})\widetilde{v}_0'' - (v'' + \tfrac{1}{2})^2 \widetilde{v}_0'' x_e \right].$$

No further simplification is possible.

13.47. We first convert all wavelengths λ to wavenumbers $\widetilde{v} = 1/(\lambda \times 10^{-7}\ \text{cm})$. Then a multiple regression analysis using the model

$$\widetilde{v} = \widetilde{T}_e + \left[(v' + \tfrac{1}{2})\widetilde{v}_0' - (v' + \tfrac{1}{2})^2 \widetilde{v}_0' x_e \right] - \left[(v'' + \tfrac{1}{2})\widetilde{v}_0'' - (v'' + \tfrac{1}{2})^2 \widetilde{v}_0'' x_e \right]$$

yields

$$\widetilde{v}(\text{cm}^{-1}) = 15738.5 + \left[(v' + \tfrac{1}{2})131.1 - (v' + \tfrac{1}{2})^2 0.9736 \right]$$

$$- \left[(v'' + \tfrac{1}{2})221.7 - (v'' + \tfrac{1}{2})^2 2.763 \right].$$

13.48.

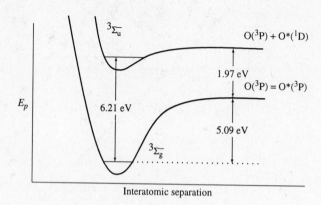

Dissociation energy of O_2 $(^3\Sigma_u^-)$ into $O(^3P) + O^*(^1D)$

$$= 5.09 + 1.97 - 6.21 = 0.85 \text{ eV} = 82.0 \text{ kJ mol}^{-1}.$$

13.49. The energy of the ground state of the molecule can be approximated as

$$\tilde{G}_0 = \frac{1}{2}\tilde{v}_0 - \frac{1}{4}\tilde{v}_0 x_e = 1481.29 \text{ cm}^{-1}$$

$$= \frac{1481.29 \text{ cm}^{-1} \times 2.998 \times 10^{10} \text{ cm s}^{-1} \times h \text{ (J s)}}{1.602 \times 10^{-19} \text{ J (eV)}^{-1}} = 0.1837 \text{ eV}.$$

Therefore,

$$D_0 = D_e - \left(\frac{1}{2}\tilde{v}_0 - \frac{1}{4}\tilde{v}_0 x_e\right) = 4.4336 \text{ eV}.$$

13.50.

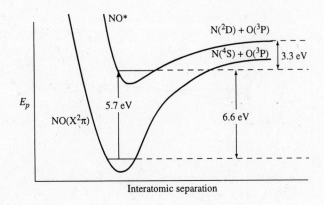

The dissociation energy into $N(^2D) + O(^3P)$ is the sum of the two values

5.7 eV + 3.3 eV = 9.0 eV.

13.51. The wavenumber of 20 302.6 cm^{-1} corresponds to a frequency of

$$\nu = 20\ 302.6\ \text{cm}^{-1} \times 2.998 \times 10^{10}\ \text{cm s}^{-1}$$

$$= 6.087 \times 10^{14}\ \text{s}^{-1}$$

and to an energy of

$$h\nu = 4.033 \times 10^{-19}\ \text{J} = 2.518\ \text{eV}$$

The wavelength 589.3 nm corresponds to a frequency of

$$\nu = \frac{2.998 \times 10^{10}\ \text{m s}^{-1}}{589.3 \times 10^{-9}\ \text{m}} = 5.087 \times 10^{14}\ \text{s}^{-1}$$

and to an energy of

$$h\nu = 3.3709 \times 10^{-19}\ \text{J} = 2.104\ \text{eV}$$

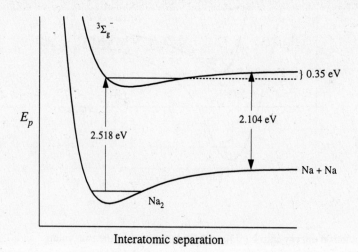

Dissociation energy of $Na_2 = 2.518 + 0.35 - 2.104$ eV

$= 0.76$ eV $= 73.7$ kJ mol^{-1}

14. SOME MODERN APPLICATIONS OF SPECTROSCOPY

■ Spectral Line Widths

14.1. From Eq. 14.4 it follows that

$$T = \left(\frac{\Delta\lambda\, c}{2\lambda}\right)^2 \frac{m}{2k_{\text{B}}}$$

The molecular mass, m, is

$$\frac{56.94 \times 10^{-3}}{6.022 \times 10^{23}}\ \text{kg} = 9.455 \times 10^{-26}\ \text{kg}$$

Therefore

$$T = \frac{(0.053\ \text{nm} \times 2.998 \times 10^8\ \text{m s}^{-1})^2 (9.455 \times 10^{-26}\ \text{kg})}{(2 \times 677.4\ \text{nm})^2\ 2 \times 1.381 \times 10^{-23}\ \text{J K}^{-1}}$$

$$= 4.7 \times 10^5\ \text{K}$$

14.2. From Eq. 14.3,

$$\Delta\tau/\text{s} = \frac{2.7 \times 10^{-12}}{\Delta\tilde{\nu}/\text{cm}^{-1}}$$

a. $\Delta\tau = 2.7 \times 10^{-10}$ s

b. $\Delta\tau = 2.7 \times 10^{-11}$ s

c. $\Delta\tau = 2.7 \times 10^{-12}$ s

d. $\Delta\tau = \dfrac{2.7 \times 10^{-12} \times 2.998 \times 10^{10}\ \text{cm s}^{-1}}{200 \times 10^6\ \text{s}^{-1}}$

$$\doteq 4.05 \times 10^{-10}\ \text{s}$$

14.3. a. In this experiment, the energy absorbed by the NO_2 molecule is used for bond dissociation, to produce $NO(\nu' = 0,1) + O(^3P)$, and the remaining energy is given to the translational motion of the two fragments. Since it takes less energy to form the NO ($\nu' = 0$) than the NO($\nu' = 1$) state, the formation of $NO(\nu' = 0)$ will correspond to higher O atom velocity (actually, higher velocities of both fragments).

b. The energy of the O atom fragment corresponding to the peak

i. at $u_O = 900$ m s^{-1}, $E_O = \frac{1}{2}m_O u_O^2 = \frac{1}{2}\left(\dfrac{0.016 \text{ kg mol}^{-1}}{6.022 \times 10^{23} \text{ mol}^{-1}}\right)(900.00)^2$
$= 1.0761 \times 10^{-20}$ J.

ii. at $u_O = 1400$ m s^{-1}, $E_O = \frac{1}{2}m_O u_O^2 = \frac{1}{2}\left(\dfrac{0.016 \text{ kg mol}^{-1}}{6.022 \times 10^{23} \text{ mol}^{-1}}\right)(1400.00)^2$
$= 2.6038 \times 10^{-20}$ J.

The energy of the NO molecule in

i. $\nu' = 0$ state: $E_0 = \frac{1}{2}hc\tilde{\nu}_0 = 1.855 \times 10^{-20}$ J.

ii. $\nu' = 1$ state: $E_1 = \frac{3}{2}hc\tilde{\nu}_0 = 5.654 \times 10^{-20}$ J.

Since the total energy available to the fragments is $E_{tot} = hc(7400 \text{ cm}^{-1}) = 1.4651 \times 10^{-19}$ J, the translational energy available to the NO fragment (since rotational energy is zero), will be

$$E_{tr}(NO) = E_{tot} - (E_{\nu'} + \tfrac{1}{2}m_O u_O^2) \text{ and } u_{NO} = \sqrt{E_{tr}(NO)/m_{NO}}.$$

	$\nu' = 1; u_O = 900$ m s^{-1}	$\nu' = 0; u_O = 1400$ m s^{-1}
$E_{tr}(NO)$	7.921×10^{-20} J	1.169×10^{-19} J
u_{NO}	1783.0 m s^{-1}	2166.0 m s^{-1}

■ Resonance Spectroscopy

14.4. A wavelength of 8.00 mm corresponds to

$$\nu = \frac{2.998 \times 10^8 \text{ m s}^{-1}}{0.008 \text{ m}} = 3.748 \times 10^{10} \text{ s}^{-1}$$

From Eq. 14.43,

$$B = \frac{h\nu}{g\mu_B}$$

$$= \frac{6.626 \times 10^{-34} \text{ J s} \times 3.748 \times 10^{10} \text{ s}^{-1}}{2.023 \times 9.274 \times 10^{-24} \text{ J T}^{-1}}$$

$$= 1.324 \text{ T}$$

14.5.   a. From Eq. 14.43,

$$g = \frac{h\nu}{B\mu_B}$$

$$= \frac{6.626 \times 10^{-34} \text{ J s} \times 10.42 \times 10^9 \text{ s}^{-1}}{9.274 \times 10^{-24} \text{ J T}^{-1} \times 0.371\ 75 \text{ T}}$$

$$= 2.0026$$

b. $$B = \frac{h\nu}{g\mu_B}$$

$$= \frac{6.626 \times 10^{-34} \text{ J s} \times 9.488 \times 10^9 \text{ s}^{-1}}{2.0026 \times 9.274 \times 10^{-24} \text{ J T}^{-1}}$$

$$= 0.3385 \text{ T}$$

14.6.

	^{2}H	^{19}F	^{35}Cl	^{37}Cl
Spin, I	1	$\frac{1}{2}$	$\frac{3}{2}$	$\frac{3}{2}$
M_I values	1, 0, –1	$+\frac{1}{2}, -\frac{1}{2}$	$+\frac{3}{2}, +\frac{1}{2}, -\frac{1}{2}, -\frac{3}{2}$	$+\frac{3}{2}, +\frac{1}{2}, -\frac{1}{2}, -\frac{3}{2}$
Number of lines	3	2	4	4
g_N	0.857	5.257	0.547	0.456
$\mu_N/10^{-27}$ J T^{-1}	4.329	13.28	4.144	3.455

14.7.   For spin $I = 1$, 3 lines are expected corresponding to $M_I = 1, 0, -1$.

14.8.   The frequency of 60×10^6 s^{-1} corresponds to an energy of

$$6.626 \times 10^{-34} \times 60 \times 10^6 \text{ J} = 3.976 \times 10^{-26} \text{ J}$$

Resonance could be observed with the proton and with ^{35}Cl. For the proton,

$$\Delta E = g_N \mu_N B = 5.586 \times 5.0508 \times 10^{-27} \times B \text{ (J T}^{-1})$$

Resonance would thus be observed at

$$B = \frac{3.976 \times 10^{-26}}{5.586 \times 5.0508 \times 10^{-27}} \text{ T}$$

$$= 1.41 \text{ T}$$

For ^{35}Cl, ($\Delta m = \pm 1$, leads to transitions from $\frac{3}{2} \rightarrow \frac{1}{2}, \frac{1}{2} \rightarrow -\frac{1}{2}, -\frac{1}{2} \rightarrow -\frac{3}{2}$ in absorption)

$$\Delta E = 0.5479 \times 5.0508 \times 10^{-27} B$$

Resonance would thus be found at

$$B = \frac{3.976 \times 10^{-26}}{0.5473 \times 5.0508 \times 10^{-27}} \text{ T}$$

$$= 14.4 \text{ T}$$

(This is outside the usual instrumental range.)

14.9.　From Eq. 14.25,

$$B_{\text{eff}} = (1 - \sigma)B$$

Thus

$$\Delta B_{\text{eff}} = -\Delta\sigma B$$

$$= -(9.80 - 2.20) \times 10^{-6} \times 1.5 \text{ T}$$

$$= -1.14 \times 10^{-5} \text{ T} = -11.4 \text{ } \mu\text{T}$$

$$\Delta\sigma = (9.80 - 2.20) \times 10^{-6} = 7.60 \times 10^{-6}$$

and this gives a frequency splitting of

$$7.60 \times 10^{-6} \times 60 \text{ MHz} = 456 \text{ Hz}$$

14.10.　There are four orientations, corresponding to

$$M_I = +\frac{3}{2}, +\frac{1}{2}, -\frac{1}{2}, -\frac{3}{2}$$

The resonance corresponds to $\Delta M_I = 1$ and is at an energy given by Eq. 14.24:

$$\Delta E = g_N \mu_N B$$

$$= 0.2606 \times 5.0508 \times 10^{-27} \times 1.0 \text{ J} = 1.316 \times 10^{-27} \text{ J}$$

This corresponds to a frequency of

$$\nu = \frac{1.316 \times 10^{-27}}{6.626 \times 10^{-34}} = 1.986 \times 10^6 \text{ s}^{-1} = 1.99 \text{ MHz}$$

14.11.　The frequency 60 MHz corresponds to an energy of

$$60 \times 10^6 \times 6.626 \times 10^{-34} = 3.976 \times 10^{-26} \text{ J}$$

Resonance is thus found at a field of

$$B = \frac{3.976 \times 10^{-26}}{1.792 \times 5.0508 \times 10^{-27}} \text{ T} = 4.39 \text{ T}$$

14.12.　Using the fact that $e^{i\omega t} = \cos(\omega t) + i \sin(\omega t)$, we obtain

$$I(\omega) = A \text{ Re}\left[\int_0^\infty \cos\omega_1 t(\cos\omega t + i \sin\omega t)e^{-t/T}dt + \int_0^\infty \cos\omega_2 t(\cos\omega t + i \sin\omega t)e^{-t/T}dt\right]$$

or

$$I(\omega) = A\left[\int_0^\infty \cos\omega_1 t \cos\omega t e^{-t/T} dt + \int_0^\infty \cos\omega_2 t \cos\omega t e^{-t/T} dt\right].$$

Assigning $a = \omega_i$ and $b = \omega$ in the given integral, we get

$$I(\omega) = T\left\{\frac{1+(\omega_1^2 + \omega^2)T^2}{[1+(\omega_1-\omega)^2 T^2][1+(\omega_2+\omega)^2 T^2]} + \frac{1+(\omega_2^2 + \omega^2)T^2}{[1+(\omega_2-\omega)^2 T^2][1+(\omega_2+\omega)^2 T^2]}\right\}.$$

We assign $\omega_1 = 199.00$ and $\omega_2 = 202.00$, and $T = 8.0$ (arbitrary units) and plot $I(\omega)$ in the range $195 \le \omega \le 205$. A plot of the full $I(\omega)$ function as well as the two terms individually are shown below.

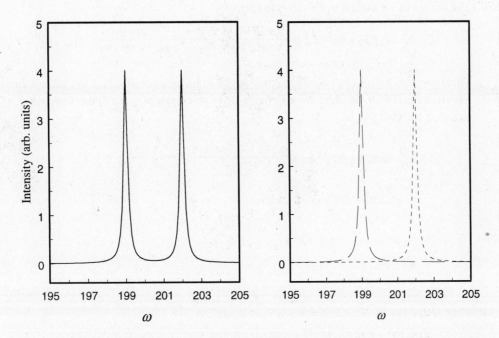

Thus, we see that $I(\omega)$ has peaks when $\omega \approx \omega_i$ and is small at all other values of the frequency. Therefore, Fourier transform of a function of time with two frequency components results in a function of frequency with two peaks centered at those frequencies.

14.13. From Eq. 14.2,

$$\Delta E = \frac{h}{4\pi\Delta\tau} = \frac{6.626 \times 10^{-34} \text{ J s}}{4\pi \times 2 \times 10^{-7} \text{ s}}$$

$$= 2.636 \times 10^{-28} \text{ J}$$

The uncertainty in the frequency is thus,

$$\Delta v = \frac{2.636 \times 10^{-28}\,\text{J}}{6.626 \times 10^{-34}\,\text{J s}} = 3.98 \times 10^5\,\text{s}^{-1}$$

The uncertainty in the wavenumber is

$$\Delta \tilde{v} = \frac{3.98 \times 10^5\,\text{s}^{-1}}{2.998 \times 10^{10}\,\text{cm s}^{-1}} = 1.33 \times 10^{-5}\,\text{cm}^{-1}$$

14.14. The planar configuration must arise from sp^2 hybridization; the point group is D_{3h}. Since planar CH$_3$ has no dipole moment, it shows no microwave spectrum.

The vibrations are as shown in Figure 13.25 for BF$_3$. As discussed in the text (p. 685), all normal modes except the completely symmetric one (a_1') give an infrared spectrum; i.e., an infrared spectrum is given by the a_2'' and e' vibrations.

14.15. The energy of the radiation with $\lambda = 58.4$ nm is

$$E = h\frac{c}{\lambda} = 6.626 \times 10^{-34}\,\text{J s} \times \frac{2.998 \times 10^8\,\text{m s}^{-1}}{58.4 \times 10^{-9}\,\text{m}} \times \frac{1.00\,\text{eV}}{1.602 \times 10^{-19}\,\text{J}} = 21.2\,\text{eV}.$$

Therefore, this source can be used to study peaks with energy below 21.2 eV, i.e., all the peaks shown in Fig. 14.45. In order to study peaks up to an energy of 410 eV, the wavelength of the radiation required will be

$$\lambda = h\frac{c}{E} = 6.626 \times 10^{-34}\,\text{J s} \times \frac{2.998 \times 10^8\,\text{m s}^{-1}}{410\,\text{eV}} \times \frac{1.00\,\text{eV}}{1.602 \times 10^{-19}\,\text{J}}$$

$$= 3.02 \times 10^{-9}\,\text{m} = 3.02\,\text{nm}.$$

15. MOLECULAR STATISTICS

■ Thermodynamic Quantities from Partition Functions

15.1. $$C_P = \left(\frac{\partial H}{\partial T}\right)_P$$

$$= \left\{\frac{\partial}{\partial T}\left[k_B T^2\left(\frac{\partial \ln Q}{\partial T}\right)_V\right]\right\}_P + Nk_B$$

15.2.   a. $Q = q^N$; $\ln Q = N \ln q$

$$U - U_0 = Nk_B T^2\left(\frac{\partial \ln q}{\partial T}\right)_V$$

$$S = Nk_B T\left(\frac{\partial \ln q}{\partial T}\right)_V + Nk_B \ln q$$

$$A - U_0 = -Nk_B T \ln q$$

$$H - U_0 = Nk_B T^2\left(\frac{\partial \ln q}{\partial T}\right)_V + Nk_B T$$

$$G - U_0 = -Nk_B T \ln q + Nk_B T = -Nk_B T \ln(q/e)$$

   b. $Q = q^N/N!$; $\ln Q = \ln q^N - \ln N!$

$$= N \ln q - N \ln N + N$$

$$U - U_0 = Nk_B T^2\left(\frac{\partial \ln q}{\partial T}\right)_V$$

$$S = Nk_B T\left(\frac{\partial \ln q}{\partial T}\right)_V + Nk_B \ln q - Nk_B \ln N + Nk_B$$

$$= Nk_B T(\partial \ln q/\partial T)_V + Nk_B \ln(qe/N)$$

$$A - U_0 = -Nk_B T \ln q + Nk_B T \ln N - Nk_B T$$

$$= -Nk_B T \ln(qe/N)$$

$$H - U_0 = Nk_BT^2\left(\frac{\partial \ln q}{\partial T}\right)_V + Nk_BT$$

$$G - U_0 = -Nk_BT \ln q + Nk_BT \ln N - Nk_BT + Nk_BT$$

$$= -Nk_BT \ln(q/N)$$

15.3.   From Eq. 3.118,

$$P = -\left(\frac{\partial A}{\partial V}\right)_T = k_BT\left(\frac{\partial \ln Q}{\partial V}\right)_T = nRT\left(\frac{\partial \ln q}{\partial V}\right)_T$$

a. $\ln Q = N \ln q$

$$P = k_BT\left(\frac{\partial \ln Q}{\partial V}\right)_T$$

$$= nRT\left(\frac{\partial \ln q}{\partial V}\right)_T$$

b. $\ln q = N \ln q - N \ln N + N$

This gives the same expression for P as above.

15.4. $S = k_BT\left(\frac{\partial \ln Q}{\partial T}\right)_V + k_B \ln Q$

For distinguishable molecules,

$$\ln Q = N \ln q \quad \text{and} \quad S_{dist} = Nk_BT\left(\frac{\partial \ln q}{\partial T}\right)_V + Nk_B \ln q$$

For indistinguishable molecules,

$$\ln Q = N \ln q - N \ln N + N$$

and $S_{indis} = Nk_BT\left(\frac{\partial \ln q}{\partial T}\right)_V + Nk_B \ln q - Nk_B \ln N + Nk_B$

Thus $S_{dist} - S_{indis} = Nk_B \ln N - Nk_B$

For 1 mol, $N = 6.022 \times 10^{23}$ and $k_B = 1.381 \times 10^{-23}$ J K^{-1}

and $Nk_B = R = 8.3145$ J

Then $(S_{dist} - S_{indis})_m = 8.3145 \ln (6.022 \times 10^{23}) - 8.3145$

$$= 446.9 \text{ J K}^{-1} \text{ mol}^{-1}$$

15.5.   The system partition function for a two-dimensional monatomic gas is

$$Q = \left(\frac{2\pi mk_BT}{h^2}\right)^N \text{ where } q = \frac{2\pi mk_BT}{h^2}$$

Then from Eq. 15.53, for 1 mol,

$$U_m - U_{0,m} = RT^2\left(\frac{\partial \ln Q}{\partial T}\right)_V = RT$$

If $T = 25.0°C = 298.15$ K,

$$U_m - U_{0,m} = 8.3145 \times 298.15 \text{ J mol}^{-1} = 2480 \text{ J mol}^{-1}$$

$$= 2.48 \text{ kJ mol}^{-1}$$

15.6. In the limit as $T \to \infty$ the vibrational partition function becomes

$$\frac{1}{1 - (1 - h\nu/k_BT)} = \frac{k_BT}{h\nu}$$

Then, as in the preceding problem,

$$U_m - U_{0,m} = RT$$

and

$$C_{V,m} = R = 8.3145 \text{ J K}^{-1} \text{ mol}^{-1}$$

15.7. The partition function is

$$Q = q^N = \left(1 + e^{-\epsilon/k_BT}\right)^N$$

a. The molar internal energy is, from Table 15.2,

$$U_m - U_{m,0} = k_BT^2\left(\frac{\partial \ln Q}{\partial T}\right)_V$$

$$\left(\frac{\partial \ln Q}{\partial T}\right)_V = \frac{1}{Q}\left(\frac{\partial Q}{\partial T}\right)_V = \frac{N\epsilon}{Qk_BT^2} e^{-\epsilon/k_BT}$$

For 1 mol,

$$= \frac{L\epsilon}{QRT^2} e^{-\epsilon/k_BT}$$

Then

$$U_m - U_{m,0} = L\epsilon \times \frac{e^{-\epsilon/k_BT}}{1 + e^{-\epsilon/k_BT}}$$

$$= \frac{1}{2} L\epsilon \quad \text{when} \quad T \to \infty$$

(Under these conditions half of the molecules are in the ground state and half are in the state of energy ϵ.)

b. The molar entropy is (see Eq. 15.63)

$$S_m = \frac{U_m - U_{m,0}}{T} + k_B \ln Q$$

When $T \to \infty$, the first term approaches zero and $Q \to 2$; thus

$$S_m(T \to \infty) = R \ln 2$$

c. $H - U_0 = U - U_0 + Nk_BT = \dfrac{1}{2} N\epsilon + Nk_BT$

$$H_m - U_{0,m} = RT$$

when T is very large.

d. $G - U_0 = -Nk_BT \ln q + Nk_BT$

$$G_m - U_{0,m} = -R \ln 2 + RT = RT$$

when T is very large.

15.8. The reduced mass of $^{16}O^1H$ is

$$\mu = \frac{1.0078 \times 15.9949}{1.0078 + 15.9949} \times \quad = 1.574 \times 10^{-27} \, \text{kg}.$$

The moment of inertia is

$$I = 1.574 \times 10^{-27} \quad \times (0.96966 \times 10^{-10} \, \text{m})^2 = 1.480 \times 10^{-47} \, \text{kg m}^2.$$

The rotational partition function is

$$q_r = \frac{2Ik_BT}{\hbar^2} = 10.95 \text{ at } T = 298 \text{ K}.$$

Now, using Equation 15.40 with $g_J = 2J + 1$, and $\epsilon_J = J(J+1)\hbar^2/2I$, we write

$$\frac{n_J}{N} = \frac{2J+1}{q_r} e^{-J(J+1)\hbar^2/(2Ik_BT)}.$$

A plot of this fraction as a function of J along with the experimental distribution is shown in the accompanying figure.

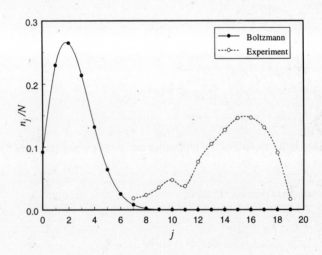

■ Partition Functions for Some Special Cases

15.9. $q = \dfrac{(2\pi m k_B T)^{3/2} V}{h^3}$

For N indistinguishable molecules,

$\ln Q = N \ln q - N \ln N + N$

$U - U_0 = k_B T^2 \left(\dfrac{\partial \ln Q}{\partial T}\right)_V = N k_B T^2 \left(\dfrac{\partial \ln Q}{\partial T}\right)_V$

$\qquad = N k_B T^2 \cdot \dfrac{3}{2} \cdot \dfrac{1}{T} = \dfrac{3}{2} N k_B T$

For 1 mol,

$U_m - U_{0,m} = \dfrac{3}{2} RT$

15.10. The molecular partition function with $V = 1 \text{ m}^3$ and $T = 300$ K is

$q_t = \dfrac{(2\pi \times 1.381 \times 10^{-23} \times 300)^{3/2}}{(6.626 \times 10^{-34})^3} (m/\text{kg})^{3/2}$

$\quad = 1.444 \times 10^{70} (m/\text{kg})^{3/2}$

a. For N_2, $m = 2 \times 14.0067 \times 10^{-3}/6.022 \times 10^{23}$

$\qquad = 4.652 \times 10^{-26}$ kg

$\qquad q_t = 1.449 \times 10^{32}$

b. For H_2O, $m = 18.015 \times 10^{-3}/6.022 \times 10^{23}$

$\qquad = 2.992 \times 10^{-26}$ kg

$\qquad q_t = 7.47 \times 10^{31}$

c. For C_6H_6, $m = 78.114 \times 10^{-3}/6.022 \times 10^{23}$

$\qquad = 1.297 \times 10^{-25}$ kg

$\qquad q_t = 6.746 \times 10^{32}$

$\ln Q = N \ln q_t - N \ln N + N$

For the molar translational partition function $Q_{t,m}$:

a. $\ln Q_{t,m} = 4.4595 \times 10^{25} - 3.297 \times 10^{25} + 6.022 \times 10^{23} = 1.223 \times 10^{25}$

b. $\ln Q_{t,m} = 4.420 \times 10^{25} - 3.297 \times 10^{25} + 6.022 \times 10^{23} = 1.183 \times 10^{25}$

c. $\ln Q_{t,m} = 4.552 \times 10^{25} - 3.297 \times 10^{25} + 6.022 \times 10^{23} = 1.315 \times 10^{25}$

15.11. Mass of N atom $= 14.0067 \times 10^{-3}/6.022 \times 10^{23}$

$\qquad = 2.326 \times 10^{-26}$ kg

Moment of inertia of N_2

$$I = \frac{1}{2} \times 2.326 \times 10^{-26} \times (0.1095 \times 10^{-9})^2$$

$$= 1.394 \times 10^{-46} \text{ kg m}^2$$

The symmetry number is 2.

$$q_r = \frac{8\pi^2 \times 1.394 \times 10^{-46} \times 1.381 \times 10^{-23} \times 300}{2 \times (6.626 \times 10^{-34})^2} = 51.9$$

$$\ln Q_r = L \ln q_r = 2.38 \times 10^{24}$$

15.12. The molecular vibrational partition function is

$$q_v = \frac{1}{1 - e^{-1890/T}} \times \frac{1}{1 - e^{-3360/T}} \left(\frac{1}{1 - e^{-954/T}} \right)^2$$

a. $q_v = 1.002 \times 1.000 \times 1.04^2 = 1.09$

b. $q_v = 2.139 \times 1.484 \times 3.671^2 = 42.8$

15.13. The Sackur-Tetrode equation is based on Eq. 15.86 for the translational partition function. This expression was obtained by replacing a summation (Eq. 15.83) by an integration (Eq. 15.84), a procedure that is valid only if the spacing between the translation levels is much smaller than $k_B T$.

This approximation is not valid at extremely low temperatures, and the Sackur-Tetrode equation then is inapplicable.

15.14. From the Sackur-Tetrode equation (Eq. 15.88),

$$S/\text{J K}^{-1} \text{ mol}^{-1} = 108.74 + 12.47 \ln M_r$$

For argon, $M_r = 39.948$ and therefore,

$$S/\text{J K}^{-1} \text{ mol}^{-1} = 108.74 + 45.98 = 154.7$$

$$S = 154.7 \text{ J K}^{-1} \text{ mol}^{-1}$$

15.15. The value of θ_v is 470 K and thus,

$$q_v = \frac{1}{1 - e^{-470 \text{ K}/T}}$$

a. At $T = 300$ K, $q_v = 1.26$

b. At $T = 3000$ K, $q_v = 6.90$

15.16. C_3O_2: $O{=}C{=}C{=}C{=}O$ $\sigma = 2$

CH_4

$$
\begin{array}{c}
\text{H} \\
| \\
\text{C} \\
\text{H} \diagup \,|\, \diagdown \text{H} \\
\text{H}
\end{array}
$$

$\sigma = 12$

C_2H_6		(Staggered)	$\sigma = 6$
C_2H_4			$\sigma = 2$
C_2H_6		(Eclipsed)	$\sigma = 6$
$CHCl_3$			$\sigma = 3$
C_3H_6			$\sigma = 6$
C_6H_6			$\sigma = 12$
NH_2D			$\sigma = 1$
CH_2Cl_2			$\sigma = 2$

15.17. The rotational constant is

$$B = \frac{h}{8\pi^2 I}$$

$$q_r = \frac{8\pi^2 I\, k_B T}{\sigma h^2} = \frac{k_B T}{\sigma B h}$$

15.18.   We use Eq. 15.88 with $V_m = Lk_BT/P$:

$$S_m = \frac{5}{2} R + R \ln \left[\left(\frac{2\pi mk_BT}{h^2} \right)^{3/2} \frac{k_BT}{P} \right]$$

For Cl_2, $m = \dfrac{2 \times 35.45 \times 10^{-3}}{6.022 \times 10^{23}} = 1.177 \times 10^{-25}$ kg

$P = 0.1$ bar $= 1.00 \times 10^4$ Pa

$$S_m = 2.5 \times 8.3145 + 8.3145 \ln \left[\left(\frac{2\pi \times 1.77 \times 10^{-25}}{(6.626 \times 10^{-34})^2} \right)^{3/2} \frac{(1.381 \times 10^{-23} \times 298.15)^{5/2}}{1.000 \times 10^4} \right]$$

$$= 20.79 + 8.3145 \ln (4.032 \times 10^{63} \times 1.088 \times 10^{-55})$$

$$= 20.79 + 165.45 = 186.2 \text{ J K}^{-1} \text{ mol}^{-1}$$

15.19.   The translational entropy is given by the Sackur-Tetrode equation (Eq. 15.90)

$$S_{t,m}/\text{J K}^{-1} \text{ mol}^{-1} = 108.74 + 12.47 \ln (28.01)$$

$$= 150.3 \text{ J K}^{-1} \text{ mol}^{-1}$$

The rotational partition function is

$$q_r = \frac{8\pi^2 I \, k_BT}{h^2}$$

$$= \frac{8\pi^2 \times 1.45 \times 10^{-46} \times 1.381 \times 10^{-23} \times 298.15}{(6.626 \times 10^{-34})^2}$$

$$= 107.4$$

The rotational entropy is

$$S_r = k_BT \left(\frac{\partial \ln Q_r}{\partial T} \right)_V + k_B \ln Q$$

$$S_{r,m} = RT \left(\frac{\partial \ln q_r}{\partial T} \right)_V + R \ln q_r \quad \text{since} \quad Q_r = q_r^N$$

$$= R + R \ln q_r$$

$$S_{r,m}/\text{J K}^{-1} \text{ mol}^{-1} = 8.3145 + 8.3145 \ln 107.4 = 47.2$$

$$S_{r,m} = 47.2 \text{ J K}^{-1} \text{ mol}^{-1}$$

Since $\nu = 6.50 \times 10^{13}$ s^{-1}, the spacing between vibrational energy levels is

$$h\nu = 6.626 \times 10^{-34} \times 6.50 \times 10^{13} = 4.31 \times 10^{-20} \text{ J}$$

$$\frac{h\nu}{k_BT} = 10.47 \quad q_\nu = \frac{1}{1 - e^{-10.47}} \approx 1$$

The vibrational entropy is therefore negligible.

15.20.   The fraction of molecules in the i-th level is

$$\frac{e^{-\epsilon_i/k_BT}}{1 + e^{-\epsilon_1/k_BT} + e^{-\epsilon_2/k_BT} + e^{-\epsilon_3/k_BT} + \ldots}$$

$$= \frac{e^{-\epsilon_i/k_BT}}{1 + e^{-\Delta\epsilon/k_BT} + e^{-2\Delta\epsilon/k_BT} + e^{-3\Delta\epsilon/k_BT} + \ldots}$$

$$= \left(1 - e^{-\Delta\epsilon/k_BT}\right) e^{-\epsilon_i/k_BT}$$

The limiting value of this fraction when $T \to \infty$ is zero; this is because the molecules are now distributed evenly among an infinite number of levels.

15.21. a. $S_m = \text{constant} + \dfrac{3}{2} R \ln M_r$ (1)

$$\frac{dS_m}{dM_r} = \frac{3}{2}\frac{R}{M_r} \tag{2}$$

b. The heat capacity of C_P is $(\partial H/\partial T)_P = (\partial S/\partial \ln T)_P$ and therefore does not depend on M_r. The Sackur-Tetrode equation can be written as

$$S_m = \text{constant} + \frac{5}{2} R \ln T \tag{3}$$

and therefore

$$C_{P,m} = \frac{5}{2} R \tag{4}$$

There is no dependence on M_r.

c. From Eq. (3),

$$\frac{dS_m}{dT} = \frac{5}{2} \cdot \frac{R}{T} \tag{5}$$

15.22. For two-dimensional translational motion,

$$q_t = \frac{2\pi m k_B T A}{h^2}$$

$$\ln Q = N \ln\left(\frac{q}{N}\right) + N = N \ln\left(\frac{A}{Nh^2} 2\pi m k_B T\right) + N$$

$$S + k_B T\left(\frac{\partial \ln Q}{\partial T}\right)_A + k_B \ln Q$$

$$S_m = RT \cdot \frac{1}{T} + R + R \ln\left(\frac{A}{Nh^2} \cdot 2\pi m k_B T\right)$$

$$= 2 R + R \ln\left(\frac{2\pi m k_B T A}{Nh^2}\right)$$

For Ar, $m = \dfrac{39.948 \times 10^{-3}}{6.022 \times 10^{23}} = 6.634 \times 10^{-26}$ kg

If 10^{10} molecules are absorbed on an area of 1 cm^2 at 25 °C,

$$S_m = 2 \times 8.3145 + 8.3145 \ln\left[\frac{2\pi\, 6.634 \times 10^{-26} \times 298.15 \times 1.381 \times 10^{-23} \times 10^{-4}}{10^{10} \times (6.626 \times 10^{-34})^2}\right]$$

$$= 16.63 + 8.3145 \ln (3.909 \times 10^7)$$

$$= 162.0 \text{ J K}^{-1} \text{ mol}^{-1}$$

■ Calculation of Gibbs-Energy Changes and Equilibrium Constants

15.23. The standard enthalpy change for the reaction $H_2 + \frac{1}{2}O_2 \rightarrow H_2O$ at 298.15 K is the enthalpy of formation of $H_2O(g)$:

$$\Delta H^\circ_{298} = -241.82 \text{ kJ mol}^{-1} \quad \text{(from Appendix D)}$$

$$\Delta(H^\circ_{298} - H^\circ_0) = 9.902 - 8.468 - \frac{1}{2}(8.680)$$

$$= -2.906 \text{ kJ mol}^{-1}$$

$$\Delta H^\circ_0 = -241.82 + 2.906 = -238.91 \text{ kJ mol}^{-1}$$

$$\Delta G^\circ_{298} = -238\,910 - 298.15\left[155.62 - 102.28 - \frac{1}{2}(177.02)\right]$$

$$= -228\,280 \text{ J mol}^{-1} = -228.3 \text{ kJ mol}^{-1}$$

15.24. $\Delta H^\circ_{298} = -110.53 - 241.82 + 393.51 = 41.16 \text{ kJ mol}^{-1} \quad \text{(from Table 15.5)}$

$$\Delta(H^\circ_{298} - H^\circ_0) = 8.665 + 9.902 - 8.468 - 9.360$$

$$= 0.739 \text{ kJ mol}^{-1}$$

$$\Delta H^\circ_0 = 41.16 - 0.739 = 40.42 \text{ kJ mol}^{-1}$$

$$\Delta G^\circ_{1000} = 40\,420 - 1000\,(204.18 + 196.83 - 226.54 - 137.07 = 3020 \text{ J mol}^{-1}$$

$$\ln K_P = -\frac{3020}{8.3145 \times 1000} = -0.3632$$

$$K_P = 0.695$$

15.25. $\Delta H^\circ_{f,298} = -(2 \times 46.11) = -92.22 \text{ kJ mol}^{-1} \quad \text{(from Appendix D)}$

$$\Delta(H^\circ_{298} - H^\circ_0) = (2 \times 10.045) - 8.669 - (3 \times 8.468)$$

$$= -13.983 \text{ kJ mol}^{-1}$$

$$\Delta H^\circ_0 = -92.220 + 13.983 = -78.24 \text{ kJ mol}^{-1}$$

  a. $\Delta G^\circ_{298} = -78\,240 + 298.15\,[-(159.08 \times 2) + 162.49 + (3 \times 102.28)]$

$$= -33\ 169\ \text{J mol}^{-1} = -33.17\ \text{kJ mol}^{-1}$$

$$\ln K_P = \frac{33\ 169}{8.3145 \times 298.15} = 13.38$$

$$K_P = 6.47 \times 10^5\ \text{bar}^{-1}$$

b.
$$\Delta G^{\circ}_{1000} = -78\ 240 + [-(2 \times 203.80) + 198.04 + (3 \times 137.07)]1000$$

$$= -78\ 240 - 407\ 600 + 198\ 040 + 411\ 210$$

$$= 123\ 410\ \text{J mol}{-1} = 123.41\ \text{kJ mol}^{-1}$$

$$\ln K^u_P = -\frac{123\ 410}{8.3145 \times 1000} = -14.84$$

$$K_P = 3.58 \times 10^{-7}\ \text{bar}^{-2}$$

15.26. The symmetry numbers are given below the molecules (They are 1 for atoms.) and the statistical factors are shown above and below the arrows; for simplicity ^{35}Cl is written as Cl and ^{37}Cl as Cl*.

a. $\text{Cl} \underset{1}{\overset{2}{\frac{}{2}}} \text{Cl} + \text{Cl*} \rightleftharpoons \text{Cl} - \text{Cl*} + \text{Cl}$ $\qquad K = 2$

b. $\text{Cl} \underset{2}{\frac{}{}} \text{Cl} + \text{Cl*} \underset{1}{\overset{4}{\frac{}{2}}} \text{Cl*} \rightleftharpoons 2\text{Cl} \underset{1}{\frac{}{}} \text{Cl*}$ $\qquad K = 4$

c. $\underset{12}{\text{CCl}_4} + \text{Cl*} \underset{3}{\overset{12}{\rightleftharpoons}} \underset{3}{\text{CCl*Cl}_3} + \text{Cl}$ $\qquad K = 4$

d. $\underset{3}{\text{NCl}_3} + \text{Cl*} \underset{1}{\overset{3}{\rightleftharpoons}} \underset{1}{\text{NCl*Cl}_2} + \text{Cl}$ $\qquad K = 3$

e. $\underset{2}{\text{Cl}_2\text{O}} + \text{Cl*} \underset{1}{\overset{2}{\rightleftharpoons}} \underset{1}{\text{ClCl*O}} + \text{Cl}$ $\qquad K = 2$

In each case K is $\sigma_A\sigma_B/\sigma_Y\sigma_Z$ and is l/r.

15.27. Mass of I atom,

$$m = 126.90 \times 10^{-3}/6.022 \times 10^{23}$$

$$= 2.107 \times 10^{-25}\ \text{kg}$$

Translational partition function for the I atom, with $V = 1\ \text{m}^3$, is

$$q_t(\text{I}) = \frac{(2\pi \times 2.107 \times 10^{-25} \times 1.381 \times 10^{-23})^{3/2}}{(6.626 \times 10^{-34})^3} \times (1273.15)^{3/2}$$

$$= 1.221 \times 10^{34}$$

The degeneracy of the ground state is

$$2\left(\frac{3}{2}\right) + 1 = 4$$

Thus

$$q_t(\mathrm{I}) = 4 \times 1.221 \times 10^{34} = 4.884 \times 10^{34}$$

For the iodine molecule,

$$q_t(\mathrm{I}_2) = \frac{(2\pi \times 2 \times 2.107 \times 10^{-25} \times 1.381 \times 10^{-23})^{3/2}}{(6.626 \times 10^{-34})^3} \times (1273.15)^{3/2}$$

$$= 2^{3/2} \, q_t(\mathrm{I}) = 3.453 \times 10^{34} \text{ m}^{-3}$$

$$q_r(\mathrm{I}_2) = \left[\frac{8\pi^2 \times 7.426 \times 10^{-45} \times 1.381 \times 10^{-23} \times 1273.15}{2(6.626 \times 10^{-34})^2}\right]$$

$$= 1.174 \times 10^4$$

$$q_v(\mathrm{I}_2) = \frac{1}{1 - \exp\left(\dfrac{-6.626 \times 10^{-34} \times 21\,367.0 \times 2.998}{10^{-8} \times 1.381 \times 10^{-23} \times 1273.15}\right)}$$

$$= \frac{1}{1 - 0.786} = 4.66$$

The molecular partition function for I_2 is thus,

$$q_{\mathrm{I}_2} = 3.453 \times 10^{34} \times 1.174 \times 10^4 \times 4.66$$

$$= 1.889 \times 10^{39}$$

The molecular equilibrium constant, K, is thus,

$$\frac{(4.884 \times 10^{34})^2}{1.889 \times 10^{39}} \, e^{-148\,450/8.3145 \times 1273.15}$$

$$= 1.263 \times 10^{30} \times 8.120 \times 10^{-7} = 1.026 \times 10^{24} \text{ m}^{-3}$$

Its value in molar units is

$$K_c = 1.704 \text{ mol m}^{-3} = 1.704 \times 10^{-3} \text{ mol dm}^{-3}$$

At 1273.15 K,   1 mol dm^{-3} = 105.8 bar

$$K_P = 0.180 \text{ bar}$$

(The experimental value obtained by Starck and Bodenstein in 1910 was 0.165 atm = 0.167 bar.)

15.28.　Mass of Na atom　$= 22.99 \times 10^{-3}/6.022 \times 10^{23}$

$$= 3.818 \times 10^{-26} \text{ kg}$$

The electronic partition function is $2(\frac{1}{2}) + 1 = 2$.

The molecular partition function for Na is

$$q(\mathrm{Na}) = \frac{2 \times (2\pi \times 3.818 \times 10^{-26} \times 1.381 \times 10^{-23})^{3/2}}{(6.626 \times 10^{-34})^3} \times (1000)^{3/2}$$

$$= 1.311 \times 10^{33} \text{ m}^{-3}$$

For Na$_2$,

$$q_t(\text{Na}_2)/\text{m}^{-3} = \frac{(2\pi \times 2 \times 3.818 \times 10^{-26})^{3/2}}{(6.626 \times 10^{-34})^3} \times (1.381 \times 10^{-23} \times 1000)^{3/2}$$

$$= 1.854 \times 10^{33} \quad (= q_t(\text{Na}) \times \sqrt{2})$$

Moment of inertia of Na$_2$ $= \dfrac{3.818 \times 10^{-26}}{2} \times (0.3716 \times 10^{-9})^2$

$$= 2.636 \times 10^{-45} \text{ kg m}^2$$

$$q_r(\text{Na}_2) = \frac{8\pi^2 \times 2.636 \times 10^{-45} \times 1.381 \times 10^{-23} \times 1000}{2 \times (6.626 \times 10^{-34})^2}$$

$$= 3274$$

$$q_v(\text{Na}_2) = \frac{1}{1 - \exp\left(\dfrac{-159.2 \times 2.998 \times 10^{10} \times 6.626 \times 10^{-34}}{1.381 \times 10^{-23} \times 1000}\right)}$$

$$= 4.89$$

Thus, the partition function for Na$_2$ at 1000 K

$$= 1.854 \times 10^{33} \times 3274 \times 4.89$$

$$= 2.97 \times 10^{37} \text{ m}^{-3}$$

The molecular equilibrium constant is

$$K = \frac{(1.311 \times 10^{33})^2}{2.97 \times 10^{37}} \, e^{-704\,00/(8.3145 \times 1000)}$$

$$= 5.787 \times 10^{28} \times 2.103 \times 10^{-4}$$

$$= 1.217 \times 10^{25} \text{ m}^{-3}$$

$$K_c = 20.2 \text{ mol m}^{-3} = 0.0202 \text{ mol dm}^{-3}$$

At 1000 K, 1 mol dm^{-3} = 83.1 bar; K_P = 1.68 bar

15.29. Mass of Cl atom $= 35.45 \times 10^{-3}/6.022 \times 10^{23}$ kg

$$= 5.89 \times 10^{-26} \text{ kg}$$

Translational partition function for the Cl atom with $V = 1$ m^3, is

$$Q_t(\text{Cl}) = \frac{(2\pi \times 5.89 \times 10^{-26} \times 1.381 \times 10^{-23} \times 1200)^{3/2}}{(6.626 \times 10^{-34})^3}$$

$$= 1.651 \times 10^{33}$$

The degeneracy of the $^2P_{3/2}$ state is 4; that of the $^2P_{1/2}$ state is 2; the electronic partition function is thus,

$$Q_e(\text{Cl}) = 4 + 2e^{-\epsilon/k_B T}$$

$$= 4 + 2\exp\left(\frac{-881 \times 2.998 \times 10^{10} \times 6.626 \times 10^{-34}}{1.381 \times 10^{-23} \times 1200}\right)$$

$$= 4 + 2e^{-1.056} = 4.696$$

The complete partition function for the Cl atom is thus

$$Q(Cl) = 7.753 \times 10^{33}$$

For the Cl_2 molecule,

$$q_i(Cl_2) = 2^{3/2}q_t(Cl/m^3) = 4.70 \times 10^{33}$$

The moment of inertia of Cl_2 is

$$I = \mu r^2 = \frac{1}{2}m_{Cl}r^2 = \frac{1}{2} \times 5.89 \times 10^{-26} \times (1.99 \times 10^{-10})^2 \text{ kg m}^2$$

$$= 1.167 \times 10^{-45} \text{ kg m}^2$$

The rotational partition function of Cl_2 ($\sigma = 2$) is

$$q_r(Cl_2) = \frac{8\pi^2 \times 1.167 \times 10^{-45} \times 1.381 \times 10^{-23} \times 1200}{2 \times (6.626 \times 10^{-34})^2}$$

$$= 1739$$

The vibrational partition function is

$$q_v(Cl_2) = \frac{1}{1 - \exp\left(\dfrac{-565 \times 2.998 \times 10^{10} \times 6.626 \times 10^{-34}}{1.381 \times 10^{-23} \times 1200}\right)}$$

$$= 2.033$$

The molecular partition function for Cl_2 is thus,

$$q(I_2) = 4.70 \times 10^{33} \times 1739 \times 2.033 = 1.66 \times 10^{37}$$

The molecular equilibrium constant is thus,

$$\frac{(7.753 \times 10^{33})^2}{1.66 \times 10^{37}} \; e^{-240\,000/(8.3145 \times 1200)}$$

$$= 3.62 \times 10^{30} \times 3.57 \times 10^{-11} = 1.29 \times 10^{20} \text{ m}^{-3}$$

Its value in molar units is

$$K_c = 2.14 \times 10^{-4} \text{ mol m}^{-3} = 2.14 \times 10^{-7} \text{ mol dm}^{-3}$$

At 1200 K,   1 mol dm^{-3} = 99.8 bar

$$K_P = 2.14 \times 10^{-5} \text{ bar}$$

15.30. $\delta\epsilon^2 \;\; = <\epsilon^2> - <\epsilon>^2$

$$= \sum_i P_i\,\epsilon_i^2 - \left(\sum_i P_i\,\epsilon_i\right)^2$$

$$= \frac{\sum_i \epsilon_i^2\, e^{-\beta \epsilon_i}}{q} - \frac{\left(\sum_i \epsilon_i\, e^{-\beta \epsilon_i}\right)^2}{q^2}$$

$$= \frac{\sum_i \left(\dfrac{\partial^2}{\partial \beta^2}\, e^{-\beta \epsilon_i}\right)_V}{q} - \frac{\sum_i \left(\dfrac{\partial}{\partial \beta}\, e^{-\beta \epsilon_i}\right)_V^2}{q^2}$$

$$= \frac{1}{q}\left(\frac{\partial^2 q}{\partial \beta^2}\right)_V - \frac{1}{q^2}\left(\frac{\partial q}{\partial \beta}\right)_V^2 \tag{1}$$

The partition function for a harmonic oscillator is

$$q_v = \frac{1}{1 - e^{-h\nu/k_B T}} = \frac{1}{1 - e^{-\beta h\nu}} \tag{2}$$

Then

$$\frac{dq_v}{d\beta} = + \frac{h\nu\, e^{-\beta h\nu}}{(1 - e^{-\beta h\nu})^2} \tag{3}$$

and

$$\frac{d^2 q_v}{d\beta^2} = \frac{(h\nu)^2\, e^{-\beta h\nu}(1 + e^{-\beta h\nu})}{(1 - e^{-\beta h\nu})^3} \tag{4}$$

Substitution of these two expressions into Eq. (1) gives, after some reduction

$$\delta \epsilon_v^2 = \frac{(h\nu)^2\, e^{-\beta h\nu}}{(1 - e^{-\beta h\nu})^2}$$

$$\delta \epsilon_v = \frac{h\nu\, e^{-\beta h\nu/2}}{1 - e^{-\beta h\nu}}$$

$$= \frac{h\nu}{e^{h\nu/2k_B T} - e^{-h\nu/2k_B T}}$$

As $T \to \infty$ the exponentials can be expanded and only the first term accepted:

$$\delta \epsilon_v = \frac{h\nu}{\left(1 + \dfrac{h\nu}{2k_B T}\right) - \left(1 - \dfrac{h\nu}{2k_B T}\right)}.$$

15.31. The equilibrium constant, K_H, is

$$\frac{q_H^2}{q_{H_2}}\, \exp(-E_0/RT)$$

q_H (translational only) involves $m_H^{1.5}$

q_{H_2} (translational and two degrees of rotational freedom) involves $m_H^{2.5}$. The preexponential factor in the expression for the equilibrium is thus proportional to $m^{0.5}$. The isotope ratio is thus

$$\frac{K_H}{K_D} = \left(\frac{1}{2}\right)^{0.5} \exp[-(26.1 + 18.5)\text{ kJ mol}^{-1}/RT]$$

$$= 0.707\ e^{-7600/(8.3145 \times 300)}$$

$$= 0.707 \times 21.1 = 14.9$$

■ Transition-State Theory

15.32.   At 300 K, $k_B T/h$ is 6.25×10^{12} s^{-1}, which for present purposes is rounded to 6×10^{12} s^{-1}.

a.   For the atom the partition function is 10^{33} m^{-3}. The diatomic molecule has three degrees of translational freedom (which will be written as t^3), two degrees of rotational freedom (r^2), and one of vibrational freedom (v). Its partition function is thus $10^{33} \times 10^2 = 10^{35}$ m^{-3}.

The linear triatomic activated complex has t^3, r^2, and v^3, and its partition function is

$$10^{33} \times 10^2 = 10^{35}\text{ m}^{-3}$$

The estimated preexponential factor is

$$\frac{k_B T}{h}\frac{q}{q_A q_B} = 6 \times 10^{12}\text{ s}^{-1}\frac{10^{35}\text{ m}^{-3}}{10^{33} \times 10^{35}\text{ m}^{-6}}$$

$$= 6 \times 10^{-21}\text{ m}^3\text{ s}^{-1}$$

Multiplication by 6×10^{23} mol^{-1} and by 1000 dm^3 m^{-3} gives

$$A = 4 \times 10^6\text{ dm}^3\text{ mol}^{-1}\text{ s}^{-1}$$

b.   Diatomic molecule A: $t^3 r^2 v$

$$q_A = q_B = 10^{33} \times 100 = 10^{35}\text{ m}^{-3}$$

Activated complex: $t^3 r^3 (rr) v^4$ (One vibration has been replaced by a restricted rotation [rr].)

$$q = 10^{33} \times 10^4 = 10^{37}\text{ m}^{-3}$$

Thus

$$A = 6 \times 10^{12}\frac{10^{37}\text{ m}^{-3}}{(10^{35})2} = 6 \times 10^{-21}\text{ m}^{-3}\text{ s}^{-1}$$

$$= 4 \times 10^6\text{ dm}^3\text{ mol}^{-1}\text{ s}^{-1}$$

c.   (The numbers of vibrational modes are now unspecified, but this makes no difference since they are unity.)

   Reactants A and B: $t^3 r^3$ $q = 10^{33} \times 10^3 = 10^{36}$ m^{-3}

   Activated complex: $t^3 r^3$ $q = 10^{36}$ m^{-3}

$$A = 6 \times 10^{12}\frac{10^{36}}{(10^{36})^2} = 6 \times 10^{-24}\text{ m}^{-3}\text{ s}^{-1}$$

$$= 4 \times 10^3 \text{ dm}^3 \text{ mol}^{-1} \text{ s}^{-1}$$

d. Reactants A, B, and C: $\quad t^3 r^3 \qquad q = 10^{36} \text{ m}^{-3}$

Activated complex: $\quad t^3 r^3 (rr) \qquad q = 10^{37} \text{ m}^{-3}$

$$A = 6 \times 10^{12} \frac{10^{37}}{(10^{36})^3} = 6 \times 10^{-59} \text{ m}^{-6} \text{ s}^{-1}$$

$$= 4 \times 10^{-29} \text{ dm}^3 \text{ mol}^{-1} \text{ s}^{-1}$$

15.33. The fact that there is no exponential dependence suggests that $E_0 = 0$. The temperature dependence arises entirely from the preexponential terms. For each of the three linear reactants there are three degrees of translational freedom and two of rotation, and they vary with temperature as (see Table 15.4)

$$T^{1.5} T = T^{2.5}$$

The activated complex is presumably nonlinear, and if there is no restricted rotation the temperature dependence is as

$$T^{1.5} T^{1.5} = T^3$$

The preexponential factor is thus proportional to

$$T \times T^3 / (T^{2.5})^3 = T^{-3.5}$$

To explain the dependence on T^{-3} we must allow the activated complex to have one degree of restricted rotation, so that its partition function is proportional to

$$T^{1.5} T^{1.5} T^{0.5} = T^{3.5}$$

The preexponential factor is then proportional to

$$T \times T^{3.5} / (T^{2.5})^3 = T^{-3}$$

We can postulate an activated complex of the following structure

$$
\begin{array}{ccc}
 & \text{O} \!-\! \text{O} & \\
 & \diagup \quad \diagdown & \\
\text{O} \!=\! \text{N} & & \text{N} \!=\! \text{O}
\end{array}
$$

with restricted rotation about the O—O bond.

15.34. The temperature dependencies of the partition functions are as follows:

A: $\quad t^3 \qquad : \qquad T^{1.5}$

L: $\quad t^3 r^2 \qquad : \qquad T^{1.5} T = T^{2.5}$

N: $\quad t^3 r^3 \qquad : \qquad T^{1.5} T^{1.5} = T^3$

The nonlinear activated complexes have partition functions proportional to $T^{1.5} T^{1.5} = T^3$. The temperature dependencies for the various types of reactions are therefore as follows:

A + L: $\quad T T^3 / T^{1.5} T^{2.5} = T^0$

A + N: $\quad T T^3 / T^{1.5} T^3 = T^{-0.5}$

L + L: $\quad T T^3 / T^{2.5} T^{2.5} = T^{-1}$

$$L + N: \qquad T\,T^3/T^{2.5}\,T^3 = T^{-1.5}$$

$$N + N: \qquad T\,T^3/T^3\,T^3 = T^{-2}$$

15.35. For the preceding problem it was shown that for an atom reacting with a linear molecule, the preexponential factor is proportional to T^0 if the activated complex is nonlinear. If the activated complex is linear the temperature dependence is $T\,T^{2.5}/T^{1.5}\,T^{2.5} = T^{-0.5}$. It thus appears that the activated complex is linear, and that the temperature dependence is due entirely to the pre-exponential factor.

15.36. Starting with Equation 15.152, we get

$$k = \left(\frac{k_B T}{h}\right)\frac{q^{\ddagger}}{q_A q_B}\,e^{-E_0/k_B T},$$

where

$$q^{\ddagger} = \frac{[2\pi(m_A + m_B)k_B T]^{3/2}}{h^3}\left(\frac{2Ik_B T}{\hbar^2}\right)$$

$$q_A = \frac{(2\pi m_A k_B T)^{3/2}}{h^3}\,; \qquad q_B = \frac{(2\pi m_B k_B T)^{3/2}}{h^3}.$$

Substituting the partition functions in the first equation, where $I = \left(\dfrac{m_A m_B}{m_A + m_B}\right)(r_A + r_B)^2 = \mu d_{AB}^2$, and simplifying, we get

$$k = d_{AB}^2\left(\frac{8\pi k_B T}{\mu}\right)^{1/2}e^{-E_0/k_B T},$$

which, when multiplied by the Avogadro constant (and also multiplying the numerator and denominator of the argument of the exponential term with the Avogadro constant), is essentially the same as Eq. (9.76).

15.37. The partition function for the ion N^+ is proportional to $T^{1.5}$, while that for N_2 is proportional to $T^{2.5}$. If the activated complex is nonlinear its partition function is proportional to T^3. The temperature dependence of the preexponential factor is thus, according to transition-state theory,

$$T\,T^3/T^{1.5}(T^{2.5})^2 = T^{-2.5}$$

The results can therefore be explained in terms of a nonlinear complex, with $E_0 = 0$.

15.38. The activated complexes will be assumed in all cases to be linear and to have no restricted rotation. If an activated complex is linear the temperature dependence decreases by 0.5, while each degree of restricted rotation increases the dependence by 0.5.

   a. $T\,T^3/(T^{2.5})^2 = T^{-1}$

   b. $T\,T^3/T^{2.5}T^3 = T^{-1.5}$

   c. $T\,T^3/T^3 T^{2.5} = T^{-1.5}$

   d. $T\,T^3/(T^3)^2 = T^{-2}$

   e. $T\,T^3/(T^{1.5})^3 = T^{-0.5}$

15.39. The two molecules differ only by isotopic substitution and will have different vibrational frequencies for the bond where substitution occurs. We make the reasonable assumption that the change in nuclear mass will not affect the electronic energy and therefore the force constant, k, will remain the same. The only difference occurs where the reduced mass of the two forms enter the equations. The reduced mass is given by Eq. 13.94 or 13.118 from which the reduced mass of the deuterated form is found to be smaller than that of the form with the normal hydrogen.

Then from Eq. 13.123, the smaller value of μ_D makes ν_0 for the deuterated bond greater than ν_0 for the hydrogen bond, and consequently, the bond energy of the deuterated form is greater than that of the hydrogen form. This has the effect of putting the deuterated form at a lower potential energy than the hydrogen form causing a greater expenditure of energy to promote the deuterated form to the activated state. Therefore the C—H cleavage is greater than that of the C—D bond for the same energy input and the reaction rate of the C—H form should be faster.

15.40. Since the vibrational partition functions are not given, we calculate them below:

$$q_v^{HCl} = [1 - \exp(-hc \times 2991.0/ k_BT)]^{-1} = 1.0008,$$

$$q_{v_1}^{\ddagger} = [1 - \exp(-hc \times 1407.9/ k_BT)]^{-1} = 1.0354,$$

$$q_{v_2}^{\ddagger} = [1 - \exp(-hc \times 266.8/ k_BT)]^{-1} = 2.1164.$$

Now, the quantity E_0 appearing in the exponent of Equation 15.152 has to be evaluated. This is the energy difference between the zero point energy of the reactants and that of the transition state.

$$E_0 = 45\,970 + \frac{1}{2}Lhc(1407.9 + 266.8 - 2991.0) = 38\,097 \text{ J mol}^{-1}.$$

Therefore, we calculate

$$k = \left(\frac{k_BT}{h}\right)\frac{(1.050 \times 10^{33})(1730)(1.0354 \times 2.1164)}{(1.767 \times 10^{32})(6.084 \times 10^{32})(39.4)(1.0008)} e^{-38\,097/(RT)}$$

$$= 5.659 \times 10^{-21} \text{ m}^3 \text{ s}^{-1}.$$

The translational partition functions are calculated for unit volume and hence, although strictly speaking, their units are simply m^{-3}, we may interpret this to mean "states m^{-3}" or molecules m^{-3}.

16. THE SOLID STATE

■ Crystal Lattices, Unit Cells, Density

16.1.　　a.　The end-centered lattice has $(1/8) \times 8 + (1/2) \times 2 = 2$ lattice points. Since one basis is at each lattice point, each unit cell has two basis groups.

　　　　b.　The primitive lattice has one lattice point and there is therefore only one basis group.

16.2.　　a.　A unit cell has 8 lattice points at the corners of a cube; each corner is shared with seven other unit cells. Therefore, only 1/8 of the 8 belong to a particular face-centered cubic (fcc) cell. Each face has an additional lattice point shared between two cells; there are therefore $(1/2) \times 6 = 3$ lattice points in the faces. For the unit cell: 1 (from corners) + 3 (from faces) = 4 lattice points.

　　　　b.　A body-centered cubic (bcc) lattice has 1 lattice point belonging to the unit cell plus $(1/8) \times 8$ corner points. There are thus $1 + 1 = 2$ lattice points.

16.3.　　a.　Consider the array of circles:

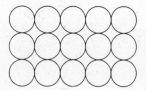

The area belonging to each circle is shown as a dotted box of area $4R^2$. The area of the circle is πR^2. The efficiency of filling space is

$$\pi R^2/4R^2 = 0.785 \quad \text{or} \quad 78.5\%$$

b. Circles on triangular lattice are shown below:

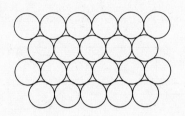

The hexagonal area belonging to a single circle is shown by the dotted lines. The hexagonal area is made up of 12 right triangles, each having an area of

$$\left(\frac{1}{2}R\right)\left(\frac{1}{\sqrt{3}}R\right) = \left(\frac{1}{2\sqrt{3}}\right) R^2$$

The total area is $12(1/2\sqrt{3})R^2 = 2\sqrt{3} R^2$, and the efficiency of filling space is given by

$$\pi R^2/2\sqrt{3} R^2 = \frac{\pi}{2\sqrt{3}} = 0.907 \quad \text{or} \quad 90.7\%$$

c. The triangular form is more efficient by a factor of

$$\frac{90.7}{78.5} = 1.16$$

16.4.   a. A *simple* cubic crystal lattice of side $2r$ contains one atom of radius r.

Free space = volume of cube − volume of atom

$$= (2r)^3 - \frac{4}{3}\pi r^3$$

$$= 8r^3 - 4.1888r^3$$

$$= 3.8112r^3$$

$$\text{Percent free space} = \frac{3.8112r^3}{8r^3} \times 100 = 47.64\%$$

A *body-centered* cube contains two atoms of radius r in the unit cell. The length of the edge of the cube is calculated using the Pythagorean theorem:

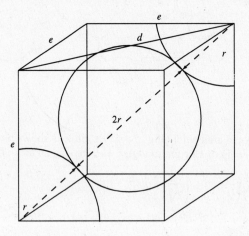

The diagonal of the cube is $4r$.

$$(4r)^2 = d^2 + e^2$$

but

$$d^2 = e^2 + e^2$$

so

$$(4r)^2 = 3e^2$$

$$e = 4r/\sqrt{3} \text{ (length of edge)}$$

$$\text{Free space } = \left(4r/\sqrt{3}\right)^3 - 2\left(\frac{4}{3}\pi r^3\right)$$

$$= 12.3168r^3 - 8.3776r^3$$

$$= 3.9392r^3$$

$$\text{Percent free space} = \frac{3.9392r^3}{12.3168r^3} \times 100 = 31.98\%$$

A *face-centered* cube contains four atoms of radius r in the unit cell. The length of the edge of the cube is calculated using the Pythagorean theorem:

$$(4r)^2 = e^2 + e^2 = 2e^2$$

$$e = 2r\sqrt{2}$$

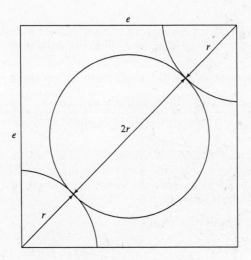

$$\text{Free space} = (2\sqrt{2}r)^3 - 4\left(\frac{4}{3}\pi r^3\right)$$

$$= 22.6274r^3 - 16.7552r^3$$

$$= 5.8722r^3$$

$$\text{Percent free space} = \frac{5.8722r^3}{22.6274r^3} \times 100 = 25.95\%$$

b. The face-centered cube has the least amount of unused space.

16.5.   The volume of a unit cell with right angles is the product abc of its edges. Since 1 mol of the crystal contains L/z unit cells, the molar volume is $V = \dfrac{abc\,L}{z}$. The molar mass M divided by V is the density D, and substitution gives $D = \dfrac{Mz}{abcL}$

16.6.   a. In a face-centered cube containing identical atoms of radius r, the edge length is $2r\sqrt{2}$. For silver, the edge length is 4.0862 Å.

$$\text{Edge length} = 4.0862 \text{ Å} = 2r\sqrt{2}$$
$$r = 1.4447 \text{ Å}$$

b. The volume of a unit cell is

$$V = (4.0862 \text{ Å} \times 10^{-8} \text{ cm/1 Å})^3$$

$$= 6.8227 \times 10^{-23} \text{ cm}^3$$

One unit cell contains four atoms, so 1 mol of Ag contains

$$\frac{6.022 \times 10^{23} \text{ atoms}}{1 \text{ mol}} \times \frac{1 \text{ unit cell}}{4 \text{ atoms}} = 1.506 \times 10^{23} \text{ unit cells mol}^{-1}$$

The mass of 1 mol of Ag is 107.8682 g mol^{-1}. The density is

$$D = \frac{m}{V} = \frac{107.8682 \text{ g mol}^{-1}}{(1506 \times 10^{23} \text{ cells mol}^{-1})(6.8227 \times 10^{23} \text{ cm}^3)} = 10.498 \text{ g cm}^{-3}$$

16.7.   a.   In a body-centered cubic unit cell, the metal atoms are in contact along the diagonal of the cube. The diagonal of the cube forms a right triangle with the unit cell edge and the diagonal of a face. Use the Pythagorean theorem to determine the length of the diagonal, d, on the face of the cube in terms of e.

$$d^2 = e^2 + e^2 = 2e^2$$
$$d = \sqrt{2}\, e$$

The diagonal of the cube is the length of four atomic radii and can be calculated by again using the Pythagorean theorem.

$$(\text{Diagonal})^2 = (4r)^2 = (2e)^2 + e^2$$
$$= 16r^2 = 3e^2$$
$$\text{Diagonal} = 4r = \sqrt{3}e$$
$$r = \frac{\sqrt{3}}{4} e = \frac{\sqrt{3}}{4}(5.025 \text{ Å}) = 2.176 \text{ Å}$$

    b.   Given a body-centered cubic structure, each unit cell contains two atoms. Use the unit cell edge length to calculate the unit cell volume and the volume occupied by each atom. Multiply to obtain the molar volume and divide the gram atomic weight by this value to obtain density (e = edge length).

$V(\text{cell}) = e^3 = (5.025 \times 10^{-8} \text{ cm})^3 = 1.26884 \times 10^{-22} \text{ cm}^3$

$V(\text{atom}) = 1.26884 \times 10^{-22} \text{ cm}^3 \text{ atom}^{-1}/2 \text{ atoms} = 6.3442 \times 10^{-23} \text{ cm}^3$

$V(\text{mole}) = 6.3442 \times 10^{-23} \times 6.022 \times 10^{23} \text{ atoms/mol} = 38.205 \text{ cm}^3$

$D(\text{Ba}) = 137.33 \text{ g}/38.205 \text{ cm}^3 = 3.595 \text{ g cm}^{-3}$

16.8.   a.   Aluminum atoms are in contact along the diagonal of the cube face:

Diagonal = 4*r*(Al)

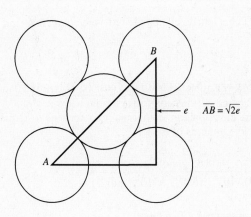

$$r(\text{Al}) = \frac{\sqrt{2}(4.0491 \text{ Å})}{4} = 1.432 \text{ Å}$$

b.   Follow the same procedure for density as used in Problem 16.7, but noting that a face-centered cubic cell contains four atoms per cell instead of two.

$$V(\text{cell}) = e^3 = (4.0491 \times 10^{-8} \text{ cm})^3 = 6.639 \times 10^{-23} \text{ cm}^3$$

$$V(\text{mol}) = 6.643 \times 10^{-23} \frac{\text{cm}^3}{\text{cell}} \times \frac{1 \text{ cell}}{4 \text{ atoms}} \times 6.022 \times 10^{23} \frac{\text{atoms}}{\text{mol}}$$

$$= 9.994 \text{ cm}^3 \text{ mol}^{-1}$$

$$D(\text{Al}) = 26.9815 \frac{\text{g}}{\text{mol}} \times \frac{1 \text{ mol}}{9.994 \text{ cm}^3} = 2.700 \text{ g cm}^{-3}$$

16.9.   From Problem 16.5:

$$D = \frac{Mz}{abcL}$$

Molar mass of $[\text{BrC}_6\text{H}_4\text{C(F)} =]_2 = 373.94 \text{ g mol}^{-1}$

$$D = \frac{4(373.94 \text{ g mol}^{-1})}{(28.32)(7.36)(6.08) \times 10^{-24} \text{cm}^3 \times 6.022 \times 10^{23}}$$

$$= 1.96 \text{ g cm}^{-3}$$

16.10. From Problem 16.5:

$$D = \frac{Mz}{abcL}$$

For a face-centered cubic lattice, all edge lengths are equal, that is, $a = b = c$.

$$z = \frac{D\,abcL}{M}$$

$$= \frac{2.328\ 99\ \text{g cm}^{-3}(5.431\ 066)^3(10^{-8}\text{cm})^3(6.022 \times 10^{23})}{28.085\ 41\ \text{g}}$$

$$= 8$$

16.11. $D = \dfrac{Mz}{abcL} = \dfrac{4(58.45\ \text{g mol}^{-1})}{(5.629)^3(10^{-8})^3(6.022 \times 10^{23}\ \text{mol}^{-1})}$

$$= \frac{233.8\text{g}}{178.4 \times 0.6022\ \text{cm}^3} = 2.176\ \text{g cm}^{-3}$$

CRC value = 2.165 g cm⁻³ at 25°C. The lower density given in the handbook may be due to voids and other imperfections in the crystal.

16.12. The structure is face-centered cubic with the hydride ions in contact along the diagonal of the face as shown in the figure.

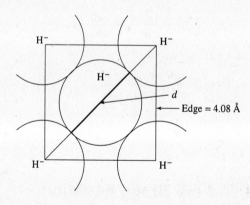

From the figure,

$$(2d_{H^-})^2 = 2(4.08\ \text{Å})^2$$

$$d_{H^-} = 1.44\ \text{Å}$$

Another way to look at this problem is to consider that lithium ions fill the space along the edge, giving an edge length of $2r_{Li^+} + 2r_{H^-}$. The radius of the hydride ion is computed from the edge length e and the reported radius of Li^+ (0.68 Å).

$$e = 4.08 \text{ Å} = 2r_{Li^+} + 2r_{H^-} = 2(0.68 \text{ Å}) + 2r_{H^-}$$

$$r_{H^-} = \frac{4.08 \text{ Å} - 1.36 \text{ Å}}{2} = 1.36 \text{ Å}$$

A recent edition of the *CRC Handbook* gives two values for Li^+: with coordination number 4, its radius is 0.56 Å; with coordination number 6, its radius is 0.76 Å. The value of 0.68 Å just cited is found in an older edition and represents an average value. Any value of r_{H^-} will depend upon the value of r_{Li^+} used in this method. It is common to consider that the anions are in contact with the cations occupying the open space. There is no requirement that the anions and cations contact. Therefore, the first method gives a better estimate of the maximum size of H^-.

16.13. The molar mass of KCl is 74.55 g mol^{-1}.

$$D = \frac{Mz}{abcL} = \frac{4(74.55 \text{ g mol}^{-1})}{(6.278 \times 10^{-8} \text{cm})^3(6.022 \times 10^{23} \text{mol}^{-1})}$$

$$= 2.001 \text{ g cm}^{-3}$$

CRC value: 1.984 g cm^{-3}

16.14. There are four Ca^{2+} ions per unit cell and eight associated F^- ions. Consequently, $z = 4$ and rearrangement of

$$D = \frac{Mz}{abcL} \quad \text{with} \quad a = b = c \quad \text{gives}$$

$$a^3 = zM/DL = \frac{4(78.08 \text{ g mol}^{-1})}{(3.18 \text{ g cm}^{-3})(6.022 \times 10^{23} \text{mol}^{-1})}$$

$$= 1.63 \times 10^{-22} \text{ cm}^3$$

$$a = 5.46 \times 10^{-8} \text{ cm} = 5.46 \text{ Å} = 546 \text{ pm}$$

■ Miller Indices and the Bragg Equation

16.15. The intercepts along the axes are spaced at a/h, b/k, c/l. For a cubic system

$$1a = \frac{a}{h}, h = 1; \quad \frac{a}{2} = \frac{a}{k}, k = 2; \quad \frac{2a}{3} = \frac{a}{l}, l = \frac{3}{2}$$

Clearing fractions we have

$$(hkl) = (243).$$

16.16. The faces are shown in the figures below:

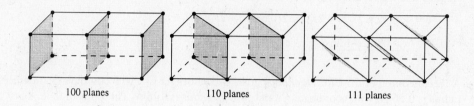

100 planes   110 planes   111 planes

The spacings are calculated from the formula

$$d_{hkl} = a/(h^2 + k^2 + l^2)^{1/2}$$

or from trigonometry as demonstrated from the 110 planes:

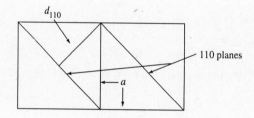

$$d_{110}^2 + d_{110}^2 = a^2$$

$$d_{110} = \frac{a}{\sqrt{2}}$$

The results are

$$d_{100} = a \quad d_{110} = a/\sqrt{2} \quad d_{111} = a/\sqrt{3}$$

16.17.

Originals	Reciprocals	Miller Indices
a. $(2a, b, 3c)$	$\frac{1}{2}, 1, \frac{1}{3}$	(362)
b. $(2a, -3b, 2c)$	$\frac{1}{2}, \frac{-1}{3}, \frac{1}{3}$	$(3\bar{2}3)$
c. $(a, b, -c)$	$1, 1, -1$	$(11\bar{1})$

16.18.   a. In the orthorhombic system all three sides are different and all angles are 90°. Therefore, all sine terms are equal to one and all cosine terms are equal to zero. From Eq. 16.19,

$$d_{hkl}^2 = \frac{1}{\frac{h^2}{a^2} + \frac{k^2}{b^2} + \frac{l^2}{c^2}}$$

or

$$d_{hkl} = \left(\frac{h^2}{a^2} + \frac{k^2}{b^2} + \frac{l^2}{c^2}\right)^{-1/2}$$

b. For the tetragonal system $a = b \neq c$ and all angles are 90°. Therefore, all sine terms are equal to one and all cosine terms are equal to zero. From Eq. 16.19,

$$d_{hkl}^2 = \frac{1}{\frac{h^2}{a^2} + \frac{k^2}{b^2} + \frac{l^2}{c^2}}$$

Then with $a = b$,

$$d_{hkl}^2 = \frac{1}{\frac{h^2 + k^2}{a^2} + \frac{l^2}{c^2}}$$

or

$$d_{hkl} = \left(\frac{1}{\frac{h^2 + k^2}{a^2} + \frac{l^2}{c^2}}\right)^{1/2}$$

16.19. $d_{hkl} = \dfrac{a}{(h^2 + k^2 + l^2)^{1/2}}$

a. $d_{100} = \dfrac{389 \text{ pm}}{1} = 389$ pm

b. $d_{111} = \dfrac{389 \text{ pm}}{(1^2 + 1^2 + 1^2)^{1/2}} = \dfrac{389 \text{ pm}}{\sqrt{3}} = 225$ pm

c. $d_{12\bar{1}} = \dfrac{389 \text{ pm}}{[1^2 + 2^2 + (-1)^2]^{1/2}} = \dfrac{389 \text{ pm}}{\sqrt{6}} = 159$ pm

16.20. $\lambda = 2d_{hkl} \sin \theta \quad \lambda = 154$ pm

From Problem 16.18,

$$\frac{1}{d^2} = \frac{h^2}{a^2} + \frac{k^2}{b^2} + \frac{l^2}{c^2}$$

$$1/d_{100}^2 = \frac{1}{a^2}; \quad d_{100} = a = 488 \text{ pm}$$

$$1/d_{010}^2 = \frac{1}{b^2}; \quad d_{010} = b = 666 \text{ pm}$$

$$1/d_{111}^2 = \frac{1}{a^2} + \frac{1}{b^2} + \frac{1}{c^2} = \frac{1}{488^2} + \frac{1}{666^2} + \frac{1}{832^2}$$

$$= 4.199 \times 10^{-6} + 2.254 \times 10^{-6} + 1.445 \times 10^{-6}$$

$$= 7.898 \times 10^{-6}$$

d_{111} $= 356$ pm

$\sin \theta$ $= \lambda/2d_{100} = 154$ pm/2(488 pm)

$= 0.1578$

θ_{100} $= 9.08°$

$\sin \theta$ $= \lambda/2d_{010} = 154$ pm/2(666 pm)

$= 0.1156$

θ_{010} $= 6.64°$

$\sin \theta$ $= \lambda/2d_{111} = 154$ pm/2(356 pm)

$= 0.2163$

θ_{111} $= 12.49°$

16.21. Assuming a first-order reflection,

$\lambda = 2d \sin \theta$

$0.154\ 18$ nm $= 2(0.400$ nm$) \sin \theta$

$\sin \theta$ $= 0.1927$

θ $= 11.1°$

16.22. $\lambda = 2d_{hkl} \sin \theta$ $\lambda = 45.5$ pm

From Eq. 16.19 or Problem 16.18,

$$\frac{1}{d_{hkl}^2} = \frac{h^2}{a^2} + \frac{k^2}{b^2} + \frac{l^2}{c^2}$$

$1/d_{100}^2 = 1/a^2;$ $d_{100} = 482$ pm

$1/d_{010}^2 = 1/b^2;$ $d_{010} = 684$ pm

$$1/d_{111}^2 = \frac{1}{a^2} + \frac{1}{b^2} + \frac{1}{c^2} = \frac{1}{482^2} + \frac{1}{684^2} + \frac{1}{867^2}$$

$$= 4.3043 \times 10^{-6} + 2.1374 \times 10^{-6} + 1.3303 \times 10^{-6}$$

$$= 7.772 \times 10^{-6}$$

d_{111} $= 359$ pm

$\sin \theta$ $= \lambda/2d_{100} = 45.5$ pm/2(482 pm)

$= 0.0472$

$$\theta_{100} = 2.71°$$

$$\sin \theta = \lambda/2d_{010} = 45.5 \text{ pm}/2(684 \text{ pm})$$

$$= 0.0333$$

$$\theta_{010} = 1.91°$$

$$\sin \theta = \lambda/2d_{111} = 45.5 \text{ pm}/2(359 \text{ pm})$$

$$= 0.0634$$

$$\theta_{111} = 3.63°$$

16.23. $\lambda = 2d_{hkl} \sin \theta$ $\lambda = 154$ pm

From Eq. 16.19 or Problem 16.18,

$$\frac{1}{d_{hkl}^2} = \frac{h^2 + k^2}{a^2} + \frac{l^2}{c^2}$$

$$1/d_{100}^2 = \frac{1}{a^2} = \frac{1}{(967 \text{ pm})^2}$$

$$d_{100} = 967 \text{ pm}$$

$$1/d_{111}^2 = \frac{1^2 + 1^2}{(967)^2} + \frac{1^2}{(892)^2}$$

$$= 2.139 \times 10^{-6} + 1.257 \times 10^{-6} = 3.396 \times 10^{-6}$$

$$d_{111} = 543 \text{ pm}$$

$$\sin \theta = \lambda/2d_{100} = 154 \text{ pm}/2(967 \text{ pm})$$

$$= 0.0796$$

$$\theta_{100} = 4.57°$$

$$\sin \theta = \lambda/2d_{111} = 154/2(543 \text{ pm})$$

$$= 0.1418$$

$$\theta_{111} = 8.15°$$

16.24. The Miller indices depend upon the way in which we draw the unit cell. Several possibilities exist, and four different ways and their corresponding values are shown. Notice that the right-hand cell has been used for convenience. The planes in the left-hand cell will have exactly the same indices.

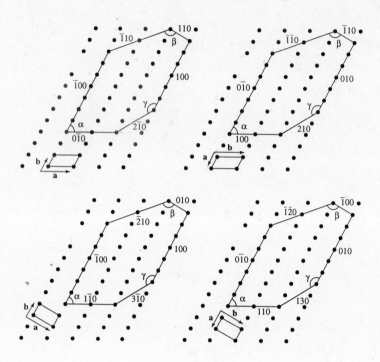

Notice that the angle between dots is not exactly 90°. This does not matter since the *a* vector that determines the reciprocal lattice will be perpendicular to the actual planes.

The indices are A, (0,1,0); B, (–1,1,0); C, (2,1,0), and D, (1,1,0).

16.25.   a.   Let $n = 1$ in the Bragg equation $n\lambda = 2d \sin \theta$

Then $\theta = \sin^{-1}(\lambda/2d)$

$$= \sin^{-1}(70.8/650) = \sin^{-1}(0.1089) = 6.25°$$

b. $\theta = \sin^{-1}(\lambda/2d) = \sin^{-1}(154/650)$

$$= \sin^{-1}(0.2369) = 13.70°$$

Notice that the shorter the wavelength, the smaller the diffraction angle.

16.26.   The kinetic energy of the electron is $\frac{1}{2} mu^2$ and is also VQ; it thus follows that

$$u = \left(\frac{2VQ}{m}\right)^{1/2}$$

$$= \left[\frac{2(40 \times 10^3 \text{ V})(1.60 \times 10^{-19} \text{ C})}{9.11 \times 10^{-31} \text{ kg}}\right]^{1/2}$$

$$= (14.5 \times 10^{15} \text{ J kg}^{-1})^{1/2}$$

Since J $= \text{kg·m}^2\text{·s}^{-2}$

$$u = 1.19 \times 10^8 \text{ m s}^{-1}$$

Substituting into the de Broglie equation (Eq. 11.56) gives

$$\lambda = \frac{6.63 \times 10^{-34} \text{ J s}}{9.11 \times 10^{-31} \text{ kg}(1.19 \times 10^8 \text{ m s}^{-1})}$$

$$= 0.0612 \times 10^{-10} \text{ m} = 6.12 \text{ pm}$$

■ Interpretation of X-Ray Data

16.27. $n\lambda = 2d \sin \theta$

$\lambda = 2(4.00 \times 10^{-8} \text{ Å}) \sin 10.40°$

$= 8.00 \times 10^{-8} (0.1805)$

$= 1.44 \text{ Å}$

16.28.   From Figure 16.28, it is determined that the face-centered cubic (fcc) system is the only one that conforms to the data. Note that the symmetry of the crystal determines which indices will appear.

16.29.   From Figure 16.28, it can only be bcc.

16.30.   a.   From Figure 16.28, it is seen that the ratio 7 is not allowed for cubic systems. The ratio must be 2, 4, 6, and so on. Consequently, the structure is bcc.

b.   For a bcc system, $a = b = c$, $z = 2$, and from Problem 16.5, $D = zMa^3L$ or

$a^3 = 2(55.85 \text{ g mol}^{-1})/(7.90 \text{ g cm}^{-3}) \times (6.022 \times 10^{23} \text{ mol}^{-1}) = 2.348 \times 10^{-23}$

$a = 2.86 \times 10^{-8} \text{ cm} = 286 \text{ pm}$

For 100 type planes, d is $a/2$ since the planes are actually (200). Therefore, from $2d \sin \theta = n\lambda$,

$$\sin \theta = \frac{154.18 \text{ pm}}{286 \text{ pm}} = 0.539$$

$$\theta = 32.6°$$

c. The body diagonal is the smallest interatomic distance and has the value $\dfrac{\sqrt{3}a}{2}$. Therefore, the radius is the distance from the center of one Fe atom to the center of the central atom divided by 2:

$$r_{Fe} = \frac{\sqrt{3}a}{2 \times 2} = \frac{1.732(286)}{4} = 123.8 \text{ pm}$$

16.31. There are two atoms in a body-centered lattice and we may write:

$$\text{Density} = \frac{(\text{number of atoms/cell}) \ (\text{atomic mass})}{La^3}$$

$$0.856 \text{ g cm}^{-3} = \frac{2(39.102 \text{ g mol}^{-1})}{(6.022 \times 10^{23} \text{mol}^{-1})a^3}$$

$$a^3 = \frac{2 \times 39.102}{6.022 \times 10^{23} \times 0.856} = 1.517 \times 10^{-22} \text{ cm}^{-3}$$

$$a = 5.333 \times 10^{-8} \text{ cm} = 0.5333 \text{ nm}$$
$$= 533.3 \text{ pm}$$

Then from the equation

$$d_{hkl} = \frac{1}{(h^2 + k^2 + l^2)^{1/2}} = \frac{533.3 \text{ nm}}{(h^2 + k^2 + l^2)^{1/2}}$$

For (200) planes, $d_{200} = 533.3/\sqrt{4} = 266.7$ pm.

For (110) planes, $d_{110} = 533.3/\sqrt{2} = 377.1$ pm.

For (222) planes, $d_{222} = 533.3/\sqrt{12} = 154.0$ pm.

16.32. First, determine the d values for the three lines and take their ratios:

$$2d_1 = \frac{\lambda}{\sin 14.18} \quad 2d_2 = \frac{\lambda}{\sin 20.25} \quad 2d_3 = \frac{\lambda}{\sin 25.10}$$

$$d_1 : d_2 : d_3 = \frac{1}{\sin 14.18} \quad : \quad \frac{1}{\sin 20.25} \quad : \quad \frac{1}{\sin 25.10}$$

$$= \quad 4.082 \quad : \quad 2.889 \quad : \quad 2.357$$

$$= \quad 1 \quad : \quad 0.7077 \quad : \quad 0.577$$

From Problem 16.16, for the cubic lattice,

$$d_{100} = a$$

$$d_{110} = \frac{\sqrt{2}}{2} a = 0.707 \, a$$

$$d_{111} = \frac{\sqrt{3}}{3} a = 0.5773 \, a$$

The ratios thus correspond to the cubic structure. To confirm the structure, if K^+ and Cl^- reflect equally $d = a/2$, the theoretical density could be compared to the experimental value.

16.33. a. $d_{111} = \dfrac{\lambda}{2 \sin \theta} = \dfrac{154.18 \text{ pm}}{2 \sin 19.076} = 235.9 \text{ pm}$

$= \dfrac{a}{(h^2 + k^2 + l^2)^{1/2}} = 2.359 \times 10^{-10} \text{ m}$

$\dfrac{a}{(1 + 1 + 1)^{1/2}} = \dfrac{a}{(3)^{1/2}} = 2.359 \times 10^{-10} \text{ m}$

$a = 4.086 \times 10^{-10} \text{ m} = 408.6 \text{ pm}$

b. The effective volume of each Ag atom is

$V_{\text{Ag}} = M/D = \dfrac{0.107\ 87 \text{ kg mol}^{-1}}{10\ 500 \text{ kg m}^{-3}\ 6.022 \times 10^{23} \text{ mol}^{-1}}$

$= 1.706 \times 10^7 \text{ pm}^3$

$V_{\text{cell}} = a^3 = (408.6 \text{ pm})^3 = 6.82 \times 10^7 \text{ pm}^3$

The number of atoms per unit cell is

$N = \dfrac{V_{\text{cell}}}{V_{\text{Ag}}} = \dfrac{6.82 \times 10^7 \text{ pm}^3}{1.706 \times 10^7 \text{ pm}^3} = 4.0$

This is an indication that Ag is fcc.

16.34. $d_{111} = \dfrac{\lambda}{2 \sin \theta} = \dfrac{154.2 \text{ pm}}{2 \sin 16.72} = 268.0 \text{ pm}$

$d_{111} = 268.0 \text{ pm} = \dfrac{a}{(h^2 + k^2 + l^2)^{1/2}} = \dfrac{a}{(3)^{1/2}}$

$a = 464.2 \text{ pm} \quad (= 4.64 \text{ Å})$

16.35. First, calculate decimal equivalent of θ values and the sin θ. From each value of θ, calculate the value of $d_{hkl} = \lambda/2 \sin \theta$.

θ		$\sin \theta$	$d_{hkl} = \dfrac{229.1 \text{ pm}}{2 \sin \theta}$	$h\ k\ l$	a
20°36′	20.600	0.3518	325.6	1 1 1	564.0
23°58′	23.967	0.4062	282.0	2 0 0	564.0
35°4′	35.067	0.5745	199.4	2 2 0	564.0
42°21′	42.350	0.6737	170.0	3 1 1	563.8
44°43′	44.717	0.7036	162.8	2 2 2	564.0
54°20′	54.333	0.8124	141.0	4 0 0	564.0
62°17′	62.283	0.8853	129.4	3 3 1	564.0
65°16′	65.267	0.9082	126.1	4 2 0	563.9

Ignore the final two columns for the time being.

Assume that NaCl has a cubic crystal structure.

Then

$$a = d_{hkl} \cdot \sqrt{h^2 + k^2 + l^2}$$

If we index the first three angles obtained from NaCl according to the (100), (110), (111) planes of the simple cubic lattice, the value of a should be the same in each case.

$$a = d_{hkl} \sqrt{1^2 + 0 + 0} = 325.6 \text{ pm } (\sqrt{1^2}) = 325.6 \text{ pm}$$

$$a = d_{hkl} \sqrt{1^2 + 1^2 + 0} = 282.0 \text{ pm } (\sqrt{2}) = 398.9 \text{ pm}$$

$$a = d_{hkl} \sqrt{1^2 + 1^2 + 1^2} = 199.4 \text{ pm } (\sqrt{3}) = 345.4 \text{ pm}$$

Since the unit cell dimension a is not the same, this does not allow indexing as a simple cubic system.

An attempt to index the first line with d_{110} gives

$$a = 325.6 \text{ pm } (\sqrt{1^2 + 1^2 + 0}) = 460.5 \text{ pm}$$

and the next set of hkl values must give the value

$$\sqrt{h^2 + k^2 + l^2} = \frac{a}{d_{hkl}} = \frac{460.5}{282.0} = 1.633$$

No set of integers will give this value.

If the first line is indexed as 111, we have

$$a = 325.6 \text{ pm } (\sqrt{1^2 + 1^2 + 1^2}) = 564.0 \text{ pm}$$

and, as in the last case, the next line must be indexed such that

$$\sqrt{h^2 + k^2 + l^2} = \frac{a}{d_{hkl}} = \frac{564.0}{282.0} = 2$$

This allows h, k, or $l = 2$ corresponding to planes (200), (020), or (002).

The rest of the planes can be indexed as listed in the final columns of the table with the corresponding values of a. These correspond to the lines expected for an fcc structure as listed in Figure 16.20.

■ Bonding in Crystals and Metals

16.36. In a closest-packed array there are two tetrahedral holes for each anion. If only half the tetrahedral holes are occupied, the numbers of anions and cations are equal. The formula for cadmium sulfide is CdS.

16.37. In a closest-packed array, there is one octahedral hole for each anion. If only half of the octahedral holes are occupied by titanium atoms, there are twice as many oxygen atoms as titanium ions, and the formula as TiO_2. With each oxygen as –2, the titanium must be +4.

16.38

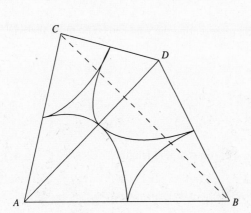

The contributions of the four atoms in contact forming the tetrahedral void may be represented at the corners A, B, C, and D. A plane through A and B and bisecting the line $\overline{CD}$ is represented as follows:

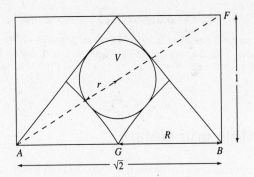

Here the atoms at A and B are shown by arcs. The right triangle AGV is similar to the right triangle ABF. Therefore,

$$\frac{AG}{AV} = \frac{AB}{AF} = \sqrt{\frac{2}{3}}$$

Then the maximum radius of the circle representing the void can be no more than $AV - R$, that is,

$$\frac{AG}{AV} = \frac{R}{R + r} = \sqrt{\frac{2}{3}}$$

$$r = \frac{\sqrt{3} - \sqrt{2}}{2} R = 0.225 R$$

16.39.   Take a section through an octahedron with sides of unit length:

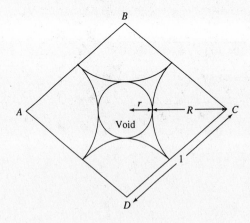

The diagonal AC is $\sqrt{2}$.

From the isosceles right triangle,

$$\frac{AC}{AB} = \frac{\sqrt{2}}{1}$$

If r is the radius of the void, then

$$\frac{2R + 2r}{2R} = \frac{\sqrt{2}}{1}$$

$$r = \sqrt{2}\,R - R = 0.414\,R$$

16.40.   From Eq. 16.58,

$$\Delta E_C = -\Delta_f H + \Delta_{sub} H + \frac{1}{2} D_0 + I - A$$

$$= \left[414 + 84 + \frac{1}{2}(192) + 397 - 318 \right] \text{ kJ mol}^{-1}$$

$$= 673 \text{ kJ mol}^{-1}$$

■ Supplementary Problems

16.41.   a. $n\lambda = 2\,d \sin \theta$

d/pm	323	309	304	284	274
θ	13.81°	14.45°	14.69°	15.75°	16.33°

b. $\dfrac{1}{d^2} = \dfrac{h^2 + k^2 + l^2}{a^2}$

For $d = 323$ pm:

$$\frac{1}{(d/\text{pm})^2} = \frac{1}{104\ 300} = \frac{h^2 + k^2 + l^2}{4\ 639\ 700}$$

$$h^2 + k^2 + l^2 = 44.48$$

$$6^2 + 2^2 + 2^2 = 36 + 4 + 4 = 44$$

For $d = 309$ pm:

$$\frac{1}{(d/\text{pm})^2} = \frac{1}{95\ 481} = \frac{h^2 + k^2 + l^2}{4\ 639\ 700}$$

$$h^2 + k^2 + l^2 = 48.63$$

$$4^2 + 4^2 + 4^2 = 48$$

For $d = 304$ pm:

$$\frac{1}{(d/\text{pm})^2} = \frac{1}{92\,400} = \frac{h^2 + k^2 + l^2}{4\,639\,700}$$

$$h^2 + k^2 + l^2 = 50.2$$

$$5^2 + 4^2 + 3^2 = 50$$

Note that other combinations of *hkl* values are possible, as well as different orders of the values given. As higher *hkl* values are used, more possibilities exist, making it more difficult to decide on the correct values. Help in this task is provided by knowledge of the lines that are forbidden for the particular crystal type, and of the extinction caused by absorption by different atoms in the crystal layers.

16.42. $n\lambda = 2\,d \sin\theta \quad d = \lambda/2 \sin\theta$

$d_{12.95°} = 344 \text{ pm}; \quad d_{13.76°} = 324 \text{ pm}; \quad d_{14.79°} = 302 \text{ pm}$

$$\frac{1}{d^2} = \frac{h^2}{a^2} + \frac{k^2}{b^2} + \frac{l^2}{c^2}$$

With the values given,

$$\frac{1}{(d/\text{pm})^2} = \frac{h^2}{822.7^2} + \frac{k^2}{1198.2^2} + \frac{l^2}{644.1^2}$$

$$= \frac{h^2}{678\,800} + \frac{k^2}{1\,436\,000} + \frac{l^2}{414\,900}$$

With $d = 344$ pm, agreement is obtained with $h = 2$, $k = 0$, and $l = 1$:

LHS $= \dfrac{1}{344^2} = 8.45 \times 10^{-6}$

RHS $= 5.90 \times 10^{-6} + 0 + 2.40 \times 10^{-6} = 8.30 \times 10^{-6}$

With $d = 324$ pm, agreement is obtained with $h = 0$, $k = 0$, and $l = 2$:

LHS $= \dfrac{1}{324^2} = 9.52 \times 10^{-6}$

RHS $= 0 + 0 + 9.60 \times 10^{-6} = 9.60 \times 10^{-6}$

With $d = 302$ pm, agreement is obtained with $h = 0$, $k = 4$, and $l = 0$:

LHS $= \dfrac{1}{302^2} = 10.96 \times 10^{-6}$

RHS $= 0 + 11.1 \times 10^{-6} + 0 = 11.1 \times 10^{-6}$

In the latter case, agreement is obtained with $h = 1$, $k = 0$, and $l = 2$ (RHS = 11.0×10^{-6}), but in view of the crystal type (040) is more likely.

16.43.　a.　From Eq. 16.38,

$$F(hkl) = \sum_{j=1}^{N} f_j \exp[2\pi i\,(hx_j + ky_j + lz_j)]$$

$$= f_{Zn}\left[\exp 2\pi i(0) + \exp 2\pi i\left(\frac{1}{2} + \frac{1}{2}\right)\right.$$

$$\left. + \exp 2\pi i\left(\frac{1}{2} + \frac{1}{2}\right) + \exp 2\pi i\left(\frac{1}{2} + \frac{1}{2}\right)\right]$$

$$+ f_S\left[\exp 2\pi i\left(\frac{1}{4} + \frac{1}{4} + \frac{1}{4}\right) + \exp 2\pi i\left(\frac{1}{4} + \frac{3}{4} + \frac{3}{4}\right)\right.$$

$$\left. + \exp 2\pi i\left(\frac{3}{4} + \frac{1}{4} + \frac{3}{4}\right) + \exp 2\pi i\left(\frac{3}{4} + \frac{3}{4} + \frac{1}{4}\right)\right] F(hkl)$$

$$= f_{Zn}(1 + 3e^{2\pi i}) + f_S(e^{3\pi i/2} + 3e^{7\pi i/2})$$

$$= 4 f_{Zn} - 4\,i\,f_S$$

b.　$a = \dfrac{\lambda}{2\sin\theta}(h^2 + k^2 + l^2)^{1/2} = \dfrac{154.18 \text{ pm } (3)^{1/2}}{2(0.247)}$

$$= 540.5 \text{ pm}$$

16.44.　The number of atoms per unit volume must be calculated

$$N/V = \frac{LD}{A}$$

where A is the atomic mass and D is the density.

$$\nu_D = \left(\frac{9N}{4\pi V}\right)^{1/3}\left(\frac{1}{c_l^3} + \frac{2}{c_t^3}\right)^{-1/3}$$

$$= \left[\frac{9}{4\pi}\frac{(19.271 \text{ g cm}^{-3})(6.022 \times 10^{23} \text{ mol}^{-1})}{183.85 \text{ g mol}^{-1}}\right]^{1/3} \times$$

$$\left[\frac{1}{(5.2496 \times 10^5 \text{ cm s}^{-1})^3} + \frac{2}{(2.9092 \times 10^5 \text{ cm s}^{-1})^3}\right]^{-1/3}$$

$$\nu_D = 8.005 \times 10^{12} \text{ s}^{-1}$$

Then　$\Theta_D = \dfrac{h\nu_D}{k_B} = \dfrac{6.6262 \times 10^{-34} \text{ J s } (8.005 \times 10^{12} \text{ s}^{-1})}{1.3807 \times 10^{-23} \text{ J K}^{-1}}$

$$= 384 \text{ K}$$

17. THE LIQUID STATE

■ Thermodynamic Properties of Liquids

17.1.   The molar mass of ethanol is 46.07 g mol^{-1}, and since the density of liquid ethanol is 0.790 g cm^{-3} the molar volume is 58.32 cm^3 mol^{-1} = 5.832 × 10^{-5} m^3 mol^{-1}. The internal pressure is therefore

$$P_i = \frac{a}{V_m^2} = \frac{1.218 \text{ Pa m}^6 \text{ mol}^{-2}}{(5.832 \times 10^{-5})^2 \text{ m}^6 \text{ mol}^{-2}} = 3.58 \times 10^8 \text{ Pa}$$

$$= 3580 \text{ bar}$$

$$E_p = -\frac{1.218 \text{ Pa m}^6 \text{ mol}^{-2}}{5.832 \times 10^{-5} \text{ m}^3 \text{ mol}^{-1}} = -20.9 \text{ kJ mol}^{-1}$$

17.2. $P_i = \left(\frac{\partial U}{\partial V}\right)_T = T\left(\frac{\partial P}{\partial T}\right)_V - P$ (Eq. 17.2)

$$= (298 \times 6.60 \times 10^6 - 100\ 000) \text{ Pa}$$

$$= 1.966 \times 10^9 \text{ Pa} = 19\ 660 \text{ bar}$$

17.3.   The molar mass is 78.11 g mol^{-1}, and the molar volume is 86.89 cm^3 mol^{-1} = 8.689 × 10^{-5} m^3 mol^{-1}. The internal pressure is therefore

$$P_i = \frac{1.824 \text{ Pa m}^6 \text{ mol}^{-2}}{(8.689 \times 10^{-5})^2 \text{ m}^6 \text{ mol}^{-2}} = 2.416 \times 10^8 \text{ Pa}$$

$$= 2416 \text{ bar}$$

$$E_p = -\frac{1.824 \text{ Pa m}^6 \text{ mol}^{-2}}{8.689 \times 10^{-5} \text{ m}^3 \text{ mol}^{-1}} = -21.0 \text{ kJ mol}^{-1}$$

17.4. $P_i = T\left(\frac{\partial P}{\partial T}\right)_V - P$ (Eq. 17.2)

$$= (298 \times 1.24 \times 10^6 - 100\ 000) \text{ Pa}$$

$$= 3.69 \times 10^8 \text{ Pa} = 3690 \text{ bar}$$

17.5. a. Hg: $P_i = (4.49 \times 10^6 \times 298 - 100\,000)$

$$= 1.34 \times 10^9 \text{ Pa} = 13\,400 \text{ bar}$$

b. n-Heptane: $P_i = (8.53 \times 10^5 \times 298 - 100\,000)$

$$= 2.54 \times 10^8 \text{ Pa} = 2540 \text{ bar}$$

c. n-Octane: $P_i = (1.01 \times 10^6 \times 298 - 100\,000)$

$$= 3.01 \times 10^8 \text{ Pa}$$

$$= 3010 \text{ bar}$$

d. Diethyl ether: $P_i = (8.06 \times 10^5 \times 298 - 100\,000)$

$$= 2.40 \times 10^8 \text{ Pa} = 2400 \text{ bar}$$

17.6. For the vapor,

$$P_i = \left(\frac{\partial U}{\partial V}\right)_T = T\left(\frac{\partial P}{\partial T}\right)_V - P \qquad \text{(Eq. 17.2)}$$

$$= (298 \times 115 - 10) \text{ Pa} = 3.43 \times 10^4 \text{ Pa} = 0.343 \text{ bar}$$

For the liquid,

$$P_i = (298 \times 1.24 \times 10^6 - 100\,000) \text{ Pa}$$

$$= 3.69 \times 10^8 \text{ Pa} = 3690 \text{ bar}$$

17.7. a. $P_i = \dfrac{\alpha T}{\kappa} - P \qquad \text{(from Eq. 17.2)}$

$$= \frac{1.06 \times 10^{-3} \times 293}{9.08 \times 10^{-10} \text{ Pa}^{-1}} - 100\,000 \text{ Pa}$$

$$= 3.42 \times 10^8 \text{ Pa} = 3420 \text{ bar}$$

b. Molar mass of acetic acid = 60.0 g mol^{-1}

Molar volume, $V_m = 60.0/1.049 - 57.2 \text{ cm}^3 \text{ mol}^{-1}$

$$= 5.72 \times 10^{-5} \text{ m}^3 \text{ mol}^{-1}$$

$$P_i = \frac{1.78}{(5.72 \times 10^{-5})^2} \text{ Pa} = 5.44 \times 10^8 \text{ Pa}$$

$$= 5440 \text{ bar}$$

17.8. a. $C_P - C_V = \left[P + \left(\dfrac{\partial U}{\partial V}\right)_T\right]\left(\dfrac{\partial V}{\partial T}\right)_P \qquad \text{(Eq. 2.117)}$

One of the thermodynamic equations of state is

$$\left(\frac{\partial U}{\partial V}\right)_T = -P + T\left(\frac{\partial P}{\partial T}\right)_V \qquad\qquad \text{(Eq. 3.128)}$$

and therefore

$$C_P - C_V = \left(\frac{\partial P}{\partial T}\right)_V \left(\frac{\partial V}{\partial T}\right)_P T$$

$$\alpha \equiv \frac{1}{V}\left(\frac{\partial V}{\partial T}\right)_P \quad \text{and} \quad \kappa \equiv -\frac{1}{V}\left(\frac{\partial V}{\partial P}\right)_T$$

$$\frac{\alpha}{\kappa} = -\left(\frac{\partial V}{\partial T}\right)_P \left(\frac{\partial P}{\partial V}\right)_T = \left(\frac{\partial P}{\partial T}\right)_V \qquad\qquad \text{(see Appendix C)}$$

Thus

$$C_P - C_V = \frac{\alpha^2 V T}{\kappa}$$

b. For CCl_4, from the data given,

$$C_{P,m} - C_{V,m} = (1.24 \times 10^{-3}\ \text{K}^{-1})^2 \times \frac{(97 \times 10^{-6}\ \text{m}^3\ \text{mol}^{-1})(298.15\ \text{K})}{(10.6 \times 10^{-5}/100\ 000)\ \text{Pa}^{-1}}$$

$$= 42.0\ \text{J K}^{-1}\ \text{mol}^{-1}$$

$$C_{P,m} = 89.5 + 42.5 = 132.0\ \text{J K}^{-1}\ \text{mol}^{-1}$$

c. For acetic acid,

$$V_m = 60.05\ \text{g mol}^{-1}/1.049\ \text{g cm}^{-3}$$

$$= 57.2 \times 10^{-6}\ \text{m}^3\ \text{mol}^{-1}$$

$$C_{P,m} - C_{V,m} = (1.06 \times 10^{-3}\ \text{K}^{-1})^2 \times \frac{(57.2 \times 10^{-6}\ \text{m}^3\ \text{mol}^{-1})(293\ \text{K})}{9.08 \times 10^{-10}\ \text{Pa}^{-1}}$$

$$= 20.7\ \text{J K}^{-1}\ \text{mol}^{-1}$$

■ Intermolecular Energies

17.9. Since the intermolecular energy is inversely proportional to the sixth power of the intermolecular distance, it is inversely proportional to the square of the volume. The volume has increased by a factor of 10^3, and the energy therefore changes by a factor of 10^{-6}.

17.10. $E_p = -\dfrac{z_A e \mu}{4\pi\varepsilon_0 r^2}$ \qquad\qquad (from Eq. 17.13)

$$= \frac{-2 \times 1.602 \times 10^{-19}\ \text{C} \times 6.18 \times 10^{-30}\ \text{C m}}{4\pi(8.854 \times 10^{-12}\ \text{C}^2\ \text{N}^{-1}\ \text{m}^{-2})(5.00 \times 10^{-10}\ \text{m})^2}$$

$$= -7.12 \times 10^{-20} \text{ J}$$

$$= -42.9 \text{ kJ mol}^{-1}$$

17.11. $E_p = \dfrac{z_A z_B e^2}{4\pi\varepsilon_0 r}$ (from Eq. 17.12)

$$= -\frac{2(1.602 \times 10^{-19} \text{ C})^2}{4\pi(8.854 \times 10^{-12} \text{ C}^2\,\text{N}^{-1}\,\text{m}^{-2})(5.00 \times 10^{-10} \text{ m})}$$

$$= -9.23 \times 10^{-19} \text{ J} = -556 \text{ kJ mol}^{-1}$$

17.12. $E_p = \dfrac{-\alpha(z_A)^2}{8\pi\varepsilon_0 r^4}$ (Eq. 17.14)

$$= -\frac{(2.0 \times 10^{-30} \text{ m}^3)(2 \times 1.602 \times 10^{-19} \text{ C})^2}{8\pi(8.854 \times 10^{-12} \text{ C}^2\,\text{N}^{-1}\,\text{m}^{-2})(5.00 \times 10^{-10} \text{ m})^4}$$

$$= -1.48 \times 10^{-20} \text{ J} = -8.89 \text{ kJ mol}^{-1}$$

17.13. $E_p = -\dfrac{\mu_A^2 \mu_B^2}{24\pi^2\varepsilon_0^2 k_B T r^6}$ (from Eq. 17.17)

$$= -\frac{(6.18 \times 10^{-30} \text{ C m})^4}{24\pi^2(8.854 \times 10^{-12} \text{ C}^2\,\text{N}^{-1}\,\text{m}^{-2})^2(1.381 \times 10^{-23} \times 298.15 \text{ J})(5.00 \times 10^{-10} \text{ m})^6}$$

$$= -1.22 \times 10^{-21} \text{ J}$$

$$= -735 \text{ J mol}^{-1}$$

17.14.   Differentiation of Eq. 17.21 gives

$$\frac{dE_p}{dr} = \frac{6A}{r^7} - \frac{12B}{r^{13}}$$

This is zero when $r = r_0$ and therefore,

$$r_0 = \left(\frac{2B}{A}\right)^{1/6}$$

$$= \left(\frac{2 \times 3.42 \times 10^{10} \text{ J pm}^{12}}{1.34 \times 10^{-5} \text{ J pm}^6}\right)^{1/6}$$

$$= 415 \text{ pm}$$

The energy at this separation is

$$E_p = \left(-\frac{1.34 \times 10^{-5}}{(415)^6} + \frac{3.42 \times 10^{10}}{(415)^{12}}\right) \text{ J}$$

$$= \left(\frac{1.34 \times 10^{-5}}{5.108 \times 10^{15}} + \frac{3.42 \times 10^{10}}{2.610 \times 10^{31}} \right) J$$

$$= (-2.62 \times 10^{-21} + 1.31 \times 10^{-21}) \, J$$

$$= -1.31 \times 10^{-21} \, J = -789 \, J \, mol^{-1}$$

17.15. The dipole-dipole energy is given by Eq. 17.17:

$$E_p = - \frac{(2.60 \times 10^{-30})^4}{24\pi^2 (8.854 \times 10^{-12})^2 (1.381 \times 10^{-23} \times 298.15) \times (5 \times 10^{-10})^6}$$

$$= -3.83 \times 10^{-23} \, J = -23.0 \, J \, mol^{-1}$$

The dipole-(induced dipole) energy is given by Eq. 17.18.

$$E_p = - \frac{3.58 \times 10^{-30} (2.60 \times 10^{-30})^2}{2\pi (8.854 \times 10^{-12}) (5 \times 10^{-10})^6}$$

$$= -2.78 \times 10^{-23} \, J = -16.8 \, J \, mol^{-1}$$

The potential energy due to the dispersion forces is given by Eq. 17.19:

$$E_p = - \frac{3(6.626 \times 10^{-34} \times 3.22 \times 10^{15} \, J) \times (3.58 \times 10^{-30} \, m^3)^2}{4(5 \times 10^{-10})^6}$$

$$= -1.31 \times 10^{-21} \, J = -790 \, J \, mol^{-1}$$

17.16. Calculate ν_0 for He and Xe from Eq. 17.19 with E_p from Table 17.3.

For H, $\nu_0 = \dfrac{-7.639 \times 10^{-24} \times 4 \times (500 \times 10^{-12})^6}{3 \times 6.626 \times 10^{-34} \times (0.205 \times 10^{-30})^2} = 5.72 \times 10^{15} \, s^{-1}$

For Xe, $\nu_0 = \dfrac{-1.411 \times 10^{-21} \times 4 \times (500 \times 10^{-12})^6}{3 \times 6.626 \times 10^{-34} \times (4.04 \times 10^{-30})^2} = 2.72 \times 10^{15} \, s^{-1}$

$$E_d = - \frac{3h\nu_0 \alpha_0^2}{4r^6}$$

$$E_d / J = - \frac{3 \times 6.626 \times 10^{-34} (\nu_0 / s^{-1})(\alpha / m^3)^2}{4 \times (500 \times 10^{-12})^6}$$

$$= -3.180 \times 10^{22} \, (\nu_0 / s^{-1}) \, (\alpha / m^3)^2$$

For He, $E_d / J = -3.180 \times 10^{22} \times 5.72 \times 10^{15} \times (0.20 \times 10^{-30})^2$

$\qquad E_d = -7.28 \times 10^{-24} \, J; \quad -4.38 \, J \, mol^{-1}$

For Ne, $E_d / J = -3.180 \times 10^{22} \times 5.21 \times 10^{15} \times (0.396 \times 10^{-30})^2$

$\qquad E_d = -2.60 \times 10^{-23} \, J; \quad -15.6 \, J \, mol^{-1}$

For Ar, E_d/J $= -3.180 \times 10^{22} \times 3.39 \times 10^{15} \times (1.63 \times 10^{-30})^2$

$\quad E_d$ $= -2.86 \times 10^{-22}$ J; $\quad -172$ J mol^{-1}

For Kr, E_d/J $= -3.180 \times 10^{22} \times 2.94 \times 10^{15} \times (2.46 \times 10^{-30})^2$

$\quad E_d$ $= -5.66 \times 10^{-22}$ J; $\quad -341$ J mol^{-1}

For Xe, E_d/J $= -3.180 \times 10^{22} \times 2.72 \times 10^{15} \times (4.04 \times 10^{-30})^2$

$\quad E_d$ $= -1.41 \times 10^{-21}$ J; $\quad -850$ J mol^{-1}

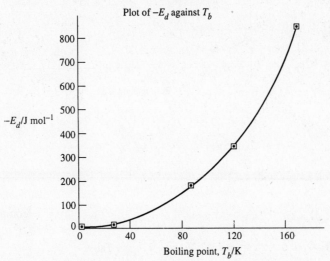

Plot of $-E_d$ against T_b

17.17. The values previously calculated (Problem 17.16) will be multiplied by the factor

$$\left(\frac{500}{r/\text{pm}}\right)^6$$

The values of $-E_d/J$ mol^{-1} are then:

	He	Ne	Ar	Kr	Xe
For 500 pm:	4.38	15.6	172	341	850
For r:	358	227	893	1300	2420

If each Ar atom has 12 nearest neighbors, the estimated enthalpy of vaporization is

$$(12/2) \times 892 \text{ J mol}^{-1} = 5.3 \text{ kJ mol}^{-1}$$

This is not bad agreement considering the simplicity of the model.

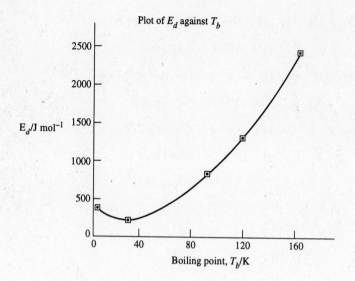

Plot of E_d against T_b

17.18. The interaction energy with $\varepsilon = 1$ is

$$E_p = -\frac{\alpha\mu^2}{4\pi\varepsilon_0 r^6}$$

(compare Eq. 17.18)

Then for Ar·H$_2$O,

$$E_p = -\frac{(1.63 \times 10^{-30}\ \text{m}^3)\ (6.18 \times 10^{-30}\ \text{C m})^2}{4\pi(8.854 \times 10^{-12}\ \text{C}^2\ \text{N}^{-1}\ \text{m}^{-2})\ (6.0 \times 10^{-10}\ \text{m})^6}$$

$$= -1.20 \times 10^{-23}\ \text{J} = -7.22\ \text{J mol}^{-1}$$

In Ag·5H$_2$O there will be in addition considerable hydrogen bonding between neighboring water molecules.

17.19. a. $E_p = -\dfrac{A}{r^6} + \dfrac{B}{r^n}$ (1)

Differentiating, with $r = r_0$ at the minimum,

$$\frac{dE_p}{dr} = \frac{6A}{r_0^{\ 7}} - \frac{nB}{r_0^{\ n+1}} = 0 \tag{2}$$

$$E_{\text{min}} = -\frac{A}{r_0^{\ 6}} + \frac{B}{r_0^{\ n}} \tag{3}$$

From (2), $B = \dfrac{6Ar_0^{\ n-6}}{n}$ (4)

Substituting Eq. 4 in Eq. 3,

$$E_{\min} = -\frac{A}{r_0^6} + \frac{6A}{nr_0^6} \tag{5}$$

$$\text{or} \quad A = \frac{nE_{\min}r_0^6}{6-n} \tag{6}$$

Insertion of Eq. 6 in Eq. 4,

$$B = \frac{6r_0^n E_{\min}}{6-n} \tag{7}$$

Then from Eq. 1,

$$\frac{E}{E_{\min}} = -\frac{n}{6-n}\left(\frac{r_0}{r}\right)^6 + \frac{6}{6-n}\left(\frac{r_0}{r}\right)^n \tag{8}$$

b. $-\dfrac{A}{(r^*)^6} + \dfrac{B}{(r^*)^n} = 0$ \hfill (9)

and thus

$$(r^*)^{n-6} = \frac{B}{A} = \frac{6r_0^{n-6}}{n} \quad \text{(from Eq. 4)}$$

Therefore

$$\left(\frac{r^*}{r_0}\right)^{n-6} = \frac{6}{n} \tag{10}$$

c. If $n = 12$, from Eq. 8,

$$E = E_{\min}\left[2\left(\frac{r_0}{r}\right)^6 - \left(\frac{r_0}{r}\right)^{12}\right] \tag{11}$$

If $n = 12$, $(r_0/r^*)^6 = 2$

and from Eq. 11,

$$E = 4E_{\min}\left[\left(\frac{r^*}{r}\right)^6 - \left(\frac{r^*}{r}\right)^{12}\right] \tag{12}$$

18. SURFACE CHEMISTRY AND COLLOIDS

Adsorption Isotherms

18.1.   a. $\theta = \dfrac{K\,[A]}{1 + K\,[A]}$

$$0.5 = \frac{K \times 1 \text{ bar}}{1 + K \times 1 \text{ bar}}$$

$$0.5 + 0.5\,K \text{ bar} = K \text{ bar}$$

$$K/\text{bar}^{-1} = 1$$

b. $0.75 = \dfrac{P \times 1 \text{ bar}^{-1}}{1 + P \times 1 \text{ bar}^{-1}} ; \quad P = \dfrac{\theta}{1 - \theta} ; \quad P = 3 \text{ bar}$

$$0.90 = \frac{P \text{ bar}^{-1}}{1 + P \text{ bar}^{-1}} ; \quad P = 9 \text{ bar}$$

$$0.99 = \frac{P \text{ bar}^{-1}}{1 + P \text{ bar}^{-1}} ; \quad P = 99 \text{ bar}$$

$$0.999 = \frac{P \text{ bar}^{-1}}{1 + P \text{ bar}^{-1}} ; \quad P = 999 \text{ bar}$$

c. $\theta = \dfrac{0.1}{1 + 0.1} = 0.091$

$$\theta = \frac{0.5}{1 + 0.5} = 0.33$$

$$\theta = \frac{1000}{1 + 1000} = 0.999$$

18.2. The Langmuir isotherm in terms of pressure P is, from Eq. 18.6,

$$\theta = \frac{KP}{1 + KP}$$

$$\theta = \frac{V}{V_0}$$

and therefore

$$\frac{V}{V_0} = \frac{KP}{1 + KP}$$

which rearranges to

$$\frac{P}{V} = \frac{1 + KP}{V_0 K} = \frac{1}{V_0 K} + \frac{P}{V_0}$$

A plot of P/V against P is therefore linear; the slope is $1/V_0$ and the intercept on the P/V_0 axis is $1/V_0 K$. The quantities V_0 and K can thus be obtained separately.

18.3. The amount x adsorbed is proportional to θ and therefore,

$$x = \frac{aK[A]}{1 + K[A]}$$

To convert atmospheres to concentrations:

$$\frac{n}{V} = \frac{P}{RT} = \frac{P/\text{atm}}{0.082\ 05 \times 293.15\ \text{dm}^3\ \text{mol}^{-1}}$$

To convert amount of gas adsorbed to moles:

$$n = \frac{PV}{RT} = \frac{1(\text{atm})V}{0.082\ 05 \times 293.15\ \text{atm dm}^3\ \text{mol}^{-1}}$$

$$= \frac{V/\text{mm}^3}{10^6 \times 0.082\ 05 \times 293.15}\ \text{mol}$$

Thus, the table becomes

[A]

Concentration mol dm^{-3}	0.116	0.166	0.249	0.391	0.711	1.39

x

Amount adsorbed 10^{-7} mol	4.99	6.28	7.90	9.94	11.7	13.7

a. A linear plot may be obtained by plotting $1/x$ against $1/[A]$:

$$\frac{1}{x} = \frac{1}{aK[A]} + \frac{1}{a}$$

$[A]^{-1}/\text{dm}^3\ \text{mol}^{-1}$	8.62	6.02	4.02	2.56	1.41	0.719
$x^{-1}/10^6\ \text{mol}^{-1}$	2.00	1.59	1.27	1.01	0.85	0.73

From a plot of x^{-1} against $[A]^{-1}$ or from linear regression,

$$a = 1.63 \times 10^{-6} \text{ mol}$$

$$K = 3.82 \text{ dm}^3 \text{ mol}^{-1}$$

b. Complete coverage corresponds to

$$1.63 \times 10^{-6} \text{ mol} = 9.82 \times 10^{17} \text{ molecules}$$

The surface area was thus about

$$10^3 \text{ cm}^2 = 10^{-1} \text{ m}^2$$

18.4.   a. From Eq. 18.7,

$$1 - \theta = \frac{1}{1 + KP} \tag{1}$$

and therefore

$$\theta = KP(1 - \theta) \tag{2}$$

Then

$$\ln \frac{\theta}{P} = \ln K + \ln (1 - \theta) \tag{3}$$

$$\approx \ln K - \theta \quad \text{if} \quad \theta \ll 1 \tag{4}$$

A plot of $\ln (\theta/P)$ against $\ln \theta$ is thus linear with a slope of -1.

b. Since $\theta = V/V_0$, Eq. (4) can be written as

$$\ln (V/P) - \ln V_0 \approx \ln K - V/V_0 \tag{5}$$

A plot of $\ln (V/P)$ against V thus has a slope of $-1/V_0$.

18.5.   The Langmuir isotherm can be used in the form

$$\frac{V}{V_0} = \frac{KP}{1 + KP} \quad \text{or} \quad \frac{1}{V} = \frac{1}{V_0 KP} + \frac{1}{V_0}$$

The data are plotted as $1/V$ against $1/P$ in the accompanying figure. The Langmuir isotherm is obeyed, with

$$V_0 = 222 \text{ cm}^3 \quad \text{and} \quad K = 7.35 \times 10^{-2} \text{ kPa}^{-1}$$

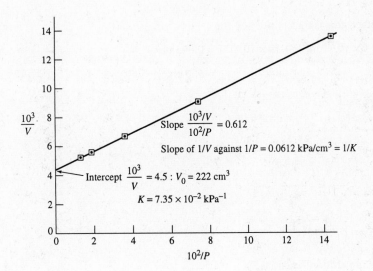

18.6.   a.   The BET isotherm can be tested in a number of ways, for example by plotting
$P/V (P_0 - P)$ against P.

   b.   If $P_0 \gg P$, the isotherm becomes

$$\frac{P}{V} = \frac{1}{V_0 K} + \frac{P}{V_0}$$

The fraction covered $\theta = V/V_0$ and therefore,

$$\frac{P}{\theta} = \frac{1}{K} + P$$

or

$$\theta = \frac{KP}{1 + KP}$$

which is the Langmuir isotherm.

18.7.   Insertion of the data into the BET equation (Eq. 18.25) gives two simultaneous equations:

$$0.796 = \frac{1}{V_0 K} + \frac{1.11}{V_0}$$

$$1.956 = \frac{1}{V_0 K} + \frac{3.08}{V_0}$$

The solution is

$$K = 4.11 \; (\text{Torr}^{-1}) \qquad V_0 = 1.70 \; (\text{cm}^3)$$

At S.T.P., 22.7 L = 22 700 cm^3 is the volume occupied by 1 mol. A volume of 1.70 cm^3 thus contains 7.49×10^{-5} mol = 4.51×10^{19} molecules.

The surface area is thus,

$$2.1 \times 10^{-21} \text{ m}^2 \times 4.51 \times 10^{19} \qquad = 0.095 \text{ m}^2$$

$$= 950 \text{ cm}^2$$

18.8.   The process is

$$A_2 + 2S \rightleftarrows 2S{-}A$$

and

$$K_c = \frac{c_a^2}{c_g c_s^2} = \frac{N_a^2}{(N_g/V)\, N_s^2}$$

$$= \left(\frac{\theta}{1-\theta}\right)^2 \frac{1}{c_g}$$

In terms of partition functions,

$$K_c = \frac{q_a^2}{q_g q_s^2}\, e^{-\Delta E_0/RT}$$

$$= \frac{h^3 b_a^2}{(2\pi m k_B T)^{3/2} b_g}\, e^{-\Delta E_0/RT} \quad (\text{if } q_s = 1)$$

Therefore

$$\frac{\theta}{1-\theta} = c_g^{1/2}\, \frac{h^{3/2} b_a}{(2\pi m k_B T)^{3/4} b_g^{1/2}}\, e^{-\Delta E_0/2RT}$$

18.9.   The adsorption centers need no longer be regarded as reactants; the equilibrium is between gas molecules and molecules forming the two-dimensional layer:

$$K_c = \frac{c_a}{c_g} = \frac{N_a/S}{N_g/V} = \frac{\dfrac{(2\pi m k_B T)}{h^2}\, b_a\, e^{-\Delta E_0/RT}}{\dfrac{(2\pi m k_B T)^{3/2}}{h^3}\, b_g}$$

Therefore

$$c_a = c_g\, \frac{h}{(2\pi m k_B T)^{1/2}}\, \frac{b_a}{b_g}\, e^{-\Delta E_0/RT}$$

■ Kinetics of Surface Reactions

18.10. a. The rate and the rate constant are both increased by a factor of 10:

$$v = 1.5 \times 10^{-3} \text{ mol dm}^{-3} \text{ s}^{-1}; \quad k = 2.0 \times 10^{-2} \text{ s}^{-1}$$

b. The rate of conversion (mol s^{-1}) remains the same, but since the volume is increased by a factor of 10 the rate is reduced by a factor of 10, as is the rate constant:

$$v = 1.5 \times 10^{-5} \text{ mol dm}^{-3} \text{ s}^{-1}; \quad k = 2.0 \times 10^{-4} \text{ s}^{-1}$$

c. Increasing the radius by a factor of 10 increases the surface area by a factor of 100 and the volume by a factor of 1000. The rate and the rate constant are thus reduced by a factor of 10:

$$v = 1.5 \times 10^{-5} \text{ mol dm}^{-3} \text{ s}^{-1}; \quad k = 2.0 \times 10^{-4} \text{ s}^{-1}$$

d. Since k is proportional to S and inversely proportional to V, the constant $k' = kV/S$ is independent of V and S.

e. Its SI unit is m s^{-1}.

18.11. The arguments are the same as in Problem 18.10, and thus,

a. (a') $k = 2.5 \times 10^{-2} \text{ mol dm}^{-3} \text{ s}^{-1}$

$v = 2.5 \times 10^{-2} \text{ s}^{-1}$

(b') $k = 2.5 \times 10^{-4} \text{ mol dm}^{-3} \text{ s}^{-1}$

$v = 2.5 \times 10^{-4} \text{ s}^{-1}$

(c') $k = 2.5 \times 10^{-4} \text{ mol dm}^{-3} \text{ s}^{-1}$

$v = 2.5 \times 10^{-4} \text{ s}^{-1}$

b. $k' = kV/S$

c. mol m^{-2} s^{-1}

18.12.

	2NH$_3$	=	N$_2$	+	3H$_2$
Initial concentrations:	a_0		0		0
Concentrations after time t:	$a_0 - \dfrac{2x}{3}$		$\dfrac{x}{3}$		x

$$\frac{dx}{dt} = k\left(a_0 - \frac{2x}{3}\right)/x$$

$$= \frac{3\, ka_0 - 2kx}{3x} = \frac{ka_0}{x} - \frac{2k}{3}$$

18.13. From Eq. 18.14, the fraction of bare surface is

$$1 - \theta = \frac{1}{1 + K^{1/2}[H_2]^{1/2}}$$

Rate of H atom formation is thus,

$$\begin{aligned} v &= k(1 - \theta)[H_2] \\ &= \frac{k[H_2]}{1 + K^{1/2}[H_2]^{1/2}} \end{aligned}$$

Kinetics are one-half order when $K^{1/2}[H_2]^{1/2} \gg 1$; i.e., at high pressure when the surface is fully covered:

$$v = \frac{k}{K^{1/2}}[H_2]^{1/2}$$

18.14. $k \propto e^{-E/RT}$, $K \propto e^{-\Delta H_A/RT}$ and $K_i \propto e^{-\Delta H_I/RT}$

   a. $v = kK[A] \propto e^{-(E+\Delta H_A)RT}$

   b. $v = k \propto e^{-E/RT}$

   c. $v = \dfrac{kK}{K_i}\dfrac{[A]}{[I]} \propto e^{-(E + \Delta H_A - \Delta H_I)/RT}$

18.15.

	A	$\rightarrow$	Y	+	Z
Initially:	a_0		0		0
At time t:	$a_0 - z$		z		z

$$\frac{dz}{dt} = k(a_0 - z) + k_s z$$

$$\frac{dz}{ka_0 + (k_s - k)z} = dt$$

Put $ka_0 - (k_s - k)z = y$; $dy = -(k_s - k)dz$

$$-\frac{1}{(k_s - k)}\int \frac{dy}{y} = \int dt$$

$$-\frac{1}{(k_s - k)} \ln[(ka_0 - k)z] = t + I$$

The boundary condition is that $z = 0$ when $t = 0$:

$$I = -\frac{1}{(k_s - k)} \ln ka_0$$

$$t = \frac{1}{(k_s - k)} \ln \frac{ka_0}{ka_0 - (k_s - k)z}$$

$$\frac{ka_0 - (k_s - k)z}{ka_0} = e^{-(k_s - k)t}$$

$$z = \frac{ka_0}{k_s - k}\left[1 - e^{-(k_s - k)t}\right]$$

18.16.　　a.　The general rate equation is Eq. 18.34. At low pressures the surface is sparsely covered and

$$v = kK[A]; \quad E_{a_{\text{observed}}} = E_{a_0} + \Delta H_{ad}$$

At high pressure it is fully covered and

$$v = k; \quad E_{a_{\text{observed}}} = E_{a_0} \quad \& \quad \Delta H_{ad} < 0$$

b.　Reaction occurs on certain surface sites on which N_2 is not adsorbed. The rate equation is Eq. 18.36.

c.　This is a Langmuir-Rideal mechanism. The general equation is Eq. 18.44; $K_{H_2}[H_2]$ is small and $K_{O_2}[O_2]$ is large, so that

$$v = k[H_2]$$

d.　The mechanism is

$$p\text{-}H_2 + -S-S- \rightleftarrows -\overset{\overset{\displaystyle H}{|}}{S}-\overset{\overset{\displaystyle H}{|}}{S}- \rightleftarrows -\overset{|}{S}-\overset{|}{S}- + o\text{-}H_2$$

with the surface fully covered. The rate is

$$v = k[H_2](1 - \theta)^2$$

where $1 - \theta$ is given by Eq. 18.15; thus

$$v = \frac{k}{K}$$

■ Surface Tension and Capillarity

18.17. According to Eq. 18.54, the capillary rise is given by

$$h = \frac{2\gamma}{\rho g r}$$

a. If $r = 10^{-3}$ m,

$$h = \frac{2 \times 7.27 \times 10^{-2} \text{ (N m}^{-1})}{998 \text{(kg m}^{-3})9.81 \text{(m s}^{-2})10^{-3} \text{ (m)}}$$

$$= 1.49 \times 10^{-2} \text{ m} = 1.49 \text{ cm}$$

b. If $r = 10^{-5}$ m,

$$h = 1.49 \text{ m}$$

18.18. From Eq. 18.55,

$$h = \frac{2\gamma \cos \theta}{\rho g r}$$

$$= \frac{2 \times 0.47 \text{ (N m}^{-1}) \text{ (}-0.766)}{1.36 \times 10^4 \text{(kg m}^{-3})9.81 \text{(m s}^{-2}) \times 5 \times 10^{-4} \text{(m)}}$$

$$= -10.8 \times 10^{-3} \text{ m} = -10.8 \text{ mm}$$

18.19. Volume of droplet $= \dfrac{10^{-12}}{0.998} = 1.002 \times 10^{-12} \text{ cm}^3$

$$= 1.002 \times 10^{-18} \text{ m}^3 = \frac{4}{3} \pi r^3$$

$$r^3 = \frac{3 \times 1.002 \times 10^{-18}}{4\pi}$$

$$= 2.392 \times 10^{-19} \text{ m}^3$$

$$r = 6.21 \times 10^{-7} \text{ m}$$

$$\ln \frac{P}{P_0} = \frac{2\gamma M}{\rho r R T}$$

$$= \frac{2 \times 7.27 \times 10^{-2} \text{ (N m}^{-1}) \times 18.02 \times 10^{-3} \text{ (kg mol}^{-1})}{0.998 \times 10^{+3} \text{(kg m}^{-3}) \times 6.21 \times 10^{-7} \text{(m)} \times 8.3145} \times \frac{1}{298.15 \text{(J mol}^{-1})}$$

$$= 0.0017$$

$$\frac{P}{P_0} = 1.0017$$

18.20. From Eq. 18.54, the height is $2\gamma/r\rho g$ and the difference in heights is

$$\Delta h = \frac{2\gamma}{\rho g}\left(\frac{1}{r_1} - \frac{1}{r_2}\right)$$

Thus

$$0.022 \text{ m} = \frac{2\gamma(2000 - 1000) \text{ m}^{-1}}{0.80 \times 10^3 (\text{kg m}^{-3})9.81(\text{m s}^{-2})}$$

$$= 0.255 \text{ (kg}^{-1} \text{ m s}^2)\gamma$$

$$\gamma = 0.086 \text{ kg s}^{-2} \equiv 0.086 \text{ N m}^{-1}$$

18.21. The rise is proportional to γ/d and is therefore the same in the second liquid, i.e., 1.5 cm.

18.22. No, the water does not flow over the edge. The meniscus will rise to the top of the tube and then the radius of curvature will decrease until the capillary pressure just balances the pressure of the column of liquid; equilibrium is then established. This will occur when the radius of curvature at the surface is half what it is in a longer tube.

18.23. The equation that applies is an extension of Eq. 18.55:

$$\gamma = \frac{rh\Delta\rho g}{2\cos\theta}$$

where $\Delta\rho$ is the difference between the two densities. Then

$$\gamma = (1.00 - 0.80) \times 10^3 \text{ kg m}^{-3} \times 9.81 \text{ m s}^{-2} \times \frac{0.040 \text{ m} \times 0.5 \times 10^{-4} \text{ m}}{2 \times 0.766}$$

$$= 2.56 \times 10^{-3} \text{ N m}^{-1} \quad (\text{N} \equiv \text{kg m s}^{-2})$$

18.24. The Gibbs energy change is the work done, which is the surface tension multiplied by the change in surface area (Eq. 18.50).

The surface area of the water in bulk can be estimated on the assumption that the liter of water was present as a sphere. The volume, 1 dm^3, is 10^{-3} m^3, and the radius r is

$$(3 \times 10^{-3} \text{ m}^3/4\pi)^{1/3} = 6.20 \times 10^{-2} \text{ m}$$

The surface area, $4\pi r^2$, is thus

$$4\pi \times (6.2 \times 10^{-2} \text{ m})^2 = 0.0483 \text{ m}^2$$

The surface area of each droplet is

$$4\pi \times (10^{-7} \text{ m})^2 = 1.257 \times 10^{-13} \text{ m}^2$$

and the volume of each droplet is

$$(4/3)\pi \ (10^{-7} \text{ m})^3 = 4.188 \times 10^{-21} \text{ m}^3$$

The number of droplets is therefore

$$10^{-3} \text{ m}^3/4.188 \times 10^{-21} \text{ m}^3 = 2.388 \times 10^{17}$$

The total surface area of the droplets is therefore

$$2.388 \times 10^{17} \times 1.257 \times 10^{-13} \text{ m}^2 = 3.00 \times 10^4 \text{ m}^2$$

This is effectively the increase in surface area, and the increase in Gibbs energy is

$$7.27 \times 10^{-2} \text{ N m}^{-1} \times 3.00 \times 10^4 \text{ m}^2 = 2182 \text{ J} = 2.182 \text{ kJ}$$

■ Surface Films

18.25. An acre is 4840 square yards $= 4840 \times (0.915)^2 \text{ m}^2 = 4052 \text{ m}^2$. Thus, half an acre is approximately 2000 m^2. The thickness of the film is thus

$$10^{-6} \text{ m}^3 / 2000 \text{ m}^2 = 5 \times 10^{-10} \text{ m} = 5 \text{ Å}$$

This is a reasonable estimate, considering the approximate nature of the area and volume. Lord Rayleigh later estimated the thickness of an oil film to be about 10 Å, and similar values were obtained in Langmuir's work.

18.26. At the three lower pressures, πA is constant within the experimental error and has a value of

$$1.11 \times 10^4 \text{ N m kg}^{-1} \ (= \text{J kg}^{-1})$$

πA is equal to RT and therefore,

$$\pi A = 8.3145 \times 288.15 = 2396 \text{ J mol}^{-1}$$

The molar mass is thus

$$\frac{2396}{11\ 100} = 0.216 \text{ kg mol}^{-1} = 216 \text{ g mol}^{-1}$$

At the highest pressure the area is 5.7 cm^2 μg^{-1}. Since 216 g $= 6.022 \times 10^{23}$ molecules,

$$1 \text{ μg} = \frac{6.022 \times 10^{23} \times 10^{-6}}{216} = 2.79 \times 10^{15} \text{ molecules}$$

Thus, 1 molecule occupies

$$\frac{5.7}{2.79 \times 10^{15}} = 2.04 \times 10^{-15} \text{ cm}^2 = 0.204 \text{ nm}^2$$

18.27. The molecular weight of 1-hexadecanol is 242.43 g mol^{-1}, and 52.0 μg therefore contains 2.14×10^{-7} mol $= 1.29 \times 10^{17}$ molecules.

Length/cm	Area/cm^2	Force/10^{-5} N	Surface Pressure/ 10^{-4} N m^{-1}	Area per molecule/nm^2
20.9	292.6	4.14	3.00	0.227
20.3	284.2	8.56	6.20	0.220
20.1	281.4	26.2	19.0	0.218
19.6	274.4	69.0	50.0	0.212
19.1	267.4	108	78.3	0.207
18.6	260.4	234	169.6	0.202
18.3	256.2	323	234.1	0.198
18.1	253.4	394	285.5	0.196
17.8	249.2	531	384.8	0.193

From the graph, area for the fully compressed layer = 0.19 nm^2.

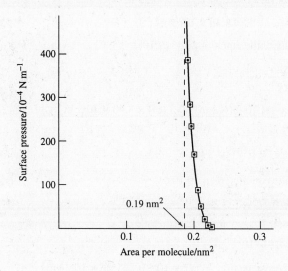

19. TRANSPORT PROPERTIES

■ Viscosity

19.1.   Area of cross section of aorta

$$= \pi(9 \times 10^{-3})^2 = 2.54 \times 10^{-4} \text{ m}^2$$

Volume rate of flow $= 2.54 \times 10^{-4} \times 0.33 \text{ m}^3 \text{ s}^{-1}$

$$= 8.40 \times 10^{-5} \text{ m}^3 \text{ s}^{-1}$$

From the Pouisseuille equation (Eq. 19.10),

$$\Delta P = \frac{8\eta l}{\pi R^4} \cdot \frac{dV}{dt}$$

$$= \frac{8 \times 4 \times 10^{-3} \text{ (N s m}^{-2}\text{)} 0.5 \text{ (m)} \times 8.40 \times 10^{-5} \text{ m}^3 \text{ s}^{-1}}{\pi(9 \times 10^{-3})^4 \text{ m}^4}$$

$$= 65.2 \text{ Pa} = 0.49 \text{ Torr}$$

19.2.   From Poiseuille equation (Eq. 19.10),

$$\frac{dV}{dt} = \frac{\pi R^4 \Delta P}{8\eta l} = \frac{\pi(2 \times 10^{-6})^4 \times 20 \times 133.3}{8 \times 4 \times 10^{-3} \times 10^{-3}}$$

$$= 4.19 \times 10^{-15} \text{ m}^3 \text{ s}^{-1}$$

a.   Area of cross section $= \pi(2 \times 10^{-6})^2 = 1.26 \times 10^{-11} \text{ m}^2$

Linear rate of flow $= \dfrac{4.19 \times 10^{-15}}{1.26 \times 10^{-11}}$

$$= 3.33 \times 10^{-4} = 0.333 \text{ mm s}^{-1}$$

b.   Volume passing each second $= 1.26 \times 10^{-11} \text{ m}^2 \times 0.333 \times 10^{-3} \text{ m}$

$$= 4.19 \times 10^{-15} \text{ m}^3$$

c.   Number of capillaries $= \dfrac{8.40 \times 10^{-5}}{4.19 \times 10^{-15}} = 2.00 \times 10^{10}$

19.3.   From the ideal gas law,

$$\frac{n}{V} = \frac{P}{RT} = \frac{101\ 325}{8.3145 \times 298.15} = 40.874 \text{ mol m}^{-3}$$

$$\frac{N}{V} = 40.874 \times 6.022 \times 10^{23} \text{ m}^{-3} = 2.461 \times 10^{25} \text{ m}^{-3}$$

$$\bar{u} = \left(\frac{8k_\text{B}T}{\pi m}\right)^{1/2}$$

$$m = \frac{28.05}{6.022 \times 10^{23}} = 4.658 \times 10^{-23} \text{ g}$$

$$= 4.658 \times 10^{-26} \text{ kg}$$

$$\bar{u} = \left(\frac{8 \times 1.381 \times 10^{-23} \times 298.15}{\pi \times 4.658 \times 10^{-26}}\right)^{1/2}$$

$$= 474.4 \text{ m s}^{-1}$$

a.   From Eq. 19.15,

$$d = \left(\frac{mu}{2\sqrt{2\pi\eta}}\right)^{1/2}$$

$$= \left(\frac{4.658 \times 10^{-26} \times 474.4}{2\sqrt{2\pi} \times 9.33 \times 10^{-6}}\right)^{1/2}$$

$$= 5.16 \times 10^{-10} \text{ m} = 0.516 \text{ nm}$$

b.   Mean free path, $\lambda = \dfrac{V}{\sqrt{2}\pi d_\text{A}^2 N_\text{A}}$ (Eq. 1.68)

$$= \frac{1}{\sqrt{2}\pi(5.16 \times 10^{-10})^2 \times 2.462 \times 10^{25}}$$

$$= 3.43 \times 10^{-8} \text{ m}$$

$$= 34.3 \text{ nm}$$

c. $Z_\text{A} = \dfrac{\bar{u}}{\lambda} = \dfrac{474.4}{3.43 \times 10^{-8}} = 1.38 \times 10^{10} \text{ s}^{-1}$

d.   From Eqs. 1.59 and 1.61,

$$Z_\text{AA} = \frac{1}{2} Z_\text{A} N/V$$

$$= \frac{1.38 \times 10^{10} \times 2.462 \times 10^{25}}{2} = 1.70 \times 10^{35} \text{ m}^{-3} \text{ s}^{-1}$$

19.4.   a.   At 0 °C,

$$e^{-E/RT} = e^{-10\ 900/8.3145 \times 273.15} = 8.234 \times 10^{-3}$$

At 40.0 °C,

$$e^{-E/RT} = e^{-10\ 900/8.3145 \times 313.15} = 15.20 \times 10^{-3}$$

$$\text{Viscosity at 40.0 °C} = 1.33 \times 10^{-3} \times \frac{8.234}{15.20} \text{ kg m}^{-1} \text{ s}^{-1}$$

$$= 7.20 \times 10^{-4} \text{ kg m}^{-1} \text{ s}^{-1}$$

b.   At 20 °C,

$$e^{-E/RT} = e^{-18\ 000/8.3145 \times 293.15} = 6.205 \times 10^{-4}$$

At 40 °C,

$$e^{-E/RT} = e^{-18\ 000/8.3145 \times 313.15} = 9.945 \times 10^{-4}$$

$$\text{Viscosity at 40.0 °C} = 1.002 \times 10^{-3} \times \frac{6.203}{9.945} \text{ kg m}^{-1} \text{ s}^{-1} = 6.25 \times 10^{-4} \text{ kg m}^{-1} \text{ s}^{-1}$$

19.5.   a. $[\eta] = \dfrac{0.050}{5.90} \cdot \dfrac{1}{0.10 \text{ g dm}^{-3}}$

$$= 0.085 \text{ dm}^3 \text{ g}^{-1} = 0.085 \text{ m}^3 \text{ kg}^{-1}$$

b. $[\eta] = \dfrac{0.15}{5.90} \cdot \dfrac{1}{0.10 \text{ g dm}^{-3}}$

$$= 0.254 \text{ dm}^3 \text{ g}^{-1} = 0.254 \text{ m}^3 \text{ kg}^{-1}$$

c. $[\eta] = \dfrac{0.37}{5.90} \cdot \dfrac{1}{0.10 \text{ g dm}^{-3}} = 0.627 \text{ m}^3 \text{ kg}^{-1}$

19.6.   Taking logarithms:

(a) $\log_{10}[M] = 4.30$ $\log_{10}[\eta] = -1.075$

(c) $\log_{10}[M] = 4.60$ $\log_{10}[\eta] = -0.202$

(b) $\log_{10}[\eta] = -0.595$

If the Mark-Houwink equation applies, there is a linear relationship between $\log_{10}[\eta]$ and $\log_{10}[M]$.

$$\frac{-0.595 + 1.075}{-0.202 + 1.075} = 0.55$$

If x = relative molar mass for (b),

$$\log_{10}x = 0.55 \times (4.60 - 4.30) + 4.30 = 4.465$$

$$x = 29\ 200$$

19.7.   a.   At 0 °C,

$$e^{-E/RT} = e^{-12\,600/8.3145 \times 273.15} = 3.895 \times 10^{-3}$$

At 40.0 °C,

$$e^{-E/RT} = e^{-12\,600/8.3145 \times 313.15} = 7.913 \times 10^{-3}$$

Viscosity at 40.0 °C $= 7.06 \times 10^{-4} \times \dfrac{3.895}{7.913}$ kg m^{-1} s^{-1} $= 3.48 \times 10^{-4}$ kg m^{-1} s^{-1}

b. At 0 °C,

$$(T/K)^{-1.72}\, e^{543(T/K)} = (273.15)^{-1.72} \times e^{543/273.15}$$

$$= 6.447 \times 10^{-5} \times 7.30 = 4.71 \times 10^{-4}$$

At 40.0 °C,

$$(T/K)^{-1.72}\, e^{543(T/K)} = (313.15)^{-1.72} \times e^{543/313.15}$$

$$= 5.097 \times 10^{-5} \times 5.663 = 2.89 \times 10^{-4}$$

Viscosity at 40.0 °C $= 7.06 \times 10^{-4} \times \dfrac{2.89}{4.707}$ kg m^{-1} s^{-1}

$$= 4.33 \times 10^{-4} \text{ kg m}^{-1} \text{ s}^{-1}$$

The fluidity can be expressed as

$$\phi = A^{-1}(T/K)^{1.72}\, e^{-4515 \text{ J mol}^{-1}/RT}$$

The activation energy is, by definition,

$$E = RT^2 \frac{d \ln \phi}{dT}$$

$$\frac{d \ln \phi}{dT} = \frac{1.72 \text{ K}}{T} + \frac{4515 \text{ J mol}^{-1}}{RT^2}$$

$$= \frac{(1.72 \times 8.3145 \times 293.15) + 4515 \text{ J mol}^{-1}}{RT^2}$$

$$= \frac{(4192 + 4515) \text{ J mol}^{-1}}{RT^2} = \frac{8707 \text{ J mol}^{-1}}{RT^2}$$

The activation energy is thus,

$$E = 8707 \text{ J mol}^{-1} = 8.7 \text{ kJ mol}^{-1}$$

19.8. From the empirical relationship,

$$\ln \eta(20 \text{ °C}) - \ln \eta(40 \text{ °C}) = \frac{3.1556 \times 20 + 1.925 \times 10^{-3} \times 400}{149}$$

$$-6.9058 - \ln \eta(40 \text{ °C}) = (63.112 + 0.77)/149 = 0.4287$$

$$\ln \eta(40 \text{ °C}) = -0.4287 - 6.9058 = -7.3345$$

$$= 6.53 \times 10^{-4} \text{ kg m}^{-1} \text{ s}^{-1}$$

By definition,

$$E \equiv RT^2 \, d \ln \phi/dT$$

$$= -RT^2 \, d \ln \eta/dT$$

From the empirical relationship, with

$$T = t + 273.15$$

$$\ln \eta_t = \ln \eta_{20°} - \frac{a(T - 293.15) + b(T - 293.15)^2}{T - 164.15}$$

where $\quad a = 3.1556$

and $\quad b = 1.925 \times 10^{-3}$

$$-\frac{d \ln \eta}{dT} = \frac{(T - 154.15)(a + 2bT - 586.3\,b) - a(T - 293.15) - b(T - 293.15)^2}{(T - 164.15)^2}$$

a. At 20 °C = 293.15 K the value of this is

$$\frac{129.0 \times (3.1556 + 1.129 - 1.128)}{16\,641} = 0.0245$$

$$E = 0.0245 \times 8.3145 \times 293.15^2 \text{ J mol}^{-1}$$

$$= 17.5 \text{ kJ mol}^{-1}$$

b. At 100 °C = 373.15 K, the value is

$$\frac{(209.0 \times 3.46) - 252.5 - 12.32}{43\,681} = 0.010\,49$$

$$E = 0.010\,49 \times 8.3145 \times 373.15^2 \text{ J mol}^{-1}$$

$$= 12.1 \text{ kJ mol}^{-1}$$

The activation energy decreases as the temperature rises because of the breaking of hydrogen bonds between water molecules; the liquid becomes less structured with increase in temperature.

19.9.   a. In a hypothetical gas in which the molecules have no size, there are no collisions and therefore no exchanges of momentum between molecules. If there are no forces between the molecules, two layers can move past each other freely, and the viscosity is zero.

b. If the molecules have no size but attract one another, a force is required to move one layer past another. The gas will therefore have a viscosity. Increasing the temperature will increase the molecular speeds and will decrease the viscosity, as in a liquid.

c. If the molecules have no size but repel one another, a force will again be required to move one layer past another. There will again be a viscosity, which decreases with increasing temperature.

■ Diffusion

19.10.   The mass of the helium atom is

$$m = 4.0026/6.022 \times 10^{23} = 6.647 \times 10^{-24} \text{ g}$$

$$= 6.647 \times 10^{-27} \text{ kg}$$

a. From Eq. 19.16,

$$\eta = \frac{(6.647 \times 10^{-27} \times 1.381 \times 10^{-23} \times 273.15)^{1/2}}{\pi^{3/2}(0.225 \times 10^{-9})^2}$$

$$= 1.78 \times 10^{-5} \text{ kg m}^{-1} \text{ s}^{-1}$$

b. $\rho = \dfrac{mN}{V} = \dfrac{mP}{k_B T} = \dfrac{6.647 \times 10^{-27} \times 1.013\,25 \times 10^5}{1.381 \times 10^{-23} \times 273.15}$

$$= 0.1785 \text{ kg m}^{-3}$$

$$D = \frac{\eta}{\rho} = \frac{1.78 \times 10^{-5}}{0.1785} = 9.97 \times 10^{-5} \text{ m}^2 \text{ s}^{-1}$$

c. $\bar{u} = (8k_B T/\pi m)^{1/2} = \left(\dfrac{8 \times 1.381 \times 10^{-23} \times 273.15}{\pi \times 6.647 \times 10^{-27}} \right)^{1/2}$

$$= 1202 \text{ m s}^{-1}$$

d. $\dfrac{V}{\sqrt{2}} \pi d^2 N = \lambda$

$$\frac{V}{N} = \frac{1.381 \times 10^{-23} \times 273.15}{1.013\,25 \times 10^5} = 3.7229 \times 10^{-26}$$

$$\lambda = \frac{3.7229 \times 10^{-26}}{\sqrt{2}\pi(0.225 \times 10^{-9})^2} = 1.655 \times 10^{-7} \text{ m}$$

e. $Z_A = \dfrac{\bar{u}}{\lambda} = \dfrac{1202}{1.655 \times 10^{-7}} = 7.263 \times 10^9 \text{ s}^{-1}$

f. $Z_{AA} = \dfrac{1}{2} Z_A \dfrac{N}{V} = \dfrac{7.263 \times 10^9}{3.7229 \times 10^{-26} \times 2} = 9.75 \times 10^{34} \text{ m}^{-3} \text{ s}^{-1}$

19.11. From Eq. 19.48,

$$\overline{x^2} = 2tD = 2 \times 10 \times 1.005 \times 10^{-4} \text{ m}^2$$

$$(\overline{x^2})^{1/2} = 0.045 \text{ m} = 4.5 \text{ cm}$$

19.12. $t = 100 \times 24 \times 60 \times 60 = 8.64 \times 10^6 \text{ s}$

a. Glucose: $\overline{x^2} = 2 \times 6.8 \times 10^{-10} \times 8.64 \times 10^6$

$$= 0.0118 \text{ m}^2$$

$$\sqrt{\overline{x^2}} = 0.108 \text{ m} = 10.8 \text{ cm}$$

b. Tobacco mosaic virus: $\overline{x^2} = 2 \times 5.3 \times 10^{-12} \times 8.64 \times 10^6$

$$= 9.12 \times 10^{-5} \text{ m}$$

$$\sqrt{\overline{x^2}} = 9.55 \times 10^{-3} \text{ m}$$

$$= 0.96 \text{ cm}$$

19.13. $D = \dfrac{8.3145 \times 298.15}{(96\,500)^2 \times 2} \, \lambda° = 1.33 \times 10^{-7} \, \lambda° \text{ cm}^2 \text{ s}^{-1}$

For Cu^{2+}: $D_+ = 1.33 \times 10^{-7} \times 56.6 = 0.753 \times 10^{-5} \text{ cm}^2 \text{ s}^{-1}$

For SO_4^{2-}: $D_- = 1.33 \times 10^{-7} \times 80.0 = 1.065 \times 10^{-5} \text{ cm}^2 \text{ s}^{-1}$

$D = \dfrac{2 \times 0.753 \times 10^{-5} \times 1.065 \times 10^{-5}}{(0.755 + 1.065) \times 10^{-5}}$ (from Eq. 19.74)

$= 8.67 \times 10^{-6} \text{ cm}^2 \text{ s}^{-1}$

19.14. From Eq. 19.72,

$$D = \frac{RT}{QL} \, u_e$$

Charge on 1 mol of a univalent ion, QL,

$= 96\,500 \text{ C mol}^{-1}$

$D = \dfrac{(8.3145 \text{ J K}^{-1} \text{ mol}^{-1}) \, (298.15 \text{ K})}{96\,500 \text{ C mol}^{-1}} \, u_e$

$= (0.0257 \text{ V}) \, u_e$ since $J = C\,V$

For Na^+, with $u_e = 5.19 \times 10^{-4} \text{ cm}^2 \text{ V}^{-1} \text{ s}^{-1}$,
$D = 0.0257 \times 5.19 \times 10^{-4} = 1.33 \times 10^{-5} \text{ cm}^2 \text{ s}^{-1}$

For CH_3COO^-, with $u_e = 4.24 \times 10^{-4} \text{ cm}^2 \text{ V}^{-1} \text{ s}^{-1}$,

$D = 0.0257 \times 4.24 \times 10^{-4} = 1.09 \times 10^{-5} \text{ cm}^2 \text{ s}^{-1}$

For the electrolyte, using Eq. 19.74,

$D = \dfrac{2 \times 1.33 \times 10^{-5} \times 1.09 \times 10^{-5}}{1.33 \times 10^{-5} + 1.09 \times 10^{-5}} \text{ cm}^2 \text{ s}^{-1}$

$= 1.20 \times 10^{-5} \text{ cm}^2 \text{ s}^{-1}$

19.15. From Stokes's law (Eq. 19.77),

$$r = \frac{1.381 \times 10^{-23} \times 293.15}{6 \times 3.1416 \times 1.002 \times 10^{-3} \times 6.3 \times 10^{-11}}$$

$$= 3.40 \times 10^{-9} \text{ m} = 3.4 \text{ nm}$$

$$\text{Volume} = \frac{4}{3}\pi(3.40 \times 10^{-9})^3 = 1.65 \times 10^{-25} \text{ m}^3$$

$$\text{Molecular mass} = 1.65 \times 10^{-25} \times \frac{1}{0.75} \times 10^6$$

$$= 2.20 \times 10^{-19} \text{ g}$$

$$\text{Molar mass} = 2.20 \times 10^{-19} \times 6.022 \times 10^{23}$$

$$= 132\ 000 \text{ g mol}^{-1}$$

19.16. $t = \dfrac{\overline{x^2}}{2D}$

$$= \frac{(10 \times 10^{-6} \text{ m})^2}{2 \times 8.2 \times 10^{-11} \text{ m}^2 \text{ s}^{-1}} = 0.61 \text{ s}$$

19.17. Radius r of particle $= 1.5 \times 10^{-7}$ m

From Stokes's law (Eq. 19.77),

$$D = \frac{k_B T}{6\pi\eta r} = \frac{1.381 \times 10^{-23} \times 293.15 \text{ J}}{6\pi \times 1.002 \times 10^{-3} \times 1.5 \times 10^{-7} \text{ kg s}^{-1}}$$

$$= 1.43 \times 10^{-12} \text{ m}^2 \text{ s}^{-1}$$

From the Einstein random-walk equation (Eq. 19.48), $\overline{x^2} = 2Dt$, and therefore

$$t = \frac{(10^{-3} \text{ m})^2}{2 \times 1.43 \times 10^{-12} \text{ m}^2 \text{ s}^{-1}} = 3.5 \times 10^5 \text{ s}$$

■ Sedimentation and Diffusion

19.18. The following values are to be inserted into Eq. 19.93:

$$R = 8.3145 \text{ J K}^{-1} \text{ mol}^{-1} \qquad\qquad D = 5.96 \times 10^{-11} \text{ m}^2 \text{ s}^{-1}$$

$$T = 393.15 \text{ K} \qquad\qquad\qquad v_1 = 0.736 \text{ cm}^3 \text{ g}^{-1}$$

$$s = 4.6 \times 10^{-13} \text{ s} \qquad\qquad\quad \rho = 0.998 \text{ g cm}^{-3}$$

Thus

$$M = \frac{8.3145 \times 293.15 \times 4.60 \times 10^{-13}}{5.96 \times 10^{-11} (1 - 0.736 \times 0.998)}$$

$$= 70.86 \text{ J m}^{-2} \text{ s}^2 \text{ mol}^{-1}$$

$$= 70.86 \text{ kg mol}^{-1} = 70\ 860 \text{ g mol}^{-1}$$

19.19. From Eq. 19.93,

$$M = \frac{8.3145 JK^1 mol^1 \times 298.15 K \times 1.13 \times 10^{12}\, s^1}{4.2 \times 10^{11} m^2 s^1 (10.997 / 1.32}$$

$$= 272.6 \text{ kg mol}^{-1} = 272\ 600 \text{ g mol}^{-1}$$

19.20. a. From Eq. 19.86,

$$f = \frac{(1 - V_2\rho)m \ \omega^2 x}{v} = \frac{(1 - V_2\rho)m}{s}$$

where $s \equiv v/\omega^2 x$ is the sedimentation coefficient (Eq. 19.87). The molecular mass, m, is

$$\frac{60\ 000 \text{ g mol}^{-1} \cdot 10^{-3}}{6.022 \times 10^{23} \text{ mol}^{-1}} = 9.96 \times 10^{-23} \text{ kg}$$

The frictional coefficient is thus,

$$f = \frac{(1 - 0.997/1.31) \times 9.96 \times 10^{-23} \text{ kg}}{4.1 \times 10^{-13} \text{ s}}$$

$$= 5.80 \times 10^{-11} \text{ kg s}^{-1}$$

b. Volume of protein molecule is

$$\frac{9.96 \times 10^{-23} \text{ kg}}{1.31 \times 10^3 \text{ kg m}^{-3}} = 7.60 \times 10^{-26} \text{ m}^3$$

If the particle is spherical and its radius is r,

$$\tfrac{4}{3}\pi r^3 = 7.60 \times 10^{-26} \text{ m}^3$$

$$r = 2.63 \times 10^{-9} \text{ m}$$

According to Stokes's law (Eq. 19.76), $f = 6\pi\eta r$

$$= 6\pi \times 8.937 \times 10^{-4} \text{ kg m}^{-1} \text{ s}^{-1} \times 2.63 \times 10^{-9} \text{ m}$$

$$= 4.43 \times 10^{-11} \text{ kg s}^{-1}$$

The fact that the observed f is greater than that calculated using Stokes's law may be attributed to the fact that the molecule is not spherical.

19.21. Limiting rate of sedimentation is given by Eq. 19.85:

$$v = \frac{(1 - V_2\rho)mg}{6\pi r\eta}$$

The mass of the particle is

$$m = \frac{4}{3} \pi (1.5 \times 10^{-7} \text{ m})^3 \times 1.18 \times 10^3 \text{ kg m}^{-3}$$

$$= 1.668 \times 10^{-17} \text{ kg}$$

Then

$$v = \frac{(1 - 0.998/1.18) \times 1.668 \times 10^{-17} \text{ kg} \times 9.81 \text{ m s}^{-1}}{6\pi \times 1.002 \times 10^{-3} \times 1.5 \times 10^{-7} \text{ kg s}^{-1}}$$

$$= 8.908 \times 10^{-9} \text{ m s}^{-1}$$

The particle thus sediments a distance of 1 mm = 10^{-3} m in

$$\frac{10^{-3} \text{ m}}{8.908 \times 10^{-9} \text{ m s}^{-1}} = 1.12 \times 10^5 \text{ s}$$

19.22. We can use Stokes's law to estimate the radius of the particles:

$$r = \frac{k_B T}{6\pi\eta D} = \frac{1.381 \times 10^{-23} \times 298.15 \text{ J}}{6\pi \times 8.937 \times 10^{-4} \text{ kg m}^{-1} \text{ s}^{-1} \, 1.2 \times 10^{-11} \text{ m}^2 \text{ s}^{-1}}$$

$$= 2.037 \times 10^{-8} \text{ m}$$

The mass of each particle is thus,

$$m = \frac{4}{3} \pi (2.037 \times 10^{-8} \text{ m})^3 \times 1.33 \times 10^3 \text{ kg m}^{-3}$$

$$= 4.71 \times 10^{-20} \text{ kg}$$

The sedimentation coefficient is

$$s = \frac{(1 - V_2\rho)m}{f} = \frac{(1 - V_2\rho)m}{6\pi r\eta}$$

$$= \left(\frac{\left(\frac{1 - 0.997}{1.33}\right) \times 4.71 \times 10^{-20} \text{ kg}}{6\pi \times 8.937 \times 10^{-4} \text{ kg m}^{-1} \text{ s}^{-1} \times 2.036 \times 10^{-8} \text{ m}}\right) = 3.44 \times 10^{-11} \text{ s}$$

19.23. a. According to Eq. 19.93,

$$M = \frac{RTs}{D(1 - V_2\rho)}$$

$$= \frac{8.3145 \times 293.15 \text{ J mol}^{-1} \times 7.75 \times 10^{-13} \text{ s}}{4.80 \times 10^{-11} \text{ m}^2 \text{ s}^{-1} [1 - (0.998 \text{ g cm}^{-3} \times 0.739 \text{ cm}^3 \text{ g}^{-1})]}$$

$$= 149.9 \text{ kg mol}^{-1}$$

The molecular weight rounded is thus 150 000

b. According to the Stokes-Einstein equation, Eq. 19.77,

$$D = \frac{k_B T}{6\pi r \eta}$$

Therefore,

$$r = \frac{1.38 \times 10^{-23} \times 293.15 \text{ J}}{6 \times 3.1416 \times 1.002 \times 10^{-3} \text{ Pa s} \times 4.8 \times 10^{-11} \text{ m}^2 \text{ s}^{-1}}$$

$$= 4.46 \times 10^{-9} \text{ m} = 4.46 \text{ nm}$$

19.24. a. The molar mass is obtained by use of Eq. 19.93:

$$M = \frac{8.3145 \times 293.15 \text{ J K}^{-1} \times 4.48 \times 10^{-13} \text{ s}}{6.9 \times 10^{-11} \text{ m}^2 \text{ s}^{-1} \times (1 - 0.749 \times 0.998)}$$

$$= 62.7 \text{ kg mol}^{-1} = 62\ 700 \text{ g mol}^{-1}$$

The molecular weight is 62 700.

b. According to Einstein's equation, Eq. 19.48,

$$\sqrt{\overline{x^2}} = \sqrt{2 \times 6.9 \times 10^{-11} \text{ m}^2 \text{ s}^{-1} \times 60 \text{ s}}$$

$$= 9.1 \times 10^{-5} \text{ m}$$

c. In a gravitational field of $g = 9.8 \text{ m s}^{-2}$, the rate of sedimentation is

$$4.48 \times 10^{-13} \text{ s} \times 9.8 \text{ m s}^{-2} = 4.39 \times 10^{-12} \text{ m s}^{-1}$$

In 1 minute the molecule would therefore sediment

$$4.39 \times 10^{-12} \times 60 = 2.6 \times 10^{-10} \text{ m}$$

This is, of course, much less than the distance it would diffuse.

d. The speed of revolution of the centrifuge is

$$\left(\frac{15\ 000 \text{ rpm}}{60 \text{ s min}^{-1}}\right) \times 2\pi = 1571 \text{ rad s}^{-1}$$

The rate of sedimentation is given by Eq. 19.87:

$$v = s\omega^2 x = 4.48 \times 10^{-13}\ \text{s} \times (1571\ \text{s}^{-1})^2 \times 0.2\ \text{m} = 2.21 \times 10^{-7}\ \text{m s}^{-1}$$

In 1 minute the molecule would therefore sediment

$$2.21 \times 10^{-7} \times 60\ \text{m} = 1.33 \times 10^{-5}\ \text{m}$$

This is still somewhat less than the distance the molecule would diffuse.

e. From Eq. 19.77,

$$r = \frac{k_B T}{6\pi \eta D} = \frac{1.38 \times 10^{-23} \times 293.15\ \text{J}}{6\pi \times 1.003 \times 10^{-3}\ \text{kg m}^{-1}\ \text{s}^{-1} \times 6.9 \times 10^{-11}\ \text{m}^2\ \text{s}^{-1}}$$

$$= 3.10 \times 10^{-9}\ \text{m}$$

f. The mass of the molecule is

$$\frac{62\ 700\ \text{g mol}^{-1}}{6.022 \times 10^{23}\ \text{mol}^{-1}} = 1.041 \times 10^{-19}\ \text{g}$$

The density is 1.335 g cm^{-3} = 1.335 × 10^6 g m^{-3}, and the volume of the molecule is

$$\frac{1.041 \times 10^{-19}}{1.335 \times 10^6} = 7.80 \times 10^{-26}\ \text{m}^3$$

The radius is

$$\left(\frac{7.80 \times 10^{-26}}{1.33\pi}\right)^{1/3} = (1.867 \times 10^{-26})^{1/3}$$

$$= 2.66 \times 10^{-9}\ \text{m}$$

This is in reasonable agreement with the value obtained in (e), on the basis of the Stokes-Einstein equation. The molecule is not in fact spherical, and the equation is more satisfactory for larger particles.

19.25. The sedimentation coefficient in each case is calculated using Eq. 19.89. $(1 - V_2\rho)$ is

$$1 - (0.833\ \text{cm}^3\ \text{g}^{-1} \times 0.998\ \text{g cm}^{-3}) = 0.169$$

a. The mass, m, of the particle is

$$(4/3)\pi(0.001\ \text{m})^3 \times 1200\ \text{kg m}^{-3} = 5.025 \times 10^{-6}\ \text{kg}$$

Then by Eq. 19.89,

$$s = \frac{0.169 \times 5.025 \times 10^{-6}\ \text{kg}}{6\pi \times 1.022 \times 10^{-3}\ \text{kg m}^{-1}\ \text{s}^{-1}\ 10^{-3}\ \text{m}}$$

$$= 0.044\ \text{s}$$

The rate of sedimentation in a gravitational field of $g = 9.8$ m s^{-2} is

$$0.044 \times 9.8\ \text{m s}^{-1} = 0.431\ \text{m s}^{-1}$$

In 1 hour the particle will therefore sediment

$$0.431 \times 3600\ \text{m} = 1552\ \text{m}$$

b. The radius is now one-tenth that of the previous case, and the mass, m, is reduced by a factor of 1000. Since m/r appears in Eq. 19.89, the sedimentation coefficient is reduced by a factor of 100 and is 4.31×10^{-3} s. The particle will now sediment 15.52 m in 1 hour.

c. With a radius of 1 μm there is a further reduction by a factor of 10. The sedimentation coefficient is reduced by a further factor of 100, and is now 4.31×10^{-5} s. The particle will now sediment 0.155 m or 15.5 cm in 1 hour.

d. With a further reduction of the radius by a factor of 10, the sedimentation coefficient is reduced by a further factor of 100; it is now 4.31×10^{-7} s, and the particle will sediment 1.55 mm in 1 hour.

e. The radius is now reduced by a further factor of 100, and the sedimentation coefficient is reduced by a factor of 10^4. The sedimentation is now quite insignificant.

The results are summarized as follows:

Radius, r/m	10^{-3}	10^{-4}	10^{-5}	10^{-6}	10^{-8}
Distance in 1 h/m	1552	15.52	0.155	1.55×10^{-3}	1.55×10^{-7}

f. The sedimentation coefficient for the particle of radius 10 nm is 4.31×10^{-11} s. A sedimentation of 1 mm in 1 hour is a rate of sedimentation of 10^{-3} m/3600 s = 2.78×10^{-7} m s^{-1}. This is equal to $s\omega^2 x$, and with $x = 10^{-2}$ m we have

$$\omega^2 = \frac{2.78 \times 10^{-7} \text{ m s}^{-1}}{4.31 \times 10^{-11} \text{ s} \times 10^{-2} \text{ m}}$$

$$= 6.45 \times 10^5 \text{ s}^{-2}$$

$$\omega = 803 \text{ rad s}^{-1}$$

This corresponds to $803/2\pi = 127.8$ revolutions per second, or 7669 rpm, rounded to 7700 rpm.

19.26. According to the Stokes-Einstein equation, Eq. 19.77, for the particle of radius 10^{-3} m,

$$D = \frac{k_B T}{6\pi\eta r} = \frac{1.38 \times 10^{-23} \times 293.15 \text{ J}}{6\pi \times 1.002 \times 10^{-3} \text{ kg m}^{-1} \text{ s}^{-1} \times 10^{-3} \text{ m}}$$

$$= 2.142 \times 10^{-16} \text{ m}^2 \text{ s}^{-1}$$

According to the random-walk equation, Eq. 19.48, the mean distance traveled is

$$\sqrt{\overline{x^2}} = \sqrt{2Dt} = \sqrt{2 \times 2.142 \times 10^{-16} \text{ m}^2 \text{ s}^{-1} \times 3600 \text{ s}}$$

$$= 1.24 \times 10^{-6} \text{ m}$$

The diffusion coefficient is inversely proportional to the radius and the distance is inversely proportional to the square root of the radius. The values for the particles are thus:

Radius, r/m	10^{-3}	10^{-4}	10^{-5}	10^{-6}	10^{-8}
$D \times 10^{16}$/ m^2 s^{-1}	2.142	21.42	214.2	2142	2.142×10^5
$\sqrt{\overline{x^2}}$ /μm	1.24	3.9	12.4	39	390

19.27. The molar mass is given by Eq. 19.98. The value of $1 - V_2\rho$ is

$$1 - (0.78 \times 0.997) = 0.222$$

The speed of revolution is

$$\left(\frac{25\ 000\ \text{rpm}}{60\ \text{s min}^{-1}}\right) \times 2\pi = 2618\ \text{rad s}^{-1}$$

Then, by Eq. 19.98,

$$M = \frac{2 \times 8.3145 \times 298.15\ \text{J mol}^{-1}\ \ln 22.49/3.52}{0.222\ (2618\ \text{rad s}^{-1})^2\ [(8.34 \times 10^{-2})^2 - (9.12 \times 10^{-2})^2]\ \text{m}^2}$$

$$= 251.6\ \text{kg mol}^{-1}$$

The molecular weight is thus 252 000.

19.28. From Eq. 19.98,

$$\omega^2 = \frac{2RT\ \ln(c_2/c_1)}{M(1 - V_2\rho)\ (x_2^2 - x_1^2)}$$

$(1 - V_2\rho) = 0.192$ and therefore,

$$\omega^2 = \frac{2 \times 8.3145 \times 298.15\ \text{J mol}^{-1}\ \ln 20}{10^3\ \text{kg mol}^{-1} \times 0.192\ (10.00^2 - 9.00^2) \times 10^{-4}\ \text{m}^2}$$

$$= 39\ 530\ \text{rad}^2\ \text{s}^{-1}$$

$$\omega = 198.8\ \text{rad s}^{-1}$$

$$= 198.8 \times 60/2\pi = 1900\ \text{rpm}$$